Study Guide with Solutions To Selected Problems

General, Organic, and Biological Chemistry

Organic and Biological Chemistry

H. Stephen Stoker

Danny V. White

Joanne A. White

Houghton Mifflin Company Boston New York

Editor-in-Chief: Kathi Prancan
Senior Sponsoring Editor: Richard Stratton
Editorial Associate: Sarah Gessner
Senior Project Editor: Nancy Blodget
Manufacturing Coordinator: Sally Culler
Executive Marketing Manager: Andy Fisher

Printed in the U.S.A.

ISBN: 0-618-052097

123456789—PP—04 03 02 01 00

Contents

Preface

If the study of chemistry is new to you, you are about to gain a new perspective on the material world. You will never again look at the objects and substances around you in quite the same way. The invisible structure and organization of matter, the "how and why" of chemical change -- knowledge of these areas will help to demystify many occurrences in the world around you. Chemistry is not an isolated academic study. We use it throughout our lives to appreciate and understand the world and to make responsible choices in that world.

The purpose of this study guide is to help you in your study of the textbook, *General, Organic, and Biological Chemistry*, by providing summaries of the text and additional practice exercises. As you use this Study Guide, we suggest that you follow the steps below.

1. Read the overview for the chapter to get a general idea of the facts and concepts in each chapter.

2. Read the section summaries and work the practice exercises as you come to them. Write out the answers even if you are sure you understand the concepts. This will help you to check your understanding of the material. Refer to the answers at the end of the chapter as soon as you have answered each practice exercise. By checking your answers, you will know whether to review or continue to the next section.

3. When you have finished answering the practice exercises, take the self-test at the end of the chapter. Check your answers with the answer key at the end of the chapter. If there are any questions that you answer incorrectly or do not understand, refer to the chapter section numbers in the answer key and review that material. The Solutions section of this book contains answers to selected exercises and problems from the textbook along with methods for solving many of the problems.

Chemistry is a discipline of patterns and rules. Once your mind has begun to understand and accept these patterns, the time you have spent on repetition and review will be well rewarded by a deeper total picture of the world around you. As teachers, we have enjoyed preparing this study guide and hope that it will assist you in your study of chemistry.

Danny and Joanne White

Basic Concepts About Matter — Chapter 1

Chapter Overview

Why is the study of chemistry important to you? Chemistry produces many substances of practical importance to us all: building materials, foods, medicines. For anyone entering one of the life sciences, such as the health sciences, agriculture or forestry, an understanding of chemistry leads to an understanding of the many life processes.

In this chapter you will be studying some of the fundamental ideas and language of chemistry. You will characterize three states of matter, differentiate between physical properties and chemical properties, and identify two different types of mixtures. You will describe elements and compounds, and practice using symbols and formulas.

Practice Exercises

1.1 **Matter** (Sec. 1.1) exists in three physical states. Complete the following table indicating the properties of each of these states of matter.

State	Definite shape?	Definite volume?
solid (Sec. 1.2)	yes	
liquid (Sec. 1.2)		
gas (Sec. 1.2)		

1.2 The **physical properties** (Sec. 1.3) of a substance can be observed without changing the identity of the substance. **Chemical properties** (Sec. 1.3) are observed when a substance changes or resists changing to another substance. Complete the following table:

Property	Physical	Chemical	Insufficient information
liquid boils at 100°C			
solid forms a gas when heated			
metallic solid exposed to air forms a white solid			

1.3 A **physical change** (Sec. 1.4) is a change in shape or form. A **chemical change** (Sec. 1.4) produces a new substance. Classify the following processes as physical or chemical changes by marking the correct column:

Process	Physical change	Chemical change
ice cube melts		
wood block burns		
salad oil freezes solid		
sugar dissolves in tea		
wood block is split		
butter becomes rancid		

1.4 **Mixtures** (Sec. 1.5) of substances may be either **homogeneous** (Sec. 1.5), one phase, uniform throughout, or **heterogeneous** (Sec. 1.5), visibly different parts or phases. Indicate whether each of the following mixtures is homogeneous or heterogeneous:

Mixture	Homogeneous	Heterogeneous
apple juice (water, sugar, fruit juice)		
cornflakes and milk		
fruit salad		
brass (copper and zinc)		

1.5 Complete the following diagram:

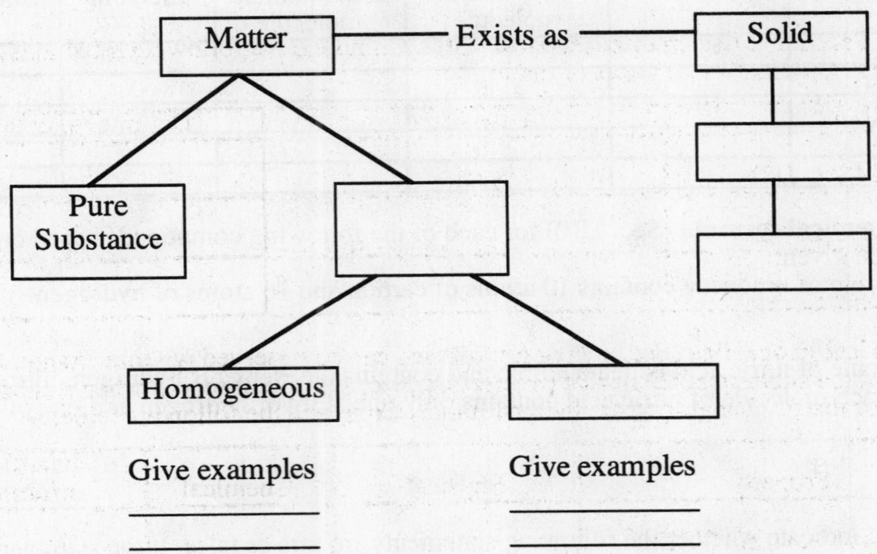

1.6 **Elements** (Sec. 1.6) are **pure substances** (Sec. 1.5) that cannot be broken down into simpler substances. **Compounds** (Sec. 1.6) can be broken down into two or more simpler substances by chemical means. Complete the following table:

	Substance is an element	Substance is a compound	Insufficient information to make a classification
Substance A reacts violently with water.			
Substance B can be broken into simpler substances by chemical processes			
Cooling substance C at 350°C turns it from a liquid to a solid.			
Substance D cannot decompose into simpler substances by chemical processes.			

1.7 In the following table, write the **chemical symbol** (Sec. 1.8) or name for each element:

Name	Symbol
calcium	
copper	
argon	
nickel	
magnesium	

Symbol	Name
C	
Ne	
Zr	
Pb	
Fe	

1.8 An **atom** (Sec. 1.9) is the smallest particle of an element that keeps the identity of that element. A **molecule** (Sec. 1.9) is a tightly-bound group of two or more atoms that functions as a unit. **Homoatomic molecules** (Sec. 1.9) are made up of atoms of one element; **heteroatomic molecules** (Sec. 1.9) contain atoms of two or more elements. Indicate whether each unit of a substance in the table below is an atom, a homoatomic molecule, or a heteroatomic molecule, and classify the substance as an element or a compound:

Unit	Atom	Homoatomic molecule	Heteroatomic molecule	Element	Compound
N_2					
CH_4					
HCN					
Au					
HF					

1.9 Write a **chemical formula** (Sec. 1.10) for each of the following compounds based on the information given:

a. A molecule of limonene contains 10 atoms of carbon and 16 atoms of hydrogen.

b. A molecule of nitric acid is pentaatomic and contains the elements hydrogen, nitrogen and oxygen. Each molecule of nitric acid contains only one atom of hydrogen and one of nitrogen.

Self-Test

True-false: Indicate whether the following statements are true or false. If the statement is false, give the word or phrase that may be substituted for the underlined portion to make the statement true.

1. Matter is anything that has volume and occupies space.
2. Gases have no definite shape or volume.
3. Liquids take the shape of the container and completely occupy the volume of the container.
4. A mixture of oil and water would be an example of a homogeneous mixture.
5. The evaporation of water from salt water is an example of a chemical change.
6. Elements cannot be broken down into simpler substances by chemical means.
7. Dissolving sugar in water is an example of a chemical change.
8. A chemical property describes the way a substance changes or resists change to form a new substance.
9. A mixture is a chemical combination of two or more pure substances.
10. Synthetic (laboratory produced) elements are all radioactive.
11. The most common (dominant) element in the universe is oxygen.
12. Two-letter symbols are always the first two letters of the element's name.
13. A compound has molecules that are homoatomic.

Multiple choice:

14. One of the three states of matter is a solid. A solid has the following characteristics:
    a) definite volume, no definite shape
    b) no definite volume, no definite shape
    c) definite volume and shape
    d) no definite volume, but definite shape
    e) none of these
15. An example of a homogeneous mixture would be:
    a) sand and water
    b) salt and water
    c) wood and water
    d) oil and water
    e) none of these
16. An example of a physical change would be:
    a) rusting of iron
    b) sugar dissolving in coffee
    c) gasoline burning in a car engine
    d) burning coal
    e) none of these
17. An example of a chemical change would be:
    a) rusting of iron
    b) burning coal
    c) gasoline burning in a car engine
    d) a, b, and c are all correct
    e) none of these
18. The symbol for the element iron is:
    a) FE b) Fe c) F d) Ir e) none of these
19. The name for the element Ne is:
    a) neodymium
    b) neon
    c) neptunium
    d) nitrogen
    e) none of these
20. $MgCO_3$ is a compound and is composed of what elements?
    a) magnesium, chlorine, iron
    b) magnesium, carbon, neon
    c) manganese, carbon, oxygen
    d) magnesium, carbon, oxygen
    e) none of these
21. The total number of atoms in one molecule of CH_4O is:
    a) 3 b) 4 c) 5 d) 6 e) none of these
22. On the basis of its formula, which of the following substances is an element?
    a) NH_3 b) Cl_2 c) CO_2 d) CO e) none of these
23. On the basis of its formula, which of the following is a compound?
    a) Fe b) Fm c) O_2 d) HI e) none of these

Answers to Practice Exercises

1.1

State	Definite shape?	Definite volume?
solid	yes	yes
liquid	no	yes
gas	no	no

1.2

Property	Physical	Chemical	Insufficient information
liquid boils at 100°C	X		
solid forms a gas when heated			X
metallic solid exposed to air forms a white solid		X	

1.3

Process	Physical change	Chemical change
ice cube melts	X	
wood block burns		X
salad oil freezes solid	X	
sugar dissolves in tea	X	
wood block is split	X	
butter becomes rancid		X

1.4

Mixture	Homogeneous	Heterogeneous
apple juice (water, sugar, fruit juice)	X	
cornflakes and milk		X
fruit salad		X
brass (copper and zinc)	X	

1.5

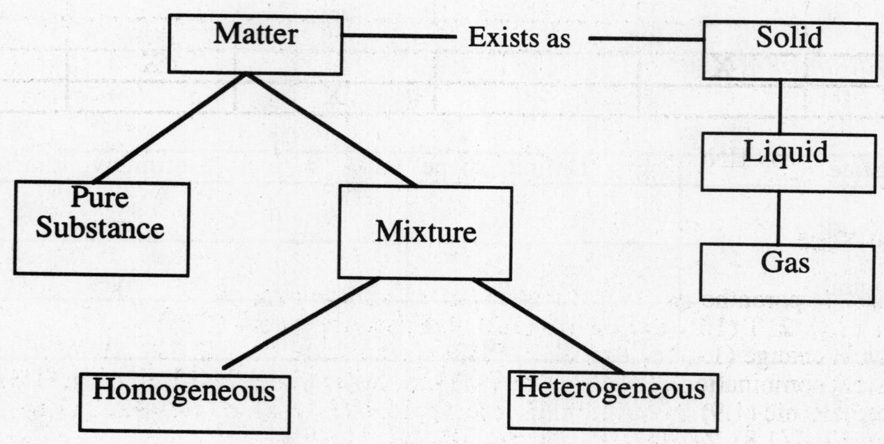

Give examples of homogeneous mixtures:

air
coffee

Give examples of heterogeneous mixtures:

smoke
concrete

(These are a few of many possible examples.)

1.6

	Substance is an element	Substance is a compound	Insufficient information to make a classification
Substance A reacts violently with water.			X
Substance B can be broken into simpler substances by chemical processes.		X	
Cooling substance C at 350°C turns it from a liquid to a solid.			X
Substance D cannot decompose into simpler substances by chemical processes.	X		

1.7

Name	Symbol
calcium	Ca
copper	Cu
argon	Ar
nickel	Ni
magnesium	Mg

Symbol	Name
C	carbon
Ne	neon
Zr	zirconium
Pb	lead
Fe	iron

1.8

Unit	Atom	Homoatomic molecule	Heteroatomic molecule	Element	Compound
N_2		X		X	
CH_4			X		X
HCN			X		X
Au	X			X	
HF			X		X

1.9 a. $C_{10}H_{16}$ b. HNO_3

Answers to Self-Test

The numbers in parentheses refer to sections in your textbook.
1. F; mass (1.1) **2**. T (1.2) **3**. F; gases (1.2) **4**. F; heterogeneous (1.5)
5. F; physical change (1.4) **6**. T (1.6) **7**. F; physical change (1.4) **8**. T (1.3)
9. F; physical combination (1.5) **10**. T (1.7) **11**. F; hydrogen (1.7) **12**. F; often (1.8)
13. F; heteroatomic (1.9) **14**. c (1.2) **15**. b (1.5) **16**. b (1.4) **17**. d (1.4) **18**. b (1.8)
19. b (1.8) **20**. d (1.8) **21**. d (1.10) **22**. b (1.10) **23**. d (1.10)

Measurements in Chemistry — Chapter 2

Chapter Overview

Measurements are very important in science. In this chapter you will study some of the common units used in measuring length, volume, mass, temperature, and heat in the modern metric (SI) system. You will solve problems involving measurement using the method of dimensional analysis, in which units associated with numbers are used as a guide in setting up the calculations.

You will learn to use scientific notation to express large and small numbers efficiently, and practice using the number of significant figures that corresponds to the accuracy of the measurements being made. You will also use equations to calculate density, heat loss or gain involved in a temperature change, and to convert temperature from one temperature scale to another.

Practice Exercises

2.1 **Significant figures** (Sec. 2.3) are the digits in any **measurement** (Sec. 2.1) that are known with certainty plus one digit that is uncertain. These guidelines will help you in determining the number of significant figures:

1. All nonzero digits are significant. (23.4 m has 3 significant figures)
2. Zeros in front of nonzero digits are not significant. (0.00025 has 2 significant figures)
3. Zeros between nonzero digits are significant. (2.005 has 4 significant figures)
4. Zeros at the end of a number are significant if a decimal point is present (1.60 has 3 significant figures), but not significant if there is no decimal point (500 has one significant figure).

In **rounding off** (Sec. 2.4) a number to a certain number of significant figures:

1. Look at the first digit to be deleted.
2. If that digit is less than 5, drop that digit and all those to the right of it. If that digit is 5 or more, increase the last significant digit by one.

Example: Round 7.3589 to three significant figures. Since 7.35 has the correct number of significant figures and 8 is greater than 5, increase 7.35 to 7.36.

In the following table state the number of significant figures in each number as written, and then round off the number to 3 significant figures.

Number	Number of significant figures	Rounded to 3 significant figures
1578		
45932		
103045		
0.00034722		
0.0345047		
0.000450700		
2984400		

2.2 When multiplying and dividing measurements, the number of significant figures in the answer is the same as the number of significant figures in the measurement that contains the fewest significant figures.

Example: 4.2 m x 3.12 m = 13.104 m^2 = 13 m^2
(Since 4.2 has fewer significant figures, the answer has two significant figures.)

Exact numbers (such as three people or twelve eggs in a dozen) do not limit the number of significant figures.

Carry out each of the mathematical operations as indicated below, and round the answer to the correct number of significant figures:

Problem	Answer before rounding off	Rounded to correct number of significant figures
160 x 0.32		
482 x 0.00358		
5 x 3.00985		
723 ÷ 4.04		
720 ÷ 1.37		
0.0485 ÷ 88.342		

2.3 When measurements are added and subtracted, the answer can have no more digits to the right of the decimal point than the measurement with the least number of decimal places.

Example: 3.58 m + 7.2 m = 10.78 m = 10.8 m

Carry out the mathematical operations as indicated below, and round the answer to the correct number of significant figures.

Problem	Answer before rounding off	Rounded to correct number of significant figures
153 + 4521		
483 + 0.223		
1.097 + 0.34		
744 – 36		
8093 – 0.566		
0.345 – 0.0221		

2.4 **Scientific notation** (Sec. 2.5) is a convenient way of expressing very large or very small numbers in a compact form. To convert a number to scientific notation:

1. Write the original number.
2. Move the decimal point to a position just to the right of the first nonzero digit.
3. Count the number of places the decimal point was moved. This number will be the exponent of 10.
4. If the decimal point is moved to the left, the exponent will be positive; if the decimal point is moved to the right the exponent will be negative.

Examples:
1. 2300 = 2.3×10^3 (decimal moved 3 places to the left)
2. 43,010,000 = 4.301×10^7 (decimal moved 7 places to the left)
3. 0.0072 = 7.2×10^{-3} (decimal moved 3 places to the right)

Note that only significant figures become a part of the coefficient.

Complete the following tables:

Decimal number	Scientific notation
4378	
783	
8400	
0.00362	
0.093200	

Scientific notation	Decimal number
6.389×10^6	
3.34×10^1	
4.55×10^{-3}	
9.08×10^{-5}	
2.0200×10^{-2}	

2.5 When numbers in exponential form (as in scientific notation) are added or subtracted, the exponents of 10 must be the same. Use the correct number of significant figures.
Example:
$(1.53 \times 10^{-3}) + (7.2 \times 10^{-4})$ (can be added in this form only with calculator)
$(1.53 \times 10^{-3}) + (0.72 \times 10^{-3}) = 2.25 \times 10^{-3}$

1. $(4.54 \times 10^{4}) + (1.804 \times 10^{4}) =$ ____________________

2. $(8.522 \times 10^{-3}) + (1.3 \times 10^{-4}) =$ ____________________

3. $(5.631 \times 10^{4}) - (1.52 \times 10^{4}) =$ ____________________

4. $(2.94 \times 10^{-4}) - (5.866 \times 10^{-5}) =$ ____________________

2.6 When multiplying numbers in exponential form, multiply the coefficients and add the exponents of 10.
Examples:
1. (two positive exponents) $(1.2 \times 10^{3}) \times (4.7 \times 10^{5}) = 5.6 \times 10^{8}$
2. (negative and positive exponents) $(2.3 \times 10^{4}) \times (1.8 \times 10^{-8}) = 4.1 \times 10^{-4}$
3. $(5.1 \times 10^{-2}) \times (7.2 \times 10^{5}) = 37 \times 10^{3} = 3.7 \times 10^{4}$

When dividing numbers in exponential form, divide the coefficients and subtract the exponents of 10.
Examples:
1. $(4.8 \times 10^{8}) \div (3.4 \times 10^{3}) = 1.4 \times 10^{5}$
2. $(6.5 \times 10^{4}) \div (3.7 \times 10^{-3}) = 1.8 \times 10^{7}$

Carry out the mathematical operations below. Express answers to the correct number of significant figures:

1. $(4.155 \times 10^{3}) \times (1.50 \times 10^{6}) =$ ____________________

2. $(7.36 \times 10^{2}) \times (4.711 \times 10^{3}) =$ ____________________

3. $(1.7 \times 10^{-3}) \times (3.363 \times 10^{-5}) =$ ____________________

4. $(4.09 \times 10^{6}) \div (2.9001 \times 10^{3}) =$ ____________________

5. $(3.1413 \times 10^{4}) \div (7.83 \times 10^{5}) =$ ____________________

6. $(5.204 \times 10^{-6}) \div (4.1 \times 10^{3}) =$ ____________________

7. $(6.61 \times 10^{-6}) \div (7.278 \times 10^{-3}) =$ ____________________

2.7 Use the method of **dimensional analysis** (Sec. 2.6) to solve the problems below. Choose the **conversion factor** (Sec. 2.6) that gives the answer in the correct units. If the wrong conversion factor is used the units will not cancel.

Example: How many millimeters (mm) are in 2.41 meters (m)?

$$2.41 \text{ m} \times \frac{1000 \text{ mm}}{1 \text{ m}} = (2.41 \times 1000)\left(\frac{\text{m} \times \text{mm}}{\text{m}}\right) = 2410 \text{ mm}$$

Complete the following table. Use the factors in Table 2.2 of your textbook for conversion between the metric and English systems.

Problem	Relationship with units	Answer with units
3.89 km = ? cm	$3.89\text{ km} \times \frac{1000\text{ m}}{1\text{ km}} \times \frac{100\text{ cm}}{1\text{ m}}$	
45.8 mm = ? m		
0.987 mm = ? km		
7.89×10^4 cm = ? m		
4.05×10^{-4} cm = ? km		
5.5×10^5 L = ? mL		
3.67 km = ? in.		
8.22 mL = ? qt		
57 fl oz = ? mL		
3.66 oz = ? kg		

2.8 **Density** (Sec. 2.7) is the ratio of the **mass** (Sec. 2.2) of an object to the volume occupied by that object.

Example: An object has a mass of 123 g and a volume of 17 cm^3. Calculate the density of the object.

$$\text{Density} = \frac{\text{mass}}{\text{volume}} = \frac{123\text{ g}}{17\text{ cm}^3} = 7.2\text{ g/cm}^3$$

Example: What mass will a cube of aluminum have if its volume is 32 cm^3? Aluminum has a density of 2.7 g/cm^3.

Using the method of dimensional analysis, we can set up an equation that uses the density as a conversion factor:

$$32\text{ cm}^3 = ?\text{ g}$$

$$32\text{ cm}^3 \times \frac{2.7\text{ g}}{1.0\text{ cm}^3} = 86\text{ g}$$

Complete the following table:

Problem	Substituted equation with units	Answer with units
1. What is the density of silver, if 38.5 cm^3 has a mass of 404 grams?		
2. Ethanol has a density of 0.789 g/mL. What is the mass of 458 mL of ethanol?		
3. Copper has a density of 8.92 g/cm^3. What would be the volume of 8.97 kg of copper?		

2.9 Convert the boiling point temperatures of the following compounds to the indicated temperature scales. Use the equations in section 2.8 of your textbook.

Boiling point	Substituted Equation	Temperature
ethyl acetate (nail polish remover) 77.0°C	°F = 9/5(77.0) + 32	°F
toluene (additive in gasoline) 111°C		°F
isopropyl alcohol (rubbing alcohol) 180°F		°C
naphthalene (moth balls) 424°F		°C
propane (fuel for camping stoves) –42°C		K
methane (natural gas) –260°F		K

2.10 The **specific heat** (Sec. 2.8) of a substance can be used to calculate how much heat is absorbed or given off when a substance changes temperature.

Example:
How many **calories** (Sec. 2.8) of heat would be needed to raise the temperature of 12.4 g of water from 21.0°C to 55.0°C?
1. The specific heat of water is 1.00 cal/g•°C
2. Heat (cal) = specific heat(cal/g•°C) x mass(g) x temperature change(°C)
3. Substitute the numerical values and solve:
   Heat (cal) = (1.00 cal/g•°C)(12.4 g)(34.0°C) = 422 cal

Calculate answers to the following problems involving specific heat.

Problem	Substituted equation with units	Answer with units
1. How much heat is required to raise the temperature of 1.00 kilogram of liquid water 25.0°C?		
2. If the temperature of a 150 g block of aluminum is raised from 22°C to 97°C, how much heat has the aluminum absorbed? (specific heat of Al = 0.21 cal/g•°C)		

Self-Test

True-false: Indicate whether the following statements are true or false. If the statement is false, give the word or phrase that may be substituted for the underlined portion to make the statement true.
1. The basic metric unit of volume is the <u>milliliter</u>.
2. The basic unit of mass in the metric system is the <u>kilogram</u>.
3. In scientific notation, a <u>coefficient</u> between 1 and 10 is multiplied by a power of ten.

4. When numbers in exponential form are multiplied or divided, the exponents of 10 must be the same.
5. In rounding numbers, the number of digits is determined by the number of significant figures in the measurement.
6. The density of an object is the ratio of its weight to its volume.
7. The calorie is a common measure of heat and is defined as the quantity of heat that raises the temperature of 1 gram of water 1°C.
8. In the metric system, the prefix micro- means one-thousandth (0.001, 10^{-3})
9. The number, 0.002010, has four significant figures.
10. On the Kelvin temperature scale, all temperature readings are positive.

Multiple choice:

11. The correct way of expressing 4174 in scientific notation is:
    a) 4.174×10^2
    b) 4.174×10^3
    c) 4.174×10^4
    d) 4.2×10^3
    e) none of these
12. The number, 0.005140, should be written in scientific notation as:
    a) 5.140×10^{-2}
    b) 514×10^{-4}
    c) 5.14×10^{-3}
    d) 5.140×10^{-3}
    e) none of these
13. The sum of 5472 plus 1946 would be written in scientific notation as:
    a) 7418
    b) 7.418×10^3
    c) 7.4×10^3
    d) 7.42×10^3
    e) none of these
14. The product of 8311 times 0.01452 would be written in scientific notation as:
    a) 120.7
    b) 1.21×10^2
    c) 1.207×10^2
    d) 1.2×10^2
    e) none of these
15. When 1.487 is rounded to three significant figures, the correct answer is:
    a) 1.480
    b) 1.490
    c) 1.48
    d) 1.49
    e) none of these
16. In converting grams to milligrams, the known quantity of grams should be multiplied by which of these conversion factors?
    a) 1 g/1000 mg
    b) 1000 mg/1 g
    c) 1 g/1000000 µg
    d) 1000000 µg/1 g
    e) none of these
17. How many milliliters would be in 25.2 kilograms of a liquid having a density of 0.833 g/mL?
    a) 30,300 mL
    b) 21.0 mL
    c) 21,000 mL
    d) 30.3 mL
    e) none of these
18. How many calories would be needed to raise 420 grams of aluminum from 120°C to 450°C? (See Table 2.4 in your text for specific heat.)
    a) 2.6×10^4 cal
    b) 2.6106×10^4 cal
    c) 2.9×10^1 cal
    d) 1.1×10^4 cal
    e) none of these
19. Which of the following measurements has three significant figures?
    a) 1.050
    b) 230 mL
    c) 0.0702 g
    d) 16.20 m
    e) none of these
20. A temperature of 22°C would have which of these values on the Fahrenheit scale?
    a) –6.6°F b) 54°F c) 97°F d) 72°F e) none of these

Answers to Practice Exercises

2.1

Number	Number of significant figures	Rounded to 3 significant figures
1578	4	1580
45932	5	45900
103045	6	103000
0.00034722	5	0.000347
0.0345047	6	0.0345
0.000450700	6	0.000451
2984400	5	2980000

2.2

Problem	Answer before rounding off	Rounded to correct number of significant figures
160 x 0.32	51.2	51
482 x 0.00358	1.72556	1.73
5 x 3.00985	15.04925	20
723 ÷ 4.04	178.96039	179
720 ÷ 1.37	525.54744	530
0.0485 ÷ 88.342	0.000549	0.000549

2.3

Problem	Answer before rounding off	Rounded to correct number of significant figures
153 + 4521	4674	4674
483 + 0.223	483.223	483
1.097 + 0.34	1.437	1.44
744 – 36	708	708
8093 – 0.566	8092.434	8092
0.345 – 0.0221	0.3229	0.323

2.4

Decimal number	Scientific notation
4378	4.378×10^3
783	7.83×10^2
8400	8.4×10^3
0.00362	3.62×10^{-3}
0.093200	9.3200×10^{-2}

Scientific notation	Decimal number
6.389×10^6	6,389,000
3.34×10^1	33.4
4.55×10^{-3}	0.00455
9.08×10^{-5}	0.0000908
2.0200×10^{-2}	0.020200

2.5 **1.** 6.34×10^4 **2.** $(8.522 \times 10^{-3}) + (0.13 \times 10^{-3}) = 8.7 \times 10^{-3}$
3. 4.11×10^4 **4.** $(2.94 \times 10^{-4}) - (0.5866 \times 10^{-4}) = 2.35 \times 10^{-4}$

2.6 **1.** 6.23×10^9 **2.** $34.7 \times 10^5 = 3.47 \times 10^6$ **3.** 5.7×10^{-8} **4.** 1.41×10^3
5. $0.401 \times 10^{-1} = 4.01 \times 10^{-2}$ **6.** 1.3×10^{-9} **7.** 9.08×10^{-4}

2.7

Problem	Relationship with units	Answer with units
3.89 km = ? cm	$3.89\text{ km} \times \frac{1000\text{ m}}{1\text{ km}} \times \frac{100\text{ cm}}{1\text{ m}}$	3.89×10^5 cm
45.8 mm = ? m	$45.8\text{ mm} \times \frac{1\text{ m}}{1000\text{ mm}}$	4.58×10^{-2} m
0.987 mm = ? km	$0.987\text{ mm} \times \frac{1\text{ m}}{1000\text{ mm}} \times \frac{1\text{ km}}{1000\text{ m}}$	9.87×10^{-7} km
7.89×10^4 cm = ? m	$7.89 \times 10^4\text{ cm} \times \frac{1\text{ m}}{100\text{ cm}}$	7.89×10^2 m
4.05×10^{-4} cm = ? km	$4.05 \times 10^{-4}\text{ cm} \times \frac{1\text{ m}}{100\text{ cm}} \times \frac{1\text{ km}}{1000\text{ m}}$	4.05×10^{-9} km
5.5×10^5 L = ? mL	$5.5 \times 10^5\text{ L} \times \frac{1000\text{ mL}}{1\text{ L}}$	5.5×10^8 mL
3.67 km = ? in.	$3.67\text{ km} \times \frac{1000\text{ m}}{1\text{ km}} \times \frac{39.4\text{ in.}}{1.00\text{ m}}$	1.45×10^5 in.
8.22 mL = ? qt	$8.22\text{ mL} \times \frac{1\text{ L}}{1000\text{ mL}} \times \frac{1.00\text{ qt}}{0.946\text{ L}}$	8.69×10^{-3} qt
57 fl oz = ? mL	$57\text{ fl oz} \times \frac{1.00\text{ mL}}{0.034\text{ fl oz}}$	1.7×10^3 mL
3.66 oz = ? kg	$3.66\text{ oz} \times \frac{28.3\text{ g}}{1.00\text{ oz}} \times \frac{1\text{ kg}}{1000\text{ g}}$	1.04×10^{-1} kg

2.8 **1.** $\text{density} = \frac{\text{mass}}{\text{volume}} = \frac{404\text{ g}}{38.5\text{ cm}^3} = 10.5\text{ g/cm}^3$

2. Use density as a conversion factor.

$$\text{mass} = \frac{0.789\text{ g}}{1\text{ mL}} \times 458\text{ mL} = 361\text{ g}$$

3. Use two conversion factors to solve this proplem.

$$\text{volume} = \frac{1.00\text{ cm}^3}{8.92\text{ g}} \times \frac{1000\text{ g}}{1\text{ kg}} \times 8.97\text{ kg} = 1.01 \times 10^3\text{ cm}^3$$

2.9

Boiling point	Substituted equation	Temperature
ethyl acetate (nail polish remover) 77°C	°F = 9/5(77.0) + 32	171°F
toluene (additive in gasoline) 111°C	°F = (9/5)111 + 32	232°F
isopropyl alcohol (rubbing alcohol) 180°F	°C = 5/9(180 – 32)	82°C
naphthalene (moth balls) 424°F	°C = 5/9(424 – 32)	218°C
propane (fuel for camping stoves) –42°C	K = (–42) + 273	231 K
methane (natural gas) –260°F	°C = 5/9(–260–32) = – 162°C K = (–162) + 273	111 K

2.10 **1**. heat absorbed(cal) = specific heat(cal/g•°C) x mass(g) x temperature change(°C)
heat absorbed = 1.00 cal/g•°C x 1000 g x 25.0°C = 2.50×10^4 cal

2. heat absorbed = 0.21 cal/g•°C x 150 g x 75°C = 2400 cal

Answers to Self-Test

The numbers in parentheses refer to sections in your textbook:
1. F; liter (2.2) **2**. T (2.2) **3**. T (2.5) **4**. F; added or subtracted (2.5) **5**. T (2.4)
6. F; its mass to its volume (2.7) **7**. T (2.8) **8**. F; milli- (2.2) **9**. T (2.3) **10**. T (2.8)
11. b (2.5) **12**. d (2.5) **13**. b (2.5) **14**. c (2.4 and 2.5) **15**. d (2.4) **16**. b (2.6)
17. a (2.7) **18**. e; 2.9×10^4 cal (2.8) **19**. c (2.3) **20**. d (2.8)

Atomic Structure and the Periodic Table — Chapter 3

Chapter Overview

All matter is made of basic building blocks called atoms. As you study the structure of atoms, you can begin to develop an understanding of how atoms bond together to form the many substances which make up our world. By studying the periodic law, you will begin to predict the properties of elements according to their positions in the periodic table.

By the end of this chapter you should be able to describe the three basic particles which make up atoms in terms of mass, charge and location, and calculate the number of each of the three types of particles in an atom using the atomic number and the mass number of that atom. You will learn to describe an isotope from its symbol, and you will calculate the average atomic mass of an element. Using the electron configuration of an element and the principle of the distinguishing electron, you will be able to classify the elements into groups with similar properties.

Practice Exercises

3.1 Early experiments indicated that atoms are made up of smaller **subatomic particles** (Sec. 3.1). Complete the following table summarizing some of these experiments.

Scientist	Experiments performed	What they showed
Goldstein		
Thomson		
Rutherford		

3.2 The **atomic number** (Sec. 3.2) of an **element** (Sec. 3.2) is the number of **protons** (Sec. 3.1) in the **nucleus** (Sec. 3.1) of that element. Since the number of protons in an atom is equal to the number of **electrons** (Sec. 3.1), the atomic number also gives us the number of electrons in a neutral atom. For example, the atomic number of oxygen is 8; therefore, the oxygen atom has 8 protons and 8 electrons.

The **mass number** (Sec. 3.2) is the total number of protons and **neutrons** (Sec. 3.1) in the nucleus. Since the atomic number is equal to the number of protons in the nucleus, and the mass number is equal to protons plus neutrons, we can find the number of neutrons by subtraction: Mass number – Atomic number = Number of neutrons

Use the relationships above and the **periodic table** (Sec. 3.3) to complete the following table:

Atomic number	Mass number	Number of protons	Number of neutrons	Number of electrons	Symbol of element
6			6		
	39	19			
			77		Xe
	64			29	
		35	45		

3.3 **Isotopes** (Sec. 3.2) are atoms that have the same number of protons but different numbers of neutrons, and therefore different mass numbers. Isotopes are usually represented as follows:

$^{14}_{6}C$ The superscript is the mass number, or A.
The subscript is the atomic number, or Z.

Complete the following table:

Isotope	A	Z	Protons	Neutrons
$^{40}_{20}Ca$				
$^{40}_{18}Ar$				
$^{23}_{11}Na$				
$^{37}_{17}Cl$				
$^{35}_{17}Cl$				

3.4 The **atomic mass** (Sec. 3.2) of an element is an average mass of the mixture of isotopes that reflects the relative abundance of the isotopes as they occur in nature. The atomic mass can be calculated by multiplying the relative mass of each isotope by its fractional abundance, and then totaling the products.

Example:
Magnesium is composed of 78.7% ^{24}Mg, 10.1% ^{25}Mg and 11.2% ^{26}Mg. To find the atomic mass for magnesium, multiply each isotope's mass by the percent abundance and add these products together:

78.7% x 23.99 amu = 18.88 amu
10.1% x 24.99 amu = 2.52 amu
11.2% x 25.98 amu = 2.91 amu
24.31 amu

An element has two common isotopes: 80.4% of the atoms have a mass of 11.01 amu and 19.6% of the atoms have a mass of 10.01 amu. In the space below, set up the equations and calculate the atomic mass for this element. Identify the element.

atomic mass =__________ element __________

3.5 According to the **periodic law** (Sec. 3.3), when elements are arranged in order of increasing atomic number, elements with similar properties occur at periodic intervals. The periodic table represents this statement graphically: elements with similar properties are found in the same **group** (Sec. 3.3) or vertical column. The horizontal rows are known as **periods** (Sec. 3.3). A steplike line in the periodic table separates the **metals** (Sec. 3.4) on the left from the **nonmetals** (Sec. 3.4) on the right.

Refer to your periodic table for information to complete the table below:

Element	Group	Period	Metal	Nonmetal
Be	IIA	2	X	
Na				
N				
Br				
O				
Sn				
K				

Which two elements in the table above would you expect to have similar chemical properties, and why?

____

3.6 The **electron configuration** (Sec. 3.6) of an atom is a statement of the number of electrons the atom has in each **electron subshell** (Sec. 3.5). A shorthand system is used to show electron configurations. For each subshell occupied by electrons, write the number and the symbol for the subshell with a superscript indicating the number of electrons in that subshell.

Example: The electron configuration for ${}_{6}C$ is $1s^22s^22p^2$. This shows that the atom has 2 electrons in the $1s$ subshell, 2 in the $2s$ subshell and 2 in the $2p$ subshell.

Write the electron configurations for the elements below. Use Figure 3-10 in your textbook to determine the order in which the **electron orbitals** (Sec. 3.5) are filled.

Element	Electron configuration
neon	
chlorine	
iron	

3.7 An **orbital diagram** (Sec. 3.6) is a statement of how many electrons an atom has in each of its orbitals. Each arrow in the diagram indicates an electron. Electron spin is denoted by the direction of the arrow.

Example: ${}_{6}C$ — $1s$ [↑↓] $2s$ [↑↓] $2p$ [↑] [↑] [ ]

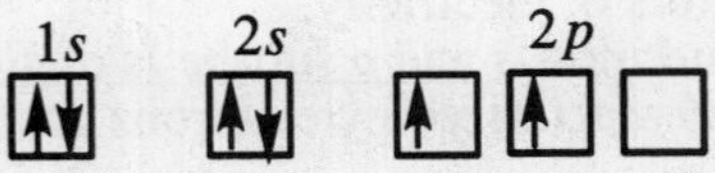

Draw orbital diagrams for the following elements:

	$1s$	$2s$	$2p$	$3s$	$3p$	$4s$
${}_{17}Cl$	[ ]	[ ]	[ ] [ ] [ ]	[ ]	[ ] [ ] [ ]	[ ]
${}_{20}Ca$	[ ]	[ ]	[ ] [ ] [ ]	[ ]	[ ] [ ] [ ]	[ ]

3.8 We can classify an element by determining the subshell of its **distinguishing electron** (Sec. 3.7), the last electron added in the electron configuration. If the distinguishing electron is added to an *s* or a *p* subshell the element is a **representative element** (Sec. 3.8), and if the *p* subshell is filled, the element is a **noble gas** (Sec. 3.8). If the distinguishing electron is added to a *d* subshell, the element is a **transition element** (Sec. 3.8); if it is added to an *f* subshell, the element is an **inner transition element** (Sec. 3.8).

Write the level of the distinguishing electron for each element below, and indicate the element's classification.

Element	Distinguishing electron	Noble gas	Representative element	Transition element	Inner transition element
Mg	$3s^2$		X		
Ti					
Ar					
Pu					
S					

Self-Test

True-false: Indicate whether the following statements are true or false. If the statement is false, give the word or phrase that may be substituted for the underlined portion to make the statement true.

1. The nucleus is the smallest particle of an atom.
2. Most of the mass of an atom is located in the nucleus.
3. The nucleus contains electrons and protons.
4. For a neutral atom, the number of protons equals the number of electrons.
5. The atomic number of an element is the number of protons and neutrons.
6. The mass number is the total number of protons and electrons in the atom.
7. Isotopes of a specific element have different numbers of neutrons in the nuclei of their atoms.
8. Electron orbitals have different shapes: *s*-orbitals are spherical.
9. The periodic law states that when elements are arranged in order of increasing atomic number, elements with similar properties occur at periodic intervals.
10. In the modern periodic table, the horizontal rows are called groups.
11. Metals are substances which have a high luster and are malleable.
12. Metals are on the left side of the periodic table.
13. Nonmetals are good conductors of electricity.
14. The noble gases have the outermost *s* and *p* subshells of electrons filled.
15. The transition elements are characterized by electrons occupying the *d* orbitals.

Multiple choice:

16. The nucleus of an atom contains these basic particles:
    a) electrons and protons
    b) neutrons and electrons
    c) protons and neutrons
    d) only neutrons
    e) none of these
17. Isotopes of a specific element vary in the following manner:
    a) electron numbers are different
    b) neutron numbers are different
    c) proton numbers are different
    d) neutron numbers will be the same
    e) none of these
18. The element, ${}^{48}_{22}Ti$, has the following electron configuration:
    a) $1s^22s^22p^63s^23p^63d^{10}4s^2$
    b) $1s^22s^22p^63s^23p^63d^4$
    c) $1s^22s^22p^63s^23p^64s^23d^2$
    d) $1s^22s^22p^63s^23p^64s^23d^4$
    e) none of these

19. In the isotope, $^{81}_{35}Br$, how many neutrons are in the nucleus?
a) 35 b) 46 c) 81 d) 116 e) none of these
20. How many nucleons are located in an atom of cesium, $^{133}_{55}Cs$?
a) 55 b) 78 c) 133 d) 188 e) none of these
21. An element has the electron configuration $1s^22s^22p^63s^23p^64s^23d^{10}4p^6$; the element is:
a) $_{10}Ne$ b) $_{54}Xe$ c) $_{18}Ar$ d) $_{36}Kr$ e) none of these
22. In the periodic table, the elements on the far left side are classified as:
a) metals
b) noble gases
c) transition metals
d) nonmetals
e) none of these
23. In the periodic table, the elements called noble gases are in:
a) Group IA
b) Group IIA
c) Group VIIA
d) Group VA
e) none of these
24. The distinguishing electron for $_{37}Rb$ would be found in what subshell?
a) 3*s* b) 2*p* c) 3*d* d) 5*s* e) none of these

Answers to Practice Exercises

3.1

Scientist	Experiments performed	What they showed
Goldstein	Discharge tubes with a grid-like cathode	Canal rays contained many different types of positive particles.
Thomson	Cathode ray experiments	Negatively charged particles (electrons) present in all matter. Model of atom: electrons embedded in a positively-charged mass.
Rutherford	Gold foil-alpha particle experiments	New atomic model: small, dense, positively-charged nucleus and lots of space containing only tiny, negatively-charged electrons.

3.2

Atomic number	Mass number	Number of protons	Number of neutrons	Number of electrons	Symbol of element
6	12	6	6	6	C
19	39	19	20	19	K
54	131	54	77	54	Xe
29	64	29	35	29	Cu
35	80	35	45	35	Br

3.3

Isotope	A	Z	Protons	Neutrons
$^{40}_{20}Ca$	40	20	20	20
$^{40}_{18}Ar$	40	18	18	22
$^{23}_{11}Na$	23	11	11	12
$^{37}_{17}Cl$	37	17	17	20
$^{35}_{17}Cl$	35	17	17	18

3.4 80.4% x 11.01 amu = 8.85 amu
19.6% x 10.01 amu = <u>1.96 amu</u>
10.81 amu

Atomic mass = <u>10.81 amu</u> element: <u>boron</u>

3.5

Element	Group	Period	Metal	Nonmetal
Be	IIA	2	X	
Na	IA	3	X	
N	VA	2		X
Br	VIIA	4		X
O	VIA	2		X
Sn	IVA	5	X	
K	IA	4	X	

The two elements with similar properties are Na and K, since both are in group IA.

3.6

Element	Electronic configuration
neon	$1s^22s^22p^6$
chlorine	$1s^22s^22p^63s^23p^5$
iron	$1s^22s^22p^63s^23p^64s^23d^6$

3.7

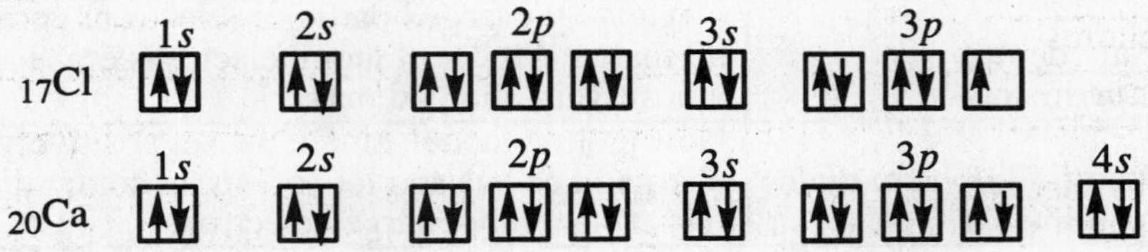

	1s	2s	2p	3s	3p	4s
$_{17}$Cl	↑↓	↑↓	↑↓ ↑↓ ↑↓	↑↓	↑↓ ↑↓ ↑	
$_{20}$Ca	↑↓	↑↓	↑↓ ↑↓ ↑↓	↑↓	↑↓ ↑↓ ↑↓	↑↓

3.8

Element	Distinguishing electron	Noble gas	Representative element	Transition element	Inner transition element
Mg	$3s^2$		X		
Ti	$3d^2$			X	
Ar	$3p^6$	X			
Pu	$5f^6$				X
S	$3p^4$		X		

Answers to Self-Test

The numbers in parentheses refer to sections in your textbook.
1. F; electron (3.1) **2**. T (3.1) **3**. F; protons and neutrons (3.1) **4**. T (3.3)
5. F; protons (3.3) **6**. F; protons and neutrons (3.3) **7**. T (3.3) **8**. T (3.6) **9**. T (3.4)
10. F; periods (3.4) **11**. T (3.5) **12**. T (3.5) **13**. F; poor (3.5) **14**. T (3.9)
15. T (3.9) **16**. c (3.1) **17**. b (3.3) **18**. c (3.7) **19**. b (3.3) **20**. c (3.3) **21**. d (3.7)
22. a (3.5) **23**. e; Group VIIIA (3.9) **24**. d (3.8)

Chemical Bonding: The Ionic Bond Model — Chapter 4

Chapter Overview

The electron configuration of the atoms of an element determines the chemical properties of that element. In this chapter you will see how electrons transfer from one atom to another to form an ionic bond.

You will identify the valence electrons of an atom using the electron configurations of the atom and the element's group number in the periodic table. You will draw the Lewis structures for atoms, and use these structures to show electron transfer in ionic bond formation. You will predict the formulas for ionic compounds, and learn to name these ionic compounds.

Practice Exercises

4.1 The **valence electrons** (Sec. 4.2) of an atom are the atoms in the outermost electron shell. Write the electron configuration and give the number of valence electrons for atoms of each of the following elements:

Element	Electron configuration	Number of valence electrons
lithium		
beryllium		
boron		
phosphorus		
sulfur		

4.2 **Lewis structures** (Sec. 4.2) are atomic symbols with one dot for each valence electron placed around the element's symbol. Give the number of valence electrons and draw the Lewis structure for each of the elements or groups below (Use the symbol X as a group symbol.)

Group	Valence electrons	Lewis structure
Group IA		
Group IVA		
Group VIIA		

Element	Valence electrons	Lewis structure
sulfur		
bromine		
magnesium		

4.3 According to the **octet rule** (Sec. 4.3), in compound formation, atoms of elements lose, gain or share electrons in such a way that their electron configurations become identical to that of the noble gas nearest them in the periodic table. Atoms on the left side of the periodic table tend to lose electrons and become positively charged **monoatomic ions** (Sec. 4.4 and 4.10) and those on the right side of the periodic table gain electrons to become negatively charged monoatomic ions. Complete the table below for ion formation.

Element	Group number	Electrons lost/gained	Ion formed	Noble gas structure
Na				
Br				
S				
Ca				
N				

4.4 Some metals have a variable ionic charge; they can form more than one type of ion. Complete the following tables for these metals with variable ionic charge:

Element	Electrons lost	Symbol for ion
tin	2	Sn^{2+}
tin	4	
cobalt	2	

Element	Electrons lost	Symbol for ion
cobalt	3	
iron	2	
iron	3	

4.5 **Ionic bonds** (Sec. 4.1) form when electrons are transferred from metal atoms to nonmetal atoms. **Binary ionic compounds** (Sec. 4.9) are named by naming the metal first, followed by the stem of the nonmetal with the ending *-ide*. Name these ionic compounds and show their formation:

Compound formula	Compound name	Formation of Lewis structure
KBr		
CaI_2		
SrS		

4.6 Formation of ionic compounds requires a charge balance: the same number of electrons must be lost as are gained. Practice balancing charges by writing chemical formulas for ionic compounds formed from the elements in the table below:

Elements	Chemical formula	Name of ionic compound
potassium and chlorine		
beryllium and iodine		
sodium and sulfur		
strontium and oxygen		
aluminum and fluorine	AlF_3	
		Cesium bromide
		Calcium oxide
		Aluminum sulfide

4.7 **Polyatomic ions** (Sec. 4.10) are **covalently bonded** (Sec. 4.1) groups of atoms having a charge. In writing the formula for an ionic compound, treat a polyatomic ion as a unit. If more than one of these ions is required for charge balance, enclose the ion in parentheses and put the number of ions outside the parentheses. Give the formulas for the ionic compounds prepared by combining the following ions:

	Bromide	Nitrate	Carbonate	Phosphate
Sodium	NaBr			
Calcium		$Ca(NO_3)_2$		
Ammonium				
Aluminum				
Lead(II)				
Tin(IV)				

4.8 In naming a compound containing a metal with a variable ionic charge, use a Roman numeral after the metal name to indicate the charge on the metal ion. Give the formulas and the names of the ionic compounds prepared by combining the following ions:

	F^-	N^{3-}	SO_4^{2-}	ClO_2^-
K^+		K_3N potassium nitride		
Pb^{2+}				
Fe^{3+}				
Sn^{4+}				

Self-Test

True-false: Indicate whether the following statements are true or false. If the statement is false, give the word or phrase that may be substituted for the underlined portion to make the statement true.

1. Lewis structures show the number of inner electrons of an atom.
2. Valence electrons determine the chemical properties of an element.
3. A negative ion is formed when an element loses an electron.
4. Metals tend to gain electrons to attain the configuration of a noble gas.
5. Bromine would accept an electron to attain the configuration of the noble gas krypton.
6. An ionic bond results from the sharing of one or more pairs of electrons between atoms.
7. The maximum number of valence electrons for any element is four.
8. The most stable electron configuration is that of the noble gases.
9. An ionic compound is formed from a metal that can donate electrons and a nonmetal that can accept electrons.
10. In binary ionic compounds, the full name of the metallic element is given first.
11. In binary ionic compounds, the fixed-charge metals are generally found in Groups VIIA and VIIIA.
12. A polyatomic ion is a group of atoms held together by ionic bonds that has acquired a charge.

Multiple choice:

13. The binary ionic compound, RbI, would be called;
    a) Rubidium(I) iodide
    b) Rubidium iodate
    c) Rubidium iodine
    d) Rubidium iodide
    e) none of these
14. The formula for the binary ionic compound silver sulfide is;
    a) SiS b) AgS c) Ag_2S d) AgS_2 e) none of these
15. In the electronic configuration for sulfur, $1s^22s^22p^63s^23p^4$, what electron shell number determines the valence electrons?
    a) 1 b) 2 c) 3 d) 2 and 3 e) none of these
16. Which representative element would have the same number of valence electrons as calcium?
    a) potassium
    b) barium
    c) scandium
    d) aluminum
    e) none of these
17. The electron configuration of a noble gas would be:
    a) $1s^22s^2$
    b) $1s^22s^22p^4$
    c) $1s^22s^22p^63s^23p^2$
    d) $1s^22s^22p^63s^23p^6$
    e) none of these
18. The electronic configuration for the ion, S^{2-}, would be:
    a) $1s^22s^22p^6$
    b) $1s^22s^22p^63s^23p^4$
    c) $1s^22s^22p^63s^23p^6$
    d) $1s^22s^22p^63s^23p^64s^2$
    e) none of these
19. In the formula, Na_3N, the total number of electrons accepted by the nitrogen is:
    a) 1 b) 2 c) 3 d) 4 e) none of these
20. In the ionic compound, calcium phosphate, how many polyatomic ions (phosphate ions) are bonded with calcium?
    a) 1 b) 2 c) 3 d) 4 e) none of these
21. The Lewis structure for a Group VA element would have dots representing the following number of electrons:
    a) two b) three c) four d) five e) none of these
22. Which of the following pairs of elements would form a binary ionic compound?
    a) sulfur and oxygen
    b) bromine and chlorine
    c) magnesium and bromine
    d) oxygen and hydrogen
    e) none of these
23. At room temperature, what type of compound would have ionic bonds?
    a) gas
    b) liquid
    c) solid
    d) gas and liquid
    e) none of these

Answers to Practice Exercises

4.1

Element	Electron configuration	Number of valence electrons
lithium	$1s^22s^1$	1
beryllium	$1s^22s^2$	2
boron	$1s^22s^22p^1$	3
phosphorus	$1s^22s^22p^63s^23p^3$	5
sulfur	$1s^22s^22p^63s^23p^4$	6

4.2

Group	Valence electrons	Lewis structure
Group IA	1	X·
Group IVA	4	·Ẋ· (4 dots)
Group VIIA	7	:Ẋ: (7 dots)

Element	Valence electrons	Lewis structure
sulfur	6	:Ṡ: (6 dots)
bromine	7	·B̈r: (7 dots)
magnesium	2	·Mg·

4.3

Element	Group number	Electrons lost/gained	Ion formed	Noble gas structure
Na	IA	1 lost	Na^{+}	Ne
Br	VIIA	1 gained	Br^{-}	Kr
S	VIA	2 gained	S^{2-}	Ar
Ca	IIA	2 lost	Ca^{2+}	Ar
N	VA	3 gained	N^{3-}	Ne

4.4

Element	Electrons lost	Symbol for ion
tin	2	Sn^{2+}
tin	4	Sn^{4+}
cobalt	2	Co^{2+}

Element	Electrons lost	Symbol for ion
cobalt	3	Co^{3+}
iron	2	Fe^{2+}
iron	3	Fe^{3+}

4.5

Compound formula	Compound name	Formation of Lewis structure
KBr	potassium bromide	K· + ·B̈r: → $[K]^{+}$ $[:\ddot{Br}:]^{-}$ → KBr
CaI_2	calcium iodide	:Ï· + ·Ca· + ·Ï: → $[Ca]^{2+}$ $[:\ddot{I}:]^{-}$ $[:\ddot{I}:]^{-}$ → CaI_2
SrS	strontium sulfide	·Sr· + :Ṡ: → $[Sr]^{2+}$ $[:\ddot{S}:]^{2-}$ → SrS

4.6

Elements	Chemical formula	Name of ionic compound
potassium and chlorine	KCl	potassium chloride
beryllium and iodine	BeI_2	beryllium iodide
sodium and sulfur	Na_2S	sodium sulfide
strontium and oxygen	SrO	strontium oxide
aluminum and fluorine	AlF_3	aluminum fluoride
cesium and bromine	$CsBr$	cesium bromide
calcium and oxygen	CaO	calcium oxide
aluminum and sulfur	Al_2S_3	aluminum sulfide

4.7

	Bromide	Nitrate	Carbonate	Phosphate
Sodium	$NaBr$	$NaNO_3$	Na_2CO_3	Na_3PO_4
Calcium	$CaBr_2$	$Ca(NO_3)_2$	$CaCO_3$	$Ca_3(PO_4)_2$
Ammonium	NH_4Br	NH_4NO_3	$(NH_4)_2CO_3$	$(NH_4)_3PO_4$
Aluminum	$AlBr_3$	$Al(NO_3)_3$	$Al_2(CO_3)_3$	$AlPO_4$
Lead(II)	$PbBr_2$	$Pb(NO_3)_2$	$PbCO_3$	$Pb_3(PO_4)_2$
Tin(IV)	$SnBr_4$	$Sn(NO_3)_4$	$Sn(CO_3)_2$	$Sn_3(PO_4)_4$

4.8

	F^-	N^{3-}	SO_4^{2-}	ClO_2^-
K^+	KF potassium fluoride	K_3N potassium nitride	K_2SO_4 potassium sulfate	$KClO_2$ potassium chlorite
Pb^{2+}	PbF_2 lead(II) fluoride	Pb_3N_2 lead(II) nitride	$PbSO_4$ lead(II) sulfate	$Pb(ClO_2)_2$ lead(II) chlorite
Fe^{3+}	FeF_3 iron(III) fluoride	FeN iron(III) nitride	$Fe_2(SO_4)_3$ iron(III) sulfate	$Fe(ClO_2)_3$ iron(III) chlorite
Sn^{4+}	SnF_4 tin(IV) fluoride	Sn_3N_4 tin(IV) nitride	$Sn(SO_4)_2$ tin(IV) sulfate	$Sn(ClO_2)_4$ tin(IV) chlorite

Answers to Self-Test

The numbers in parentheses refer to sections in your textbook.
1. F; outermost or valence electrons (4.2) **2**. T (4.2) **3**. F; gains (4.4) **4**. F; lose (4.5) **5**. T (4.5) **6**. F; covalent (4.1) **7**. F; eight (4.5) **8**. T (4.3) **9**. T (4.6) **10**. T (4.9) **11**. F; Groups IA and IIA (4.9) **12**. F; covalent bonds (4.10) **13**. d (4.9) **14**. c (4.9) **15**. c (4.2) **16**. b (4.2) **17**. d (4.3) **18**. c (4.5) **19**. c (4.7) **20**. b (4.11) **21**. d (4.2) **22**. c (4.5, 4.9) **23**. c (4.1)

Chemical Bonding: The Covalent Bond Model   Chapter 5

Chapter Overview

Covalent bonds are the result of the sharing of electrons between atoms. Covalent bonds join atoms together to form molecules.

In this chapter you will use Lewis structures to indicate the various types of covalent bonds in molecules. You will study the concept of electronegativity differences between atoms and how this determines whether a bond is ionic or covalent, polar or nonpolar. You will use VSEPR theory to predict the three-dimensional shape of molecules and determine molecular polarity.

Practice Exercises

5.1 Remember that the valence electrons of an atom, those in the atom's outermost shell, are the electrons involved in bond formation. Draw Lewis structures showing valence electrons for the following elements:

·Mg·				
Mg	C	P	Br	Ar

5.2 **Bonding electrons** (Sec. 5.2) are pairs of valence electrons that are shared between atoms in a covalent bond. **Nonbonding electrons** (Sec. 5.2) are pairs of electrons that are not involved in sharing. Molecules tend to be stable when each atom in the molecule shares in an octet of electrons. Draw the Lewis structures for one molecule of each of these covalent compounds. Circle the bonding electrons in each **single covalent bond** (Sec. 5.3) in the molecule:

H(:)Br: (with nonbonding pairs on Br)				
HBr	F_2	BrI	H_2O	CH_4

5.3 In a **double covalent bond** (Sec. 5.3), two atoms share two pairs of electrons; in a **triple covalent bond** (Sec. 5.3), two atoms share three pairs of electrons. Draw Lewis structures showing the single, double, and triple covalent bonds in these molecules, as well as the nonbonding electrons. For help in determining which atom should be the central atom of a Lewis structure, see Section 5.6 of your textbook.

H:C::C:H with H on each C (H₂C::CH₂)				
C_2H_4	CS_2	HCN	H_2CO	C_2H_2

5.4 Since a covalent bond consists of a pair of shared electrons, we would expect an atom of nitrogen, having three unpaired electrons, to share in either three single bonds, one single and one double bond, or one triple bond. According to this concept, how many bonds would you expect each of the following atoms to form?

oxygen ______________________________

carbon ______________________________

5.5 In order that all atoms in a molecule share in an octet of electrons, it is possible for one atom to supply both electrons of the shared pair of a bond. This is called a **coordinate covalent bond** (Sec. 5.5). The electron pair for this bond comes from one of the nonbonding electron pairs of one of the atoms. Draw Lewis structures for the following molecules and circle the coordinate covalent bonds:

H : C̤̈l (:) Ö̤ : HClO	 $HBrO_2$

5.6 In Lewis structures for molecules, the shared electron pairs may be represented with dashes. Rewrite the structures in exercise 5.3 by replacing the bonding electron pairs with a dash to show the covalent bond between atoms. Include the nonbonding electron pairs as dots:

H₂C=CH₂ (H and H on each C) C_2H_4	 CS_2	 HCN	 H_2CO	 C_2H_2

5.7 Polyatomic ions consist of covalently bonded atoms acting as a unit with a charge on it. The Lewis structure for a polyatomic ion is drawn in the same way as for a molecule, except that the total number of electrons is increased or decreased according to the charge on the ion. Draw Lewis structures for the following polyatomic ions:

[:Ö : N̈ : Ö:]⁻ NO_2^-	 NO_3^-	 ClO_4^-

5.8 According to **VSEPR theory** (Sec. 5.8), constantly moving electron pairs form electron clouds that electrically repel each other. The three-dimensional shape of a molecule is the shape that gives the greatest distance between electron clouds. Both bonding and nonbonding electrons are counted in this determination, and double and triple bonds each count as one electron cloud.

The shape that will give the greatest distance between electron clouds is: for two electron clouds, 180° bond angle (linear); for three electron clouds, 120° bond angle (trigonal planar or angular); for four electron clouds, 109° bond angle (tetrahedral, trigonal pyramidal, or angular). Use these guidelines to predict the shape of these molecules:

Molecular formula	Lewis structure	Name the 3-D shape of the molecule	Give the bond angle
CS_2			
H_2CO			
H_2S			

5.9 **Electronegativity** (Sec. 5.9) is a measure of the relative attraction that an atom has for the shared electrons in a bond. The electronegativity difference between the two bonded atoms is a measure of the polarity of the bond. If the difference is 2.0 or greater, the bond is ionic.

Calculate the electronegativity difference for each element pair, and indicate whether the bond formed between them will be ionic, **nonpolar covalent** (Sec. 5.10), or **polar covalent** (Sec. 5.10). (Electronegativities are in Figure 5.5 of your textbook.)

Pair of elements	Electronegativity difference	Ionic bond	Nonpolar covalent bond	Polar covalent bond
sodium and fluorine				
bromine and bromine				
sulfur and oxygen				
phosphorus and bromine				

5.10 In naming binary molecular compounds, name the element of lower electronegativity first, followed by the stem of the more electronegative nonmetal and the suffix *-ide*. Include prefixes to indicate the number of atoms of each nonmetal. Name the following molecular compounds:

a. CCl_4 ____________________

b. CS_2 ____________________

c. NCl_3 ____________________

d. N_4S_4 ____________________

Self-Test

True-false: Indicate whether the following statements are true or false. If the statement is false, give the word or phrase that may be substituted for the underlined portion to make the statement true.

1. Covalent bonds form between atoms of dissimilar elements.
2. Covalent bond formation between nonmetal atoms involves electron transfer.
3. Nonbonding electrons are pairs of valence electrons that are not shared between atoms having a covalent bond.
4. A nitrogen molecule, N_2, would have a double covalent bond between the two nitrogen atoms.
5. Carbon can form multiple covalent bonds with other nonmetallic elements.
6. A coordinate covalent bond is a covalent bond formed when both electrons are donated by one atom.
7. According to VSEPR, the electron clouds in the valence shell arrange themselves to maximize the repulsion between the electron pairs.
8. According to VSEPR, a water molecule, H_2O, would have a linear arrangement of the valence electron clouds.
9. According to VSEPR theory convention, single and triple bonds are equal.
10. Electronegativity is a measure of the relative repulsion that an atom has for the shared electrons in a bond.
11. Electronegativity values increase from left to right across periods in the Periodic Table.
12. The bond between fluorine and bromine would be a nonpolar covalent bond.
13. Nonbonding electron clouds are important in determining the shape of a molecule.

Multiple choice:

14. Which of the following pairs of elements would form a covalent bond?
    a) sulfur and oxygen
    b) potassium and iodine
    c) magnesium and bromine
    d) calcium and fluorine
    e) none of these

15. Which of the following pairs of elements would form a nonpolar covalent bond?
    a) nitrogen and oxygen
    b) fluorine and fluorine
    c) calcium and iodine
    d) potassium and bromine
    e) none of these
16. Which of the following pairs of elements would form a polar covalent bond?
    a) carbon and carbon
    b) bromine and bromine
    c) potassium and fluorine
    d) sodium and oxygen
    e) none of these
17. How many valence electrons are found in a triple covalent bond?
    a) 2 b) 3 c) 4 d) 6 e) none of these
18. An element that forms a triple covalent bond could be found in:
    a) Group VA
    b) Group IIA
    c) Group VIIA
    d) Group VIIIA
    e) none of these
19. What types of electron clouds are used in VSEPR calculations?
    a) core electrons
    b) bonding electron clouds
    c) nonbonding electron clouds
    d) both b. and c.
    e) none of these
20. How many VSEPR electron clouds are found in ammonia, NH_3?
    a) one b) two c) three d) four e) none of these
21. Which element would be more electronegative than chlorine?
    a) sulfur b) lithium c) bromine d) carbon e) none of these
22. The correct name for the binary molecular compound, SO_3 is:
    a) sulfur oxide
    b) sulfur trioxygen
    c) sulfur trioxide
    d) trioxygen sulfide
    e) none of these

Answers to Practice Exercises

5.1

·Mg·	·C·	·P·	·Br:	:Ar:

5.2

H⊙Br:	:F⊙F:	:Br⊙I:	H⊙O: (H below O)	H⊙C⊙H (H above and below C)

5.3

H₂C::CH₂	S::C::S	H:C:::N:	H:C::O (H below C)	H:C:::C:H

5.4 oxygen 2; two single or one double
carbon 4; four single, or two double, or two single, one double, or one single, one triple.

5.5

H:Cl⊙O:	:O: above Br; H:Br⊙O:
HClO	$HBrO_2$

5.6

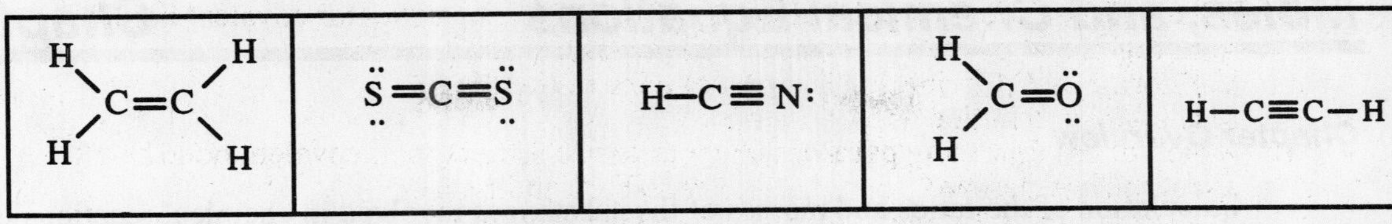

$H_2C=CH_2$ (H, H—C=C—H, H)	S=C=S	H—C≡N:	$H_2C=O$ (H, H—C=O)	H—C≡C—H

5.7

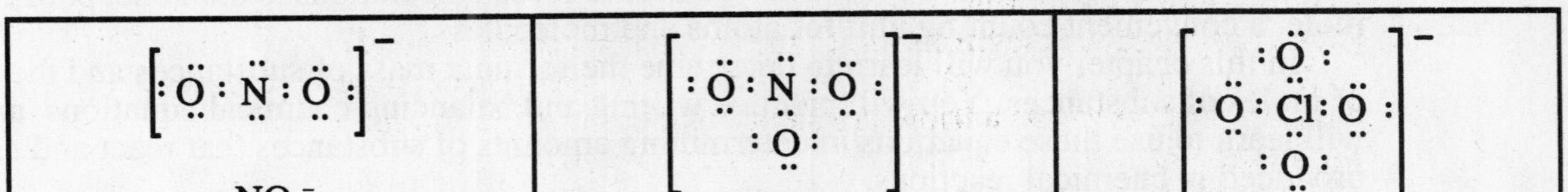

[:O:N:O:]⁻ NO_2^-	[:O:N:O: / :O:]⁻ NO_3^-	[:O: / :O:Cl:O: / :O:]⁻ ClO_4^-

5.8

Molecular formula	Lewis structure	Name the 3-D shape of the molecule	Give the bond angle
CS_2	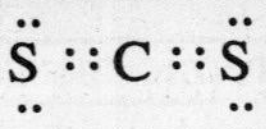 S::C::S	linear	180°
H_2CO	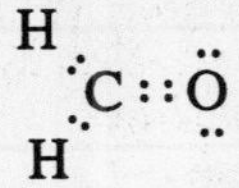 H, H :C::O	trigonal planar	120°
H_2S	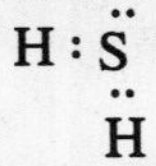 H:S: / H	angular	109°

5.9

Pair of elements	Electronegativity difference	Ionic bond	Nonpolar covalent bond	Polar covalent bond
sodium and fluorine	3.1	X		
bromine and bromine	0.0		X	
sulfur and oxygen	1.0			X
phosphorus and bromine	0.7			X

5.10 a. CCl_4 carbon tetrachloride

b. CS_2 carbon disulfide

c. NCl_3 nitrogen trichloride

d. N_4S_4 tetranitrogen tetrasulfide

Answers to Self-Test

The numbers in parentheses refer to sections in your textbook.
1. F; similar or identical (5.1) **2**. F; sharing (5.1) **3**. T (5.2) **4**. F; triple (5.3) **5**. T (5.4) **6**. T (5.5) **7**. F; minimize (5.8) **8**. F; an angular (5.8) **9**. T (5.8) **10**. F; attraction (5.9) **11**. T (5.9) **12**. F; a polar covalent (5.10) **13**. T (5.8) **14**. a (5.10) **15**. b (5.10) **16**. e (5.10) **17**. d (5.3) **18**. a (5.4) **19**. d (5.8) **20**. d (5.8) **21**. e (5.9) **22**. c (5.12)

Chemical Calculations: Formula Masses, Moles, and Chemical Equations — Chapter 6

Chapter Overview

Calculation of the ratios and masses of the substances involved in chemical reactions is very important in many chemical processes. Central to these calculations is the concept of the mole, a convenient counting unit for atoms and molecules.

In this chapter you will learn to determine the formula mass of substances and the number of moles of substances. You will practice writing and balancing chemical equations, and you will learn to use these equations in determining amounts of substances that react and are produced in chemical reactions.

Practice Exercises

6.1 The **formula mass** (Sec. 6.1) of a compound is the sum of the atomic masses of the atoms in its formula. Calculate the formula mass for these compounds:

Formula	Calculations with atomic masses	Formula mass
Na_2CO_3		
$(NH_4)_3PO_4$		

6.2 The **mole** (Sec. 6.2 and 6.3) is a useful unit for counting numbers of atoms and molecules. The number of particles in a mole is 6.02×10^{23}, known as **Avogadro's number** (Sec. 6.2). Set up and complete the calculations below:

Moles	Relationship with units	Number of atoms
2.60 moles of sodium atoms		
0.316 mole of argon atoms		

6.3 The mass of one mole, called **molar mass** (Sec. 6.3), of any substance is its formula mass expressed in grams. To express grams of a substance as moles of that substance, multiply grams by the molar mass (equal to grams/mole) of that substance. Calculate the mass in grams of the following:

Moles	Relationship with units	Mass
0.455 mole of Na_2CO_3		
3.42 moles of H_2SO_4		

6.4 The subscripts in a formula show the number of atoms of each element per formula unit. They also show the number of moles of atoms of each element in one mole of the substance (atoms/molecule = moles of atoms/moles of molecules). Determine the moles of carbon atoms in the following problems:

Moles of compound	Carbon atoms/molecule	Moles of carbon atoms
1.00 mole of $C_{10}H_{16}$ (limonene)		
13.5 moles of C_2H_6O (ethanol)		
0.705 mole of C_5H_{12} (pentane)		

6.5 Remember that the number of grams of one substance cannot be compared directly to the number of grams of another substance; however, moles can be related to moles quite easily by looking at subscripts in chemical formulas. Figure 6.3 in your textbook gives you a map of the steps to take in solving problems involving grams and moles.

a. How many molecules of KCl would be found in 0.125 g of KCl?

b. Calculate the number of fluorine atoms in 1.77 g of AlF_3.

c. Calculate the number of grams of Cl in 12.5 g of KCl.

d. Calculate the number of grams of F in 1.77 g of AlF_3.

6.6 The description of a chemical reaction can be expressed efficiently by replacing the words of the description with the formulas and symbols of a **chemical equation** (Sec. 6.6). The substances that react (the reactants) are placed on the left, and those that are produced (products) are on the right. The arrow in the chemical equation is read as "to produce"; plus signs on the left side mean "reacts with," and plus signs on the right are read as "and."

Write chemical equations for the following chemical reactions:

a. Hydrogen chloride reacts with sodium hydroxide to produce sodium chloride and water.

b. Silver nitrate and potassium bromide react with one another to produce silver bromide and potassium nitrate.

6.7 To be most useful, chemical equations must be **balanced** (Sec. 6.6); that is, the number of atoms of each element must be the same on both sides of the equation. A suggested method for balancing equations is in section 6.6 of your textbook. Remember: use the **coefficients** (Sec. 6.6) to balance equations, but do not change the subscripts within the formulas.

Balance the following equations:

a. $Mg + H_2O \rightarrow Mg(OH)_2 + H_2$

b. $HCl + Ba(OH)_2 \rightarrow BaCl_2 + H_2O$

c. $H_3PO_4 + NaOH \rightarrow H_2O + Na_3PO_4$

6.8 Balanced chemical equations are useful in telling us what amounts of products we can expect from a given amount of reactant, or how much reactant to use for a specific amount of product. The chemical equation tells us the ratio of the numbers of atoms and molecules involved in the reaction, but it also tells us the ratio of the numbers of moles of the substances involved.

The mole diagram in Figure 6.5 of your textbook will help you to determine the sequence of steps to use in solving the following types of problems.

Balanced equation: $4Na + O_2 \rightarrow 2Na_2O$

a. How many moles of Na_2O could be produced from 0.300 mole of Na?

b. How many moles of Na_2O could be produced from 2.18 g of sodium?

c. How many grams of Na_2O could be produced from 5.15 g of Na?

6.9 Balance the equation: $Al + Cl_2 \rightarrow AlCl_3$

a. How many moles of chlorine gas will react with 0.160 mole of aluminum?

b. How many grams of aluminum chloride could be produced from 5.27 moles of aluminum?

c. What is the maximum number of grams of aluminum chloride that could be produced from 14.0 g of chlorine gas?

d. How many grams of chlorine gas would be needed to react with 0.746 g of aluminum?

Self-Test

True-false: Indicate whether the following statements are true or false. If the statement is false, give the word or phrase that may be substituted for the underlined portion to make the statement true.

1. The mass of one mole of helium atoms would be <u>the same as</u> the mass of one mole of gold atoms.
2. In a balanced equation, the total number of atoms on the reactants side is <u>equal to</u> the total number of atoms on the products side.
3. Formula masses are calculated on the <u>$^{16}_{8}O$ relative-mass</u> scale.
4. In balancing a chemical equation, do not change the <u>coefficients</u> within the formulas.
5. Atomic mass and formula mass are both expressed in <u>amu</u>.
6. The term molar mass can be used only when referring to <u>ionic compounds</u>.
7. The number of <u>atoms</u> in a mole of H_2O is equal to 6.02×10^{23}.
8. One mole of glucose ($C_6H_{12}O_6$) contains <u>6 moles</u> of carbon atoms.

9. In a chemical equation, the products are the materials that are consumed.
10. A mole of nitrogen gas contains 6.02×10^{23} nitrogen atoms.

Multiple choice:

11. How many atoms are contained in 6.8 moles of calcium?
   a) 4.1×10^{24}
   b) 1.6×10^{26}
   c) 6.2×10^{22}
   d) 3.3×10^{23}
   e) none of these
12. What is the formula mass for iron(III) carbonate?
   a) 115.86 g/mole
   b) 287.57 g/mole
   c) 291.73 g/mole
   d) 171.71 g/mole
   e) none of these
13. How many grams are contained in 4.72 moles of $NaHCO_3$?
   a) 84.1
   b) 283
   c) 264
   d) 396
   e) none of these
14. A mole of butane contains four moles of carbon atoms and ten moles of hydrogen atoms. Its formula is:
   a) C_6H_6
   b) H_6C_{10}
   c) C_4H_{10}
   d) H_4C_6
   e) none of these
15. What is the total number of moles of all atoms in 6.55 moles of $(NH_4)_2CO_3$?
   a) 52.4
   b) 91.7
   c) 111
   d) 157
   e) none of these
16. The conversion factor used in changing grams of O_2 to moles of O_2 is:
   a) 16.00 g/1 mole
   b) 1 mole/16.00 g
   c) 32.00 mole/1 g
   d) 1 mole/32.00 g
   e) none of these
17. When oxygen gas and hydrogen gas combine to form water, which of the following is true? (Hint: Write the balanced equation.)
   a) 2 moles of O_2 produce 1 mole of H_2O
   b) 2 moles of H_2 react with 1 mole of H_2O
   c) 1 mole of O_2 produces 1 mole of H_2O
   d) 2 moles of H_2 produce 2 moles of H_2O
   e) none of these
18. Using the equation you wrote in question 17, find the mass of water in grams that would be produced by the complete reaction of 4.00 g of oxygen gas.
   a) 4.50 g
   b) 36.0 g
   c) 7.32 g
   d) 14.7 g
   e) none of these
19. The mass of 0.560 mole of methanol, CH_4O, would equal:
   a) 32.0 amu
   b) 32.0 g
   c) 17.9 amu
   d) 17.9 g
   e) none of these
20. 42.0 g of ethanol, C_2H_6O, would equal:
   a) 1.10 moles
   b) 0.911 mole
   c) 1.10 amu
   d) 0.911 amu
   e) none of these
21. After balancing the equation below, calculate the sum of all the coefficients in the balanced equation:

   $C_2H_6 + O_2 \rightarrow CO_2 + H_2O$

   a) 4 b) 9 c) 15 d) 19 e) none of these

Use the following balanced equation to answer questions 22 through 24 below.

$CH_4 + 2O_2 \rightarrow CO_2 + 2H_2O$

22. How many moles of water would be produced from 0.420 mole of methane (CH_4)?
    a) 2.00 moles
    b) 0.210 mole
    c) 0.420 mole
    d) 0.840 mole
    e) none of these
23. How many grams of carbon dioxide could be produced by the reaction of 6.90 g of oxygen gas?
    a) 4.74 g
    b) 6.92 g
    c) 9.55 g
    d) 19.0 g
    e) none of these
24. How many grams of methane would be used to produce 10.0 g of water?
    a) 2.22 g
    b) 4.44 g
    c) 8.88 g
    d) 10.0 g
    e) none of these

Answers to Practice Exercises

6.1

Formula	Calculations with atomic masses	Formula mass
Na_2CO_3	2(22.99) + 12.01 + 3(16.00)	105.99 amu
$(NH_4)_3PO_4$	3[14.01 + 4(1.01)] + 30.97 + 4(16.00)	149.12 amu

6.2

Moles	Relationship with units	Number of atoms
2.60 moles of sodium atoms	$2.60 \text{ moles} \times \dfrac{6.02 \times 10^{23} \text{ atoms}}{1 \text{ mole}} =$	1.57×10^{24}
0.316 mole of argon atoms	$0.316 \text{ mole} \times \dfrac{6.02 \times 10^{23} \text{ atoms}}{1 \text{ mole}} =$	1.90×10^{23}

6.3

Moles	Relationship with units	Mass
0.455 mole of Na_2CO_3	0.455 mole x 105.99 g/mole	48.2 g
3.42 moles of H_2SO_4	3.42 moles x 98.08 g/mole	335 g

6.4

Moles of compound	Carbon atoms/molecule	Moles of carbon atoms
1.00 mole of $C_{10}H_{16}$ (limonene)	10	10(1.00) = 10.0 moles
13.5 moles of C_2H_6O (ethanol)	2	2(13.5) = 27.0 moles
0.705 mole of C_5H_{12} (pentane)	5	5(0.705) = 3.53 moles

6.5 a. $0.125 \text{ g} \times \dfrac{1 \text{ mole}}{74.6 \text{ g}} \times \dfrac{6.02 \times 10^{23} \text{ molecules}}{1 \text{ mole}} = 1.01 \times 10^{21} \text{ molecules}$

b. $1.77 \text{ g } AlF_3 \times \frac{1 \text{ mole } AlF_3}{83.98 \text{ g } AlF_3} \times \frac{6.02 \times 10^{23} \text{ molecules } AlF_3}{1 \text{ mole } AlF_3} \times \frac{3 \text{ atoms F}}{1 \text{ molecule } AlF_3} = 3.81 \times 10^{22} \text{ atoms F}$

c. $12.5 \text{ g KCl} \times \frac{1 \text{ mole KCl}}{74.55 \text{ g KCl}} \times \frac{1 \text{ mole Cl}}{1 \text{ mole KCl}} \times \frac{35.45 \text{ g Cl}}{1 \text{ mole Cl}} = 5.94 \text{ g Cl}$

d. $1.77 \text{ g } AlF_3 \times \frac{1 \text{ mole } AlF_3}{83.98 \text{ g } AlF_3} \times \frac{3 \text{ moles F}}{1 \text{ mole } AlF_3} \times \frac{19.00 \text{ g F}}{1 \text{ mole F}} = 1.20 \text{ g F}$

6.6 a. $HCl + NaOH \rightarrow NaCl + H_2O$

b. $AgNO_3 + KBr \rightarrow AgBr + KNO_3$

6.7 a. $Mg + 2H_2O \rightarrow Mg(OH)_2 + H_2$

b. $2HCl + Ba(OH)_2 \rightarrow BaCl_2 + 2H_2O$

c. $H_3PO_4 + 3NaOH \rightarrow 3H_2O + Na_3PO_4$

6.8 Balanced equation: $4Na + O_2 \rightarrow 2Na_2O$

a. $0.300 \text{ mole Na} \times \frac{2 \text{ moles } Na_2O}{4 \text{ moles Na}} = 0.150 \text{ mole } Na_2O$

b. $2.18 \text{ g Na} \times \frac{1 \text{ mole Na}}{22.99 \text{ g Na}} \times \frac{2 \text{ moles } Na_2O}{4 \text{ moles Na}} = 0.0474 \text{ mole } Na_2O$

c. $5.15 \text{ g Na} \times \frac{1 \text{ mole Na}}{22.99 \text{ g Na}} \times \frac{2 \text{ moles } Na_2O}{4 \text{ moles Na}} \times \frac{61.98 \text{ g } Na_2O}{1 \text{ mole } Na_2O} = 6.94 \text{ g } Na_2O$

6.9 $2Al + 3Cl_2 \rightarrow 2AlCl_3$ (molar mass of $AlCl_3$: 133.33 g/mole)

a. $0.160 \text{ mole Al} \times \frac{3 \text{ moles } Cl_2}{2 \text{ moles Al}} = 0.240 \text{ mole } Cl_2$

b. $5.27 \text{ moles Al} \times \frac{2 \text{ moles } AlCl_3}{2 \text{ moles Al}} \times \frac{133.33 \text{ g } AlCl_3}{1 \text{ mole } AlCl_3} = 703 \text{ g } AlCl_3$

c. $14.0 \text{ g } Cl_2 \times \frac{1 \text{ mole } Cl_2}{70.90 \text{ g}} \times \frac{2 \text{ moles } AlCl_3}{3 \text{ moles } Cl_2} \times \frac{133.33 \text{ g } AlCl_3}{1 \text{ mole } AlCl_3} = 17.6 \text{ g } AlCl_3$

d. $0.746 \text{ g Al} \times \frac{1 \text{ mole Al}}{26.98 \text{ g Al}} \times \frac{3 \text{ moles } Cl_2}{2 \text{ moles Al}} \times \frac{70.90 \text{ g } Cl_2}{1 \text{ mole } Cl_2} = 2.94 \text{ g } Cl_2$

Answer to Self-Test

1. F; different than (6.3) **2**. T (6.6) **3**. F; $^{12}_{6}C$ relative mass (6.1) **4**. F; subscripts (6.6) **5**. T (6.1) **6**. F; molecular compounds (6.3) **7**. F; molecules (6.4) **8**. T (6.4) **9**. F; produced (6.6) **10**. F; molecules (6.4) **11**. a (6.4) **12**. c (6.1) **13**. d (6.4) **14**. c (6.4) **15**. b (6.4) **16**. d (6.5) **17**. d (6.6) **18**. a (6.8) **19**. d (6.3) **20**. b (6.3) **21**. d (6.6) **22**. d (6.8) **23**. a (6.8) **24**. b (6.8)

Gases, Liquids, and Solids — Chapter 7

Chapter Overview

The physical states of matter and the behavior of matter in these states are determined by the behavior of the particles (atoms, molecules, ions) of which matter is made. The movements and interactions of these particles are described by the kinetic molecular theory of matter.

In this chapter you will study the five statements of the **kinetic molecular theory** (Sec. 7.1) and the ways that these statements explain the physical behavior of matter. You will use the gas laws to describe quantitatively various changes in the conditions of pressure, temperature, and volume of matter in the gaseous state. You will study three types of **intermolecular forces** (Sec. 7.13) that affect liquids and solids and their changes of state.

Practice Exercises

7.1 **Solids, liquids, and gases** (Sec. 7.2) have different physical properties. Describe the states of matter in terms of each of the properties listed below (one or two words for each property).

State	Shape	Volume	Density	Compressibility	Thermal Expansion
gas					
liquid					
solid					

7.2 Gases can be described by simple quantitative relationships called **gas laws** (7.3). According to **Boyle's law** (7.4), the volume of a gas is inversely proportional to the **pressure** (Sec. 7.3) applied to it if the temperature is held constant: $P_1 \times V_1 = P_2 \times V_2$

Complete the following table using Boyle's law. Rearrange the equation to solve for the needed variable.

Problem	Relationship with units	Answer with units
a. The pressure on 2.45 L of helium is changed from 2340 mm Hg to 3580 mm Hg at 50.5°C. What is the new volume?		
b. The pressure on 12.5 L of nitrogen gas is doubled from 1.00 atm to 2.00 atm. What is the new volume of the nitrogen gas?		
c. The volume of 8.24 L of gas at 3630 mm Hg is increased to 16.4 L. If the temperature is held constant, what is the new pressure?		

7.3 According to **Charles's law** (Sec. 7.5), the volume of a gas at constant pressure is proportional to the Kelvin temperature of the gas:

$$V_1/T_1 = V_2/T_2$$

Complete the following table using Charles's law. Rearrange the equation to solve for the needed variable.

Problem	Relationship with units	Answer with units
a. The temperature of 4.71 L of gas is reduced from 278°C to 122°C. What is the new volume of the gas?		
b. The volume of 14.5 L of a gas at 345 K was increased to 20.5 L. What is the new temperature of the gas?		

7.4 The gas laws can be combined into a single equation called the **combined gas law** (Sec. 7.6):

$$\frac{P_1 \times V_1}{T_1} = \frac{P_2 \times V_2}{T_2}$$

Rearrange this equation to solve for the correct variable in completing the following combined gas law problems:

a. The volume of a gas is 5.72 L at 30°C and 1.25 atm. If the gas is heated to 50°C and compressed to a volume of 4.50 L, what will be the new pressure?

b. A gas at 514 K and 338 mm Hg was heated to 311°C and 507 mm Hg. What would be the final volume of the gas, if the initial volume was 14.2 L?

7.5 Avogadro's law states that the volume of a gas, at constant temperature and pressure, is directly proportional to the number of moles of gas present: $V_1/n_1 = V_2/n_2$

Complete the following table using Avogadro's law:

Problem	Relationship with units	Answer with units
a. What is the new volume of a gas if 1.00 mole is removed from 2.63 moles of CH_4 gas having a volume of 18.3 L. Pressure and temperature remain constant.		
b. How many moles of helium would be added to a balloon containing 0.354 mole of helium to increase the volume from 1.55 L to 2.34 L?		

7.6 Combining the three gas laws gives an equation that describes the state of a gas at a single set of conditions. This equation, $PV = nRT$, is called the **ideal gas equation** (Sec. 7.7). T is measured on the Kelvin scale, and the value of R (the ideal gas constant) is 0.0821 atm·L/mole·K.

Rearrange the ideal gas equation to solve for the correct variable in completing the following combined gas law problems.

a. What is the volume of 1.49 moles of helium with a pressure of 1.21 atm at 224°C?

b. What would be the temperature of neon gas, if 0.339 mole of neon gas is in a 5.72 liter tank and the pressure gauge reads 2.53 atm?

7.7 **Dalton's law of partial pressures** (Sec. 7.8) states that the total pressure exerted by a mixture of gases is the sum of the **partial pressures** (Sec. 7.8) of the individual gases: $P_T = P_1 + P_2 + P_3 + \ldots$ Using Dalton's law of partial pressures, complete the following problems:

a. What is the total pressure exerted by a mixture of helium and argon? The partial pressures of helium and argon are: P_{He} = 270 mm Hg and P_{Ar} = 400 mm Hg.

b. What is the partial pressure of oxygen gas in a mixture of O_2, CO_2(P_{CO_2} = 341 mm Hg) and CO (P_{CO} = 114 mm Hg) if the total pressure of the mixture is 744 mm Hg?

7.8 A **change of state** (Sec. 7.9) is a process in which matter changes from one state to another. If heat is absorbed, the change is **endothermic** (Sec. 7.9); if heat is released during the process, it is **exothermic** (Sec. 7.9).

a. Complete the following table with the correct term for the physical change involved.

	To solid	To liquid	To gas
From solid	XXXXXXXX		
From liquid		XXXXXXXX	
From gas			XXXXXXXX

b. In the table above, indicate for each physical change whether it is endothermic(A) or exothermic(B).

7.9 **Hydrogen bonds** (Sec. 7.13) are strong **dipole-dipole interactions** (Sec. 7.13) that occur when hydrogen is bonded to fluorine, oxygen, or nitrogen. The hydrogen atom in this case is almost a "bare" nucleus, and it is very strongly attracted to a pair of electrons on an electronegative atom of another molecule.

Complete the following table by indicating for each substance whether hydrogen bonding can occur between individual molecules or with a water molecule.

Molecule	Hydrogen bonding between individual molecules	Hydrogen bonding with water molecules
HF		
HI		
NH_3		
CH_4		
CO		
$CH_3CH_2\ddot{O}H$		
$H_3C—\ddot{O}—CH_3$		

Self-Test

True-false: Indicate whether the following statements are true or false. If the statement is false, give the word or phrase that may be substituted for the underlined portion to make the statement true.

1. Boyle's law states for a given mass of gas at constant temperature, the volume of a gas varies directly with pressure.
2. Charles's law states for a given mass of gas at constant pressure, the volume of a gas varies directly with temperature.
3. Gases cool when they are compressed.
4. Doubling the volume of a gas will double the pressure of the gas. Assume the temperature and number of moles of gas remain constant.
5. One atmosphere is the pressure required to support 760 mm of Hg.
6. The total pressure exerted by a mixture of gases is equal to the sum of the partial pressures.
7. The vapor pressure of a liquid decreases as temperature increases.
8. As temperature increases, the velocity of molecules in a liquid decreases.
9. The energy resulting from the attractions and repulsions between particles in matter is a part of that matter's potential energy.
10. Liquids are very compressible because there is a lot of empty space between particles.
11. A balloon filled with helium and left to warm in the sun will decrease in volume.
12. A volatile liquid is one that has a high vapor pressure.
13. A state of equilibrium may exist between a liquid and a gas in a closed container.

Multiple choice:

14. Oxygen gas with a volume of 5.00 L at 1 atm and 273 K was heated to 402 K and the pressure was doubled. What was the new volume of the oxygen gas?
    a) 3.68 L
    b) 5.00 L
    c) 1.71 L
    d) 14.7 L
    e) none of these
15. Two gases, nitrogen and oxygen, in the same container, have a total pressure of 600 mm Hg. If the partial pressure of oxygen equals the partial pressure of nitrogen, what is the partial pressure of oxygen?
    a) 600 mm Hg
    b) 400 mm Hg
    c) 300 mm Hg
    d) 200 mm Hg
    e) none of these
16. How many moles of helium are in a 28.4 L balloon at 45°C and 1.03 atm?
    a) 0.271 mole
    b) 0.0453 mole
    c) 6.98 moles
    d) 1.12 moles
    e) none of these
17. The strongest intermolecular forces between water molecules are:
    a) ionic bonds
    b) covalent bonds
    c) hydrogen bonds
    d) London forces
    e) none of these
18. Hydrogen bonding would not occur between two molecules of which of these compounds?
    a) HF
    b) CH_4
    c) CH_3NH_2
    d) H_2O
    e) CH_3OH
19. London forces would be the strongest attractive forces between two molecules of which of these substances?
    a) HF
    b) BrCl
    c) F_2
    d) H_2O
    e) none of these

20. Liquids that have significant hydrogen bonding, compared to liquids that have no hydrogen bonding, would have:
    a) higher vapor pressure
    b) higher boiling point
    c) lower condensation temperature
    d) greater tendency to evaporate
    e) none of these
21. The pressure on 526 mL of gas is increased from 755 mm Hg to 974 mm Hg. If temperature remains constant, what is the new volume?
    a) 408 mL d) 387 mL
    b) 633 mL e) none of these
    c) 215 mL
22. Which of the following changes is endothermic?
    a) condensation d) deposition
    b) freezing e) none of these
    c) sublimation
23. What will increase the pressure of a gas in a closed container?
    a) decreasing the temperature of the gas
    b) adding more gas to the container
    c) increasing the volume of the container
    d) replacing the gas with the same number of moles of a different gas
    e) both b) and c)

Answers to Practice Exercises

7.1

	Shape	Volume	Density	Compressibility	Thermal Expansion
Gas	indefinite	indefinite	low	large	moderate
Liquid	indefinite	definite	high	small	small
Solid	definite	definite	high	small	very small

7.2

Relationship with units	Answer with units
a. $V_2 = \frac{P_1 \text{ x } V_1}{P_2} = \frac{2340 \text{ mm Hg x } 2.45 \text{ L}}{3580 \text{ mm Hg}} =$	1.60 L
b. $V_2 = \frac{P_1 \text{ x } V_1}{P_2} = \frac{1.00 \text{ atm x } 12.5 \text{ L}}{2.00 \text{ atm}} =$	6.25 L
c. $P_2 = \frac{P_1 \text{ x } V_1}{V_2} = \frac{3630 \text{ mm Hg x } 8.24 \text{ L}}{16.4 \text{ L}} =$	1820 mm Hg

7.3

Relationship with units	Answer with units
a. $V_2 = \frac{V_1 \text{ x } T_2}{T_1} = \frac{4.71 \text{ L x } 395 \text{ K}}{551 \text{ K}}$	3.38 L
b. $T_2 = \frac{V_2 \text{ x } T_1}{V_1} = \frac{20.5 \text{ L x } 345 \text{ K}}{14.5 \text{ L}}$	488 K

7.4 a. $P_2 = \dfrac{P_1 \text{ x } V_1 \text{ x } T_2}{T_1 \text{ x } V_2} = \dfrac{1.25 \text{ atm x } 5.72 \text{ L x } 323 \text{ K}}{4.50 \text{ L x } 303 \text{ K}} = 1.69 \text{ atm}$

b. $V_2 = \dfrac{P_1 \text{ x } V_1 \text{ x } T_2}{P_2 \text{ x } T_1} = \dfrac{338 \text{ mm Hg x } 14.2 \text{ L x } 584 \text{ K}}{507 \text{ mm Hg x } 514 \text{ K}} = 10.8 \text{ L}$

7.5

Relationship with units	Answer with units
a. $V_2 = \dfrac{V_1 \text{ x } n_2}{n_1} = \dfrac{18.3 \text{ L x } 1.63 \text{ moles}}{2.63 \text{ mole}} =$	11.3 L
b. $n_2 = \dfrac{V_2 \text{ x } n_1}{V_1} = \dfrac{2.34 \text{ L x } 0.354 \text{ mole}}{1.55 \text{ L}} =$	$n_2 = 0.534$ mole moles added = (0.534 – 0.354) mole = 0.180 mole

7.6 a. $V = \dfrac{n \text{ x } R \text{ x } T}{P} = \dfrac{1.49 \text{ moles x } 0.0821 \text{ atm L/mole K x } 497 \text{ K}}{1.21 \text{ atm}} = 50.2 \text{ L}$

b. $T = \dfrac{P \text{ x } V}{n \text{ x } K} = \dfrac{2.53 \text{ atm x } 5.72 \text{ L}}{0.339 \text{ mole x } 0.0821 \text{ atm L/mole K}} = 5.20 \text{ x } 10^2 \text{ K}$

7.7 a. $P_{total} = P_{He} + P_{Ar} = 270$ mm Hg + 400 mm Hg = 670 mm Hg

b. $P_{total} = P_{O_2} + P_{CO_2} + P_{CO}$

$P_{O_2} = P_{total} - P_{CO_2} - P_{CO} = 744$ mm Hg – 114 mm Hg – 341 mm Hg = 289 mm Hg

7.8

	To solid	To liquid	To gas
From solid	XXXXXX	melting A	sublimation A
From liquid	freezing B	XXXXXX	evaporation (and boiling) A
From gas	deposition B	condensation B	XXXXXX

7.9

Molecule	Hydrogen bonding between individual molecules	Hydrogen bonding with water molecules
HF	yes	yes
HI	no	no
NH_3	yes	yes
CH_4	no	no
CO	no	yes
$CH_3CH_2\ddot{\underset{..}{O}}H$	yes	yes
$H_3C—\ddot{\underset{..}{O}}—CH_3$	no	yes

Answers to Self-Test

The numbers in parentheses refer to sections in your textbook:
1. F; varies inversely (7.4) **2**. T (7.5) **3**. F; become hotter (7.5) **4**. F; halve (7.4) **5**. T (7.3) **6**. T (7.9) **7**. F; increases (7.12) **8**. F; increases (7.1) **9**. T (7.1) **10**. F; gases (7.2) **11**. F; increase (7.5) **12**. T (7.12) **13**. T (7.12) **14**. a (7.7) **15**. c (7.9) **16**. d (7.8) **17**. c (7.14) **18**. b (7.14) **19**. c (7.14) **20**. b (7.14) **21**. a (7.4) **22**. c (7.10) **23**. b (7.8)

Solutions — Chapter 8

Chapter Overview

Many chemical reactions take place in **solutions** (Sec. 8.1), particularly in water solutions. The properties of water make it a vital part of all living systems.

In this chapter you will define terms associated with solutions, study how solutions form, and calculate the concentrations of solutions using various units. You will study osmotic pressure and the factors that control the important process of osmosis.

Practice Exercises

8.1 **Solubility** (Sec. 8.2) of a substance can be predicted to some extent by the generalization that substances of like polarity tend to be more soluble in each other than substances that differ in polarity: "like dissolves like." However, the solubility of ionic compounds is more complex. Table 8.3 in your textbook gives water solubilities of ionic compounds.

Indicate the solubility of each of the substances below in the two **solvents** (Sec. 8.1), water and benzene.

Substance	Water (polar)	Benzene (nonpolar)
$CaCO_3$ (ionic solid)		
Ag_2S (ionic solid)		
$NaNO_3$ (ionic solid)		
K_2SO_4 (ionic solid)		
petroleum jelly (nonpolar solid)		
butane (nonpolar liquid)		
acetone (polar liquid)		

8.2 The **concentration** (Sec. 8.5) of a solution is the amount of **solute** (Sec. 8.1) present in a specified amount of solution. One way of expressing concentration is **percent by mass** (Sec. 8.5), the mass of solute divided by the mass of solution multiplied by 100.

$$\%(\text{m/m}) = \frac{\text{mass of solute}}{\text{mass of solution}} \times 100$$

a. What is the percent by mass, %(m/m), concentration of NaCl in a solution prepared by dissolving 14.8 g of NaCl in 122 g of water?

b. How many grams of NaCl were added to 225 g of water to prepare a 7.52%(m/m) NaCl solution?

8.3 Another way of expressing concentration is **mass-volume percent** (Sec. 8.2), the mass of solute divided by the volume of solution:

$$\%(\text{m/v}) = \frac{\text{mass of solute (g)}}{\text{volume of solution (mL)}} \times 100$$

How many grams of KCl must be added to 250.0 mL of water to prepare a 9.82%(m/v) solution?

8.4 The **molarity** (Sec. 8.5) of a solution is a ratio giving the number of moles of solute per liter of solution:

$$\text{Molarity(M)} = \frac{\text{moles of solute}}{\text{liters of solution}}$$

a. What is the molarity of a solution that contains 1.44 moles of NaCl in 2.50 L of solution?

b. A solution with a volume of 425 mL is prepared by dissolving 2.64 moles of $CaCl_2$ in water. What is the molarity?

c. If 7.21 g of KCl is dissolved in water to prepare 0.333 L of solution, what is the molarity?

8.5 The molarity of a solution can be used as a conversion factor to relate liters of solution to moles of solute. Use dimensional analysis to solve for the correct variable in the following problems.

a. How many moles of $CaCl_2$ were dissolved in 3.55 L of a 1.47 M solution?

b. How many grams of $CaCl_2$ were used to prepare 0.250 L of a 0.143 M $CaCl_2$ solution?

c. How many liters of solution would be needed to produce 21.5 g of $CaCl_2$ from a 0.842 M solution?

8.6 **Dilution** (Sec. 8.5) is the process in which more solvent is added to a solution in order to lower its concentration. The simple relationship used for dilution is: the concentration of the stock solution times the volume of the stock solution is equal to the concentration of the **diluted solution** (Sec. 8.2) times the volume of the diluted solution.

$$C_s \times V_s = C_d \times V_d$$

a. If 255 mL of water is added to 325 mL of a 0.477 M solution, what is the molarity of the new solution?

b. How many milliliters of water would have to be added to 250.0 mL of a 2.33 M solution to prepare a 0.551 M solution?

8.7 **Osmosis** (Sec. 8.8) is the movement of water across a **semipermeable membrane** (Sec. 8.8) from a more dilute solution to a more **concentrated solution** (Sec. 8.2). **Osmotic pressure** (Sec. 8.8), the amount of pressure necessary to stop this flow of water, depends on the number of particles of solute in the solutions. **Osmolarity** (Sec. 8.8) is the product of the molarity of the solution and the number of particles produced when the solute dissociates: Osmolarity = molarity x i

Complete the following table on osmolarity.

Molarity of solutions	Osmolarity
3 M KCl	
2 M KCl + 1 M glucose	
3 M $CaBr_2$ + 2 M glucose	
2 M $CaBr_2$ + 1 M KBr	

8.8 Water flows through the semipermeable membrane of a cell from a solution of higher solute concentration to one of lower solute concentration. Indicate which way water will flow across the cell membranes of red blood cells under each of these conditions:

Condition	Water flows into cells	Water flows out of cells
Cells immersed in concentrated NaCl solution		
Solution around cells is **hypotonic** (Sec. 8.8)		
Hypertonic (Sec. 8.8) solution surrounds cells		
Solute concentration around cells is decreased		
Cells immersed in pure water		

Self-Test

True-false: Indicate whether the following statements are true or false. If the statement is false, give the word or phrase that may be substituted for the underlined portion to make the statement true.

1. A saturated solution contains the maximum amount of <u>solute</u> that will dissolve in the solution.
2. Colligative properties are properties that depend on the <u>amount of solute</u> dissolved in a given mass of solution.
3. Osmotic semipermeable membranes permit only <u>dissolved salt</u> to flow through.
4. Hypertonic solutions contain a <u>smaller</u> number of solute molecules than the intracellular fluid.
5. In a salt water solution, the <u>solute</u> is water.
6. Fog is a <u>solution</u> of water droplets in air.
7. Undissolved solute is in equilibrium with dissolved solute in a <u>saturated solution</u>.
8. An unsaturated solution is <u>always</u> a dilute solution.
9. Carbon dioxide is <u>more</u> soluble in water when pressure increases.
10. A polar gas, such as NO_2, is <u>insoluble</u> in water.
11. Percent by mass (% m/m) is mass of solute divided by mass of <u>solvent</u>, x 100.
12. Red blood cells in a hypotonic solution may undergo <u>hemolysis</u>.
13. <u>The same number of moles</u> of NaCl are in 225 mL of a 1.55 M NaCl solution as are in 450 mL of a 1.55 M NaCl solution.
14. Dialysis can be used to remove <u>dissolved ions</u> from solutions containing large molecules.

Multiple choice:

15. Which of the following compounds would not dissolve in water?
    a) NaCl d) HCl
    b) CCl_4 e) all would dissolve
    c) $CaCl_2$
16. A solution containing 10.0 g of NaCl in 0.500 L of solution would have what molarity?
    a) 0.342 M d) 0.174 M
    b) 0.200 M e) none of these
    c) 0.500 M
17. How many milliliters of 4.57 M potassium bromide solution would be needed to prepare 1.00 L of 2.08 M potassium bromide solution?
    a) 155 mL d) 695 mL
    b) 325 mL e) none of these
    c) 455 mL
18. How much solute is present in 215 mL of a 0.500 M solution of HCl in water?
    a) 1.12 moles d) 0.108 mole
    b) 0.0566 mole e) none of these
    c) 0.752 mole
19. If 53.0 mL of a 3.00 M NaCl solution is diluted to give a solution whose molarity is 0.150 M, what is the volume of the new solution?
    a) 0.520 L d) 626 mL
    b) 1.06 L e) none of these
    c) 835 mL
20. The osmolarity of a solution that is 2 M $CaCl_2$ and 2 M glucose is:
    a) 4 M d) 10 M
    b) 6 M e) none of these
    c) 8 M
21. Which of the following solutions would be isotonic with 0.1 M NaCl?
    a) 0.5 M $CaCl_2$ d) 0.05 $Ca(NO_3)_2$
    b) 0.2 M glucose e) none of these
    c) 0.1 M sucrose
22. Water flows out of red blood cells placed in which of the following solutions?
    a) hypotonic d) both a) and c)
    b) isotonic e) none of these
    c) hypertonic
23. What mass-volume percent %(m/v) would result from dissolving 5.00 g of NaCl in enough water to form 50.0 mL of saline solution?
    a) 5.00%(m/v) d) 20.0%(m/v)
    b) 9.09%(m/v) e) none of these
    c) 10.0%(m/v)
24. Dissolving 7.5 g of NaCl in 50.3 g of water would yield a solution that is what percent by mass, %(m/m)?
    a) 7.50%(m/m) d) 74.6%(m/m)
    b) 14.9%(m/m) e) none of these
    c) 13.0%(m/m)

Answers to Practice Exercises

8.1

Substance	Water (polar)	Benzene (nonpolar)
$CaCO_3$ (ionic solid)	soluble	insoluble
Ag_2S (ionic solid)	insoluble	insoluble
$NaNO_3$ (ionic solid)	soluble	insoluble
K_2SO_4 (ionic solid)	soluble	insoluble
petroleum jelly (nonpolar solid)	insoluble	soluble
butane (nonpolar liquid)	insoluble	soluble
acetone (polar liquid)	soluble	soluble

8.2 a. $\%(m/m) = \dfrac{\text{mass of solute}}{\text{mass of solution}} \times 100 = \dfrac{14.8 \text{ g NaCl}}{14.8 \text{ g NaCl} + 122 \text{ g } H_2O} \times 100 = 10.8\%(m/m)$

b. 100 g solution – 7.52 g NaCl = 92.48 g H_2O

$$225 \text{ g } H_2O \times \frac{7.52 \text{ g NaCl}}{92.48 \text{ g } H_2O} = 18.3 \text{ g NaCl}$$

8.3 $\%(m/v) = \dfrac{\text{mass of solute (g)}}{\text{volume of solution (mL)}} \times 100 = 250.0 \text{ mL solution} \times \dfrac{9.82 \text{ g KCl}}{100 \text{ mL solution}} = 24.6 \text{ g KCl}$

8.4 a. $\text{Molarity}(M) = \dfrac{\text{moles of solute}}{\text{liters of solution}} = \dfrac{1.44 \text{ moles NaCl}}{2.50 \text{ L}} = 0.576 \text{ M}$

b. $425 \text{ mL} \times \dfrac{1 \text{ L}}{1000 \text{ mL}} = 0.425 \text{ L}$

$$\text{Molarity}(M) = \frac{\text{moles of solute}}{\text{liters of solution}} = \frac{2.64 \text{ moles } CaCl_2}{0.425 \text{ L}} = 6.21 \text{ M}$$

c. $7.21 \text{ g} \times \dfrac{1.00 \text{ mole}}{74.6 \text{ g}} = 9.66 \times 10^{-2} \text{ moles}$

$$M = \frac{\text{moles of solute}}{\text{liters of solution}} = \frac{9.66 \times 10^{-2} \text{ moles}}{0.333 \text{ L}} = 0.290 \text{ M}$$

8.5 a. $\dfrac{1.47 \text{ moles}}{1 \text{ L}} \times 3.55 \text{ L} = 5.22 \text{ moles } CaCl_2$

b. $\dfrac{0.143 \text{ mole}}{1 \text{ L}} \times \dfrac{111 \text{ g}}{1 \text{ mole}} \times 0.250 \text{ L} = 3.97 \text{ g } CaCl_2$

c. $\text{liters of solution} = \text{moles} \times \dfrac{1}{M} = \text{moles} \times \dfrac{\text{liters of solution}}{\text{moles}}$

$$\text{liters} = 21.5 \text{ g } CaCl_2 \times \frac{1 \text{ mole}}{111 \text{ g } CaCl_2} \times \frac{1 \text{ L}}{0.842 \text{ mole}} = 0.230 \text{ L}$$

8.6 a. $(C_s \times V_s = C_d \times V_d)$

$$C_d = \frac{C_s \times V_s}{V_d} = \frac{0.477\ M \times 325\ mL}{325\ mL + 255\ mL} = 0.267\ M$$

b. $$V_d = \frac{C_s \times V_s}{C_d} = \frac{2.33\ M \times 250.0\ mL}{0.551\ M} = 1060\ mL$$

$$V_d - V_s = 1060\ mL - 250.0\ mL = 810\ mL \text{ water added}$$

8.7

Molarity of solutions	Osmolarity
3 M KCl	6
2 M KCl + 1 M glucose	5
3 M $CaBr_2$ + 2 M glucose	11
2 M $CaBr_2$ + 1 M KBr	8

8.8

Condition	Water flows into cells	Water flows out of cells
Cells immersed in concentrated NaCl solution		X
Solution around cells is hypotonic	X	
Hypertonic solution surrounds cells		X
Solute concentration around cells is decreased	X	
Cells immersed in pure water	X	

Answers to Self-Test

The numbers in parentheses refer to sections in your textbook:
1. T (8.2) **2**. F; number of particles (8.7) **3**. F; ions and small molecules (8.8) **4**. F; larger (8.8) **5**. F; solvent (8.1) **6**. F; suspension (8.1) **7**. T (8.2) **8**. F; sometimes (8.2) **9**. T (8.2) **10**. F; soluble (8.4) **11**. F; solution (8.5) **12**. T (8.8) **13**. F; fewer moles (8.5) **14**. T (8.9) **15**. b (8.4) **16**. a (8.5) **17**. c (8.5) **18**. d (8.5) **19**. b (8.5) **20**. c (8.8) **21**. b (8.8) **22**. c (8.8) **23**. c (8.5) **24**. c (8.5)

Chemical Reactions — Chapter 9

Chapter Overview

Chemical reactions are the means by which new substances are formed. The concepts of collision theory explain how and under what conditions reactions take place.

In this chapter you will learn to recognize four basic types of chemical reactions. You will identify oxidizing agents and reducing agents in redox reactions. You will study factors that affect the rate of a chemical reaction. Not all chemical reactions go to completion; you will learn to calculate the concentrations of reactants and products in an equilibrium state.

Practice Exercises

9.1 In a **chemical reaction** (Sec. 9.1) at least one new substance is produced as the result of chemical change. Chemical reactions can be classified by the numbers of reactants and products. Classify the following reactions as **combination, decomposition, single replacement, or double replacement reactions** (Sec. 9.1):

Reaction	Classification
a. $2\,NaNO_3 \rightarrow 2\,NaNO_2 + O_2$	
b. $H_2 + Cl_2 \rightarrow 2\,HCl$	
c. $AgNO_3 + KBr \rightarrow AgBr + KNO_3$	
d. $Cu + 2\,AgNO_3 \rightarrow 2\,Ag + Cu(NO_3)_2$	

9.2 The **oxidation number** (Sec. 9.2) of an atom represents the charge that the atom would have if all the electrons in each of its bonds were transferred to the more electronegative atom of the two atoms in the bond.

Assign oxidation numbers for all the atoms in the following substances, using the rules for determining oxidation numbers found in your textbook in section 9.2.

Substance	Oxidation number
Fe	
Ne	
Br_2	
KBr	
MgO	

Substance	Oxidation number
NO_2	
NO_2^-	
PO_4^{3-}	
SO_4^{2-}	
NH_4^+	

9.3 In **oxidation-reduction (redox) reactions** (Sec. 9.2) electrons are transferred from one reactant to another reactant. Electrons are lost by the substance being **oxidized** (Sec. 9.3), so its oxidation number is increased. A substance being **reduced** (Sec. 9.3) gains electrons, and its oxidation number decreases.

In the following equations assign oxidation numbers to each atom in the reactants and products. Looking at oxidation number changes, classify the reaction as a redox or nonredox reaction.

Reaction	Redox or nonredox
a. $KOH + HBr \rightarrow KBr + HOH$	
b. $2NaNO_3 \rightarrow 2NaNO_2 + O_2$	
c. $Cu + 2AgNO_3 \rightarrow 2Ag + Cu(NO_3)_2$	

9.4 In redox reactions an **oxidizing agent** (Sec. 9.3) accepts electrons and is reduced. A **reducing agent** (Sec. 9.3) loses electrons and is oxidized. For the equations below, first assign the oxidation numbers, and then identify the oxidizing and reducing agents and the substances oxidized and reduced.

Equation	Substance oxidized	Substance reduced	Oxidizing agent	Reducing agent
a. $4Na(s) + O_2(g) \rightarrow 2Na_2O(s)$				
b. $Ca(s) + S(s) \rightarrow CaS(s)$				
c. $Mg(ClO_3)_2 \rightarrow MgCl_2 + 3O_2$				

9.5 According to **collision theory** (Sec. 9.4), a chemical reaction takes place when two reactant particles collide with a certain minimum amount of energy, called **activation energy** (Sec. 9.4), and the proper orientation. In an energy diagram, the activation energy is the energy difference between the top of the energy "hill" and the energy of the reactants. Some of this energy is regained during the reaction, so that the energy absorbed during the reaction is the difference in the energies of the reactants and the products.

Sketch an energy diagram representing an endothermic reaction and label these parts on the diagram: a. average energy of reactants, b. average energy of products, c. energy absorbed during the reaction, and d. activation energy.

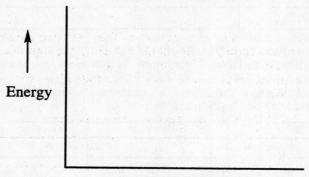

9.6 The **rate of a chemical reaction** (Sec. 9.6) is the rate at which reactants are consumed or products produced in a given time period. Various factors affect the rate of a reaction: the physical nature of reactants, reactant concentrations, reaction temperature, the presence of a **catalyst** (Sec. 9.6).

Indicate whether the listed conditions would increase or decrease the rate of the following reaction:

A(solid) + B $\rightarrow$ C + D (heat)

Condition	Rate of reaction	Explanation
Decreasing the concentration of reactants		
Decreasing the temperature of the reaction		
Introduction of an effective catalyst for this reaction		
Increasing the surface area of the solid reactant by dividing the solid into smaller particles		

9.7 In a system at **chemical equilibrium** (Sec. 9.7) the concentrations of the reactants and products remain constant. An **equilibrium constant** (Sec. 9.8) that describes numerically the extent of the reaction can be obtained by writing an equilibrium constant expression and evaluating it numerically.

Write the equilibrium constant expression for each of the equations below. Rules for writing these expressions are found in section 9.8 of your textbook.

a. $2P + 3I_2 \rightleftharpoons 2PI_3$

b. $CH_4(g) + 2O_2(g) \rightleftharpoons CO_2(g) + 2H_2O(g)$

9.8 If the concentrations of reactants and products are known for a given reaction at equilibrium, the equilibrium constant can be evaluated. Write the equilibrium constant expression for the following equation and substitute the given molarities to calculate a numerical value for Keq.

$2HI(g) \rightleftharpoons H_2(g) + I_2(g)$ 2.3 moles HI, 0.45 mole H_2 and 0.24 mole I_2

9.9 A system at chemical equilibrium can be disturbed by outside forces. There are two possible results of this disruption: either more products form (the equilibrium shifts to the right) or more reactants form (the equilibrium shifts to the left).

Indicate what effect each of the conditions below would have on the following **exothermic reaction** (Sec. 9.5) at equilibrium:

$CH_4(g) + 2O_2(g) \rightarrow CO_2(g) + 2H_2O(g) + \text{heat}$ (exothermic reaction)

Condition	Change in equilibrium	Explanation
Increasing the concentration of reactants		
Increasing the temperature of the reaction		
Introduction of an effective catalyst for this reaction		
Increasing the pressure exerted on the reaction		

Self-Test

True-false: Indicate whether the following statements are true or false. If the statement is false, give the word or phrase that may be substituted for the underlined portion to make the statement true.

1. The reaction $2CuO \rightarrow 2Cu + O_2$ is an example of a single-replacement reaction.
2. The oxidation number of a metal in its elemental state is always positive.
3. The oxidation number of oxygen in most compounds is –2.
4. A substance that is oxidized loses electrons.
5. A reducing agent gains electrons.
6. Adding heat to an endothermic reaction helps the reaction to go toward the products side.

7. The addition of a catalyst will not change the equilibrium position of a reaction.
8. The rate of a reaction is not affected by the addition of a catalyst.
9. Increasing the concentration of products in an equilibrium reaction shifts the equilibrium toward the product side of the reaction.
10. In an equilibrium constant expression the concentrations of the reactants are found in the numerator.
11. A large equilibrium constant indicates that the equilibrium position is to the right side of the equation.
12. In writing equilibrium constants we consider that concentrations of pure solids and pure liquids remain constant.

Multiple choice:

13. The equation $X + YZ \rightarrow Y + XZ$ is a general equation for which type of reaction?
    a) combination
    b) single-displacement
    c) double-displacement
    d) decomposition
    e) none of these
14. The oxidation number of chromium in the compound $K_2Cr_2O_7$ is:
    a) +6 b) –7 c) +5 d) –3 e) none of these
15. In the reaction $Zn + Cu(NO_3)_2 \rightarrow Zn(NO_3)_2 + Cu$ the oxidizing agent is:
    a) Zn
    b) $Cu(NO_3)_2$
    c) $Zn(NO_3)_2$
    d) Cu
    e) none of these
16. Which of these factors does **not** affect the rate of a reaction?
    a) the frequency of the collisions
    b) the energy of the collisions
    c) the orientation of the collisions
    d) the product of the collisions
    e) all of these affect rate
17. For a reaction at equilibrium, the concentrations of the products:
    a) increase rapidly
    b) increase slowly
    c) remain the same
    d) decrease slowly
    e) none of these

Answer questions 18 through 21 using the general equilibrium equation:

$$A + B \rightleftharpoons C + D + \text{heat}$$

18. What is the equilibrium constant for this reaction, if the following concentrations are measured at equilibrium: [A] = 0.20 M; [B] = 1.5 M; [C] = 5.2 M; [D] = 3.7 M
    a) 64
    b) 0.016
    c) 5.7
    d) 0.25
    e) none of these
19. The rate of the forward reaction could be increased by:
    a) decreasing the concentration of B
    b) increasing the concentration of A
    c) increasing the concentration of C
    d) both b) and c)
    e) none of these
20. The value of the equilibrium constant could be increased by:
    a) increasing the temperature of the reaction mixture
    b) increasing the concentration of A
    c) decreasing the concentration of B
    d) decreasing the temperature of the reaction mixture
    e) none of these

21. If more A is added to the reaction mixture at equilibrium:
    a) the amount of C will increase
    b) the amount of B will increase
    c) the amount of B will decrease
    d) both a) and c)
    e) none of these
22. In the reaction $2\ Mg + O_2 \rightarrow 2\ MgO$ the magnesium is:
    a) reduced and is the oxidizing agent
    b) reduced and is the reducing agent
    c) oxidized and is the oxidizing agent
    d) oxidized and is the reducing agent
    e) none of these

Answers to Practice Exercises

9.1

Reaction	Classification
a. $2\ NaNO_3 \rightarrow 2\ NaNO_2 + O_2$	decomposition
b. $H_2 + Cl_2 \rightarrow 2\ HCl$	combination
c. $AgNO_3 + KBr \rightarrow AgBr + KNO_3$	double-replacement
d. $Cu + 2\ AgNO_3 \rightarrow 2\ Ag + Cu(NO_3)_2$	single-replacement

9.2

Substance	Oxidation number
Fe	0
Ne	0
Br_2	0
KBr	K(+1), Br(–1)
MgO	Mg(+2), O(–2)

Substance	Oxidation number
NO_2	N(+4), O(–2)
NO_2^-	N(+3), O(–2)
PO_4^{3-}	P(+5), 0(–2)
SO_4^{2-}	S(+6), O(–2)
NH_4^+	N(–3), H(+1)

9.3

Oxidation numbers for all atoms	Redox or nonredox
a. $KOH + HBr \rightarrow KBr + H_2O$ +1,–2,+1 +1,–1 +1,–1 +1,–2	nonredox, no oxidation numbers change.
b. $2NaNO_3 \rightarrow 2NaNO_2 + O_2$ +1,+5,–2 +1,+3,–2 0	redox, N (+5 → +3) O (–2 → 0)
c. $Cu + 2AgNO_3 \rightarrow 2Ag + Cu(NO_3)_2$ 0 +1,+5,–2 0 +2,+5,–2	redox Cu (0 → +2) Ag (+1 → 0)

9.4

Equation	Substance oxidized	Substance reduced	Oxidizing agent	Reducing agent
a. $4Na(s) + O_2(g) \rightarrow 2Na_2O(s)$ 0 0 +1,–2	sodium	oxygen	oxygen	sodium
b. $Ca(s) + S(s) \rightarrow CaS(s)$ 0 0 +2,–2	calcium	sulfur	sulfur	calcium
c. $Mg(ClO_3)_2 \rightarrow MgCl_2 + 3O_2$ +2,+5,–2 +2,–1 0	oxygen	chlorine	chlorine	oxygen

9.5 a. average energy of reactants, b. average energy of products, c. energy absorbed during the reaction, and d. activation energy.

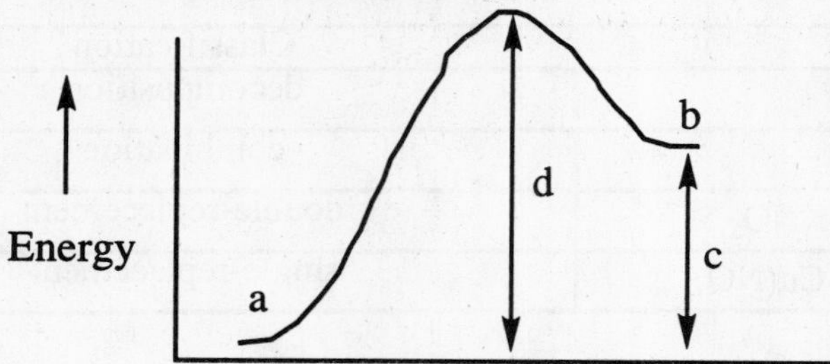

9.6 Indicate if the listed conditions would increase or decrease the rate of the following reaction: A(solid) + B → C + D (heat):

Rate of reaction	Explanation
Decrease	Fewer molecules collide, fewer molecules react.
Decrease	Lower kinetic energy, lower collision energy, so fewer collisions are effective.
Increase	Catalysts provide alternative reaction pathways that have lower energies of activation.
Increase	Larger surface area of solid provides more chances for collision.

9.7 a. $K_{eq} = \dfrac{\text{Products}}{\text{Reactants}} = \dfrac{[PI_3]^2}{[P]^2[I_2]^3}$

b. $K_{eq} = \dfrac{\text{Products}}{\text{Reactants}} = \dfrac{[CO_2][H_2O]^2}{[CH_4][O_2]^2}$

9.8 $K_{eq} = \dfrac{[H_2][I_2]}{[HI]^2} = \dfrac{(0.45\text{ mole/L}) \times (0.24\text{ mole/L})}{(2.3\text{ moles/L})^2} = 0.020$

9.9

Change in equilibrium	Explanation
Shift to right	An increase in concentration of a reactant produces more products.
Shift to left	In an exothermic reaction the equilibrium will shift to the reactant side to decrease the amount of heat produced.
No change	A catalyst cannot change the position of the equilibrium because it only lowers the energy of activation.
No change	Since the moles of gas on each side of the equation are the same, pressure changes have no effect.

Answers to Self-Test

The numbers in parentheses refer to sections in your textbook:
1. F; decomposition (9.1) **2**. F; zero (9.2) **3**. T (9.2) **4**. T (9.3) **5**. F; loses (9.3)
6. T (9.5) **7**. T (9.5) **8**. F; increased (9.6) **9**. F; reactants (9.9)
10. F; denominator (9.8) **11**. T (9.8) **12**. T (9.8) **13**. b (9.1) **14**. a (9.2)
15. b (9.3) **16**. d (9.4) **17**. c (9.7) **18**. a (9.8) **19**. b (9.6) **20**. d (9.8)
21. d (9.9) **22**. d (9.3)

Acids, Bases, and Salts — Chapter 10

Chapter Overview

Acids, bases, and salts play a central role in much of the chemistry that affects our daily lives. Learning the terms and concepts associated with these compounds will give you a greater understanding of the chemistry of the human body, and of the ways in which chemicals are manufactured.

In this chapter you will learn to identify acids and bases according to the Arrhenius and Brφnsted-Lowry definitions, write equations for acid and base dissociations in water, and calculate pH, a measure of acidity. You will write equations for the hydrolysis of salts of a weak acid or a weak base, and you will study the actions of buffers.

Practice Exercises

10.1 According to the **Brφnsted-Lowry** (Sec. 10.2) definitions, an acid is a proton donor and a base is a proton acceptor. The **conjugate base** (Sec. 10.2) of an acid is the species that remains when an acid loses a proton. The **conjugate acid** (Sec. 10.2) of a base is the species formed when a base accepts a proton.

Give the formula of the conjugate acid or base for the following substances:

Base	Conjugate acid
NH_3	
BrO_3^-	
HCO_3^-	

Acid	Conjugate base
HCO_3^-	
$HClO_2$	
HNO_3	

10.2 Identify the acid and base in the reactants in the following equations. Identify the conjugate acid and conjugate base in the products.

a. $HClO_3 + H_2O \rightarrow ClO_3^- + H_3O^+$

b. $HNO_2 + OH^- \rightarrow NO_2^- + H_2O$

10.3 An **acid ionization constant** (Sec. 10.5), K_a, is the equilibrium constant corresponding to the **ionization** (Sec. 10.1) of an acid. It gives a measure of the **strength** (Sec. 10.4) of the acid. The **base ionization constant** (Sec. 10.5), K_b, gives a measure of the strength of a base.

a. Write the ionization equation and the ionization constant expression for the ionization of nitrous acid, HNO_2, in water.

b. Write the ionization equation and the ionization constant expression (K_b) for the ionization of ethylamine, $C_2H_5NH_2$, in water. (The nitrogen atom accepts a proton.)

10.4 The acid ionization constant can be calculated for an acid if its concentration and percent ionization are known.

A 0.0150 M solution of an acid, HA, is 22% ionized at equilibrium. Determine the individual ion concentrations, and then calculate the K_a for this acid.

10.5 **Salts** (Sec. 10.6) are compounds made up of positive metal or polyatomic ions, and negative nonmetal or polyatomic (except hydroxide) ions. Identify each of the following compounds as a salt, base, or acid.

Compound	Acid	Base	Salt
HCl			
NaCl			
H_2SO_4			
NaOH			
$CaBr_2$			
$Ba(OH)_2$			

10.6 Salts dissolved in water are completely dissociated into ions in solution. Write a balanced equation for the **dissociation** (Sec. 10.1) of the following soluble ionic compounds in water.

a. KI

b. Na_3PO_4

c. CaI_2

d. Na_2CO_3

10.7 **Neutralization** (Sec. 10.7) is the reaction between an acid and a hydroxide base to form a salt and water. Complete the following neutralization equations by adding the missing products or reactants. Under each reactant molecule, write acid or base, and under each product molecule, write salt or water.

a. HCl + NaOH →

b. → $CaCl_2$ + $2H_2O$

c. → $Sr_3(PO_4)_2$ + $3H_2O$

10.8 In pure water an extremely small number of molecules transfer protons to form the ions H_3O^+ and OH^-. The equilibrium constant for this self-ionization is called the **ion product constant** (Sec. 10.8): $[H_3O^+][OH^-] = 1.00 \times 10^{-14}$. Using the ion product constant for water, determine the following concentrations:

Given concentrations	Substituted equation	Answer
$[H_3O^+] = 2.4 \times 10^{-6}$		$[OH^-] =$
$[OH^-] = 3.2 \times 10^{-8}$		$[H_3O^+] =$

10.9 Since the hydronium ion concentrations in aqueous solutions have a very large range of values, a more practical way to handle them is by using the **pH scale** (Sec. 10.9), defined as follows: $pH = -\log[H_3O^+]$

Using your calculator, complete the following pH relationships:

$[H_3O^+]$	pH
1.0×10^{-4}	
1.0×10^{-9}	
2.8×10^{-3}	
7.9×10^{-8}	

pH	$[H_3O^+]$
5.00	
2.00	
3.80	
10.40	

10.10 The neutralization reaction between an acid and a base forms a salt and water. The properties of a salt in water depend on the strength or weakness of the acid and base from which it is formed.

Complete the following reactions. Classify each acid and base as strong or weak. Classify each salt formed as strong acid-strong base salt, strong acid-weak base salt, weak acid-strong base salt, or weak acid-weak base salt.

a. $HBr + KOH \rightarrow$

b. $\rightarrow Al_2(SO_4)_3 + 6H_2O$

c. $HNO_2 + Mg(OH)_2 \rightarrow$

10.11 **Hydrolysis** (Sec. 10.10) is the reaction of a substance with water to produce hydronium ion or hydroxide ion or both. Some salts hydrolyze, depending on their "parentage," the acids and bases from which they are formed.

Write the aqueous hydrolysis equations for each of the following ions:

a. HCO_3^- (proton acceptor)

b. NH_4^+ (proton donor)

c. NO_2^- (proton acceptor)

10.12 A **buffer** (Sec. 10.11) solution is a solution that resists a change in pH when small amounts of acid or base are added to it. Buffers (the solutes in buffer solutions) consist of one of the following combinations: a weak acid and the salt of its conjugate base or a weak base and the salt of its conjugate acid. These are known as **conjugate acid-base pairs** (Sec. 10.2).

Predict whether each of the following pairs of substances could function as a buffer in an aqueous solution. Explain your answer.

Pairs of substances	Explanation
KOH, KCl	
HI, NaI	
NH_4I, NH_3	
NaH_2PO_4, H_3PO_4	

10.13 **Buffers contain a substance that reacts with and removes added base and a substance that reacts with and removes added acid. Write an equation to show the buffering action in each of the following aqueous solutions.**

a. NH_4I/NH_3 in a basic solution, OH^-

b. NaH_2PO_4/H_3PO_4 in an acidic solution, H_3O^+

10.14 **The pH of a buffered solution may be determined by using the Henderson-Hasselbalch equation:**

$$pH = pKa + \frac{\log[A^-]}{\log[HA]}$$

Calculate the pH of each of these buffer solutions:

Given values	Substituted equation	Answer
[HA] = 0.34 M [A⁻] = 0.51 M pKa = 5.48		pH =
[HA] = 0.27 M [A⁻] = 0.55 M Ka = 8.4×10^{-5}		pH =
[HA] = 0.25 M [A⁻] = 0.37 M Ka = 6.2×10^{-7}		pH =

10.15 **An electrolyte (Sec. 10.12) is a substance that forms a solution in water that conducts electricity. A strong electrolyte (Sec. 10.12) is a substance that completely dissociates into ions in aqueous solution. Salts and strong acids and strong bases are strong electrolytes. A weak electrolyte (Sec. 10.12) is a substance that is only partially ionized in aqueous solution. Weak acids and weak bases are weak electrolytes.**

Classify each of the substances below, dissolved in aqueous solution, as a weak electrolyte or a strong electrolyte.

Formula	Weak electrolyte	Strong electrolyte
H_2SO_4		
NH_4OH		
NH_4Cl		
MgI_2		

10.16 **Acid-base titration** (Sec. 10.13) is a procedure used to determine the concentration of an acid or base solution. A measured volume of an acid or a base of known concentration is exactly reacted with a measured volume of a base or an acid of unknown concentration. The unknown concentration can be calculated using dimensional analysis.

a. Determine the molarity of an unknown HCl solution, if 21.9 mL of 0.338 M NaOH was needed to neutralize 41.6 mL of the HCl solution. Hint: Write the balanced neutralization reaction equation.

b. What is the molarity of an unknown sulfuric acid solution, if 33.2 mL of 0.225 M NaOH is needed to neutralize 13.8 mL of the H_2SO_4 solution? Hint: Write the balanced neutralization reaction equation.

Self-Test

True-false: Indicate whether the following statements are true or false. If the statement is false, give the word or phrase that may be substituted for the underlined portion to make the statement true.

1. The value of the ion product constant of water is always <u>1×10^{-10}</u>.
2. Arrhenius defined a base as a substance that, in water, produces <u>hydroxide ions</u>.
3. According the Brϕnsted-Lowry theory, NH_3 is <u>an acid</u>.
4. Aqueous solutions of acids have a hydronium ion concentration <u>less than</u> 1×10^{-7} moles per liter.
5. A diprotic acid can transfer <u>two protons</u> per molecule during an acid-base reaction.
6. The pH of an acid is the negative logarithm of the <u>hydronium ion concentration</u>.
7. A neutralization reaction produces a salt and <u>a base</u>.
8. Hydrolysis of the salt of a weak acid and a strong base produces a solution that is <u>basic</u>.
9. A solution whose hydronium ion concentration is 1.0×10^{-4} has a pH of <u>–4</u>.
10. A buffer is a weak acid plus the salt of its conjugate <u>base</u>.
11. An amphoteric substance can function as an acid or a <u>salt</u>.

Multiple choice:

12. The pH of a solution in which $[H_3O^+] = 1.0 \times 10^{-5}$ is:
 a) –5.0 b) –1.5 c) 5.0 d) 1.5 e) none of these
13. The pH of a solution in which $[OH^-] = 1.0 \times 10^{-8}$ is:
 a) 6.0 b) 8.0 c) –8.0 d) –1.8 e) none of these
14. If the pH of a solution is 4.8, the hydronium ion concentration is:
 a) 4.0×10^{-8} M
 b) 8.3×10^{-4} M
 c) 3.2×10^{-12} M
 d) 1.6×10^{-5} M
 e) none of these
15. If the pH of a solution is 9.2, the hydroxide ion concentration is:
 a) 1.6×10^{-5} M
 b) 2.1×10^{-9} M
 c) 9.3×10^{-2} M
 d) 7.5×10^{-10}
 e) none of these
16. How many milliliters of 0.512 M HCl would be required to neutralize 35.8 mL of 1.50 M KOH?
 a) 11.3 mL
 b) 35.8 mL
 c) 105 mL
 d) 183 mL
 e) none of these

17. The conjugate base of the acid HNO_2 is:
   a) H_2O
   b) H_3O^+
   c) NO_2^-
   d) OH^-
   e) none of these
18. A 0.400 M solution of an acid is 9.0% ionized. The value of K_a for this acid is:
   a) 3.6×10^{-3}
   b) 9.8×10^{-2}
   c) 2.1×10^{-4}
   d) 8.9×10^{-4}
   e) none of these
19. The salt K_2CO_3 is produced by the reaction of:
   a) a weak acid with a weak base
   b) a weak acid with a strong base
   c) a strong acid with a weak base
   d) a strong acid with a strong base
   e) none of these
20. Which of the following is a weak electrolyte?
   a) K_3PO_4
   b) HNO_3
   c) $Ba(OH)_2$
   d) H_2CO_3
   e) none of these
21. If a small amount of hydroxide ion is added to the buffer H_2CO_3/HCO_3^-, the reaction will produce more:
   a) H_3O^+
   b) HCO_3^-
   c) H_2CO_3
   d) H_2
   e) none of these

Answers to Practice Exercises

10.1

Base	Conjugate acid
NH_3	NH_4^+
BrO_3^-	$HBrO_3$
HCO_3^-	H_2CO_3

Acid	Conjugate base
HCO_3^-	CO_3^{2-}
$HClO_2$	ClO_2^-
HNO_3	NO_3^-

10.2 a. $HClO_3 + H_2O \rightarrow ClO_3^- + H_3O^+$
acid, base, conjugate base, conjugate acid

b. $HNO_2 + OH^- \rightarrow NO_2^- + H_2O$
acid, base, conjugate base, conjugate acid

10.3 a. $HNO_2 + H_2O \rightarrow NO_2^- + H_3O^+$ $\quad K_a = \dfrac{[H_3O^+][NO_2^-]}{[HNO_2]}$

b. $C_2H_5NH_2 + H_2O \rightarrow C_2H_5NH_3^+ + OH^-$ $\quad K_b = \dfrac{[C_2H_5NH_3^+][OH^-]}{[C_2H_5NH_2]}$

10.4 $HA + H_2O \rightarrow H_3O^+ + A^-$

$$K_a = \frac{[H_3O^+][A^-]}{[HA]} = \frac{[0.0150 \times 0.22][0.0150 \times 0.22]}{[(0.0150)-(0.0150 \times 0.22)]} = \frac{[0.0033][0.0033]}{[0.0117]} = 9.3 \times 10^{-4}$$

10.5

Compound	Acid	Base	Salt
HCl	X		
NaCl			X
H_2SO_4	X		
NaOH		X	
$CaBr_2$			X
$Ba(OH)_2$		X	

10.6 a. $KI \rightarrow K^+ + I^-$

b. $Na_3PO_4 \rightarrow 3Na^+ + PO_4^{3-}$

c. $CaI_2 \rightarrow Ca^{2+} + 2I^-$

d. $Na_2CO_3 \rightarrow 2Na^+ + CO_3^{2-}$

10.7 a. $HCl + NaOH \rightarrow NaCl + H_2O$
acid base salt water

b. $2HCl + Ca(OH)_2 \rightarrow CaCl_2 + 2H_2O$
acid base salt water

c. $2H_3PO_4 + 3Sr(OH)_2 \rightarrow Sr_3(PO_4)_2 + 3H_2O$
acid base salt water

10.8

Given concentrations	Substituted equation	Answer
$[H_3O^+] = 2.4 \times 10^{-6}$	$[OH^-] = \dfrac{1.00 \times 10^{-14}}{[2.4 \times 10^{-6}]} =$	$[OH^-] = 4.2 \times 10^{-9}$
$[OH^-] = 3.2 \times 10^{-8}$	$[H_3O^+] = \dfrac{1.00 \times 10^{-14}}{[3.2 \times 10^{-8}]} =$	$[H_3O^+] = 3.1 \times 10^{-7}$

10.9

$[H_3O^+]$	pH
1.0×10^{-4}	4.00
1.0×10^{-9}	9.00
2.8×10^{-3}	2.55
7.9×10^{-8}	7.10

pH	$[H_3O^+]$
5.00	1.0×10^{-5}
2.00	1.0×10^{-2}
3.80	1.6×10^{-4}
10.40	4.0×10^{-11}

10.10 a. $HBr + KOH \rightarrow KBr + H_2O$ strong acid-strong base salt
strong weak

b. $3H_2SO_4 + 2Al(OH)_3 \rightarrow Al_2(SO_4)_3 + 6H_2O$ strong acid-weak base salt
strong weak

c. $2HNO_2 + Mg(OH)_2 \rightarrow Mg(NO_2)_2 + 2H_2O$ weak acid-weak base salt
weak weak

10.11 a. $HCO_3^- + H_2O \rightarrow H_2CO_3 + OH^-$

b. $NH_4^+ + H_2O \rightarrow H_3O^+ + NH_3$

c. $NO_2^- + H_2O \rightarrow HNO_2 + OH^-$

10.12

Pairs of substances	Explanation
KOH, KCl	No. Strong base and strong acid-strong base salt
HI, NaI	No. Strong acid and strong acid-strong base salt
NH_4I, NH_3	Yes. Weak base and strong acid-weak base salt
NaH_2PO_4, H_3PO_4	Yes. Weak acid and weak acid-strong base salt

10.13 a. NH_4I/NH_3 in a basic solution, OH^-

$$NH_4^+ + OH^- \rightarrow NH_3 + H_2O$$

b. NaH_2PO_4/H_3PO_4 in an acidic solution, H_3O^+

$$H_2PO_4^- + H_3O^+ \rightarrow H_3PO_4 + H_2O$$

10.14

Given values	Substituted equation	Answer
[HA] = 0.34 M $[A^-]$ = 0.51 M pKa = 5.48	$\text{pH} = 5.48 + \frac{\log\left[3.4 \times 10^{-1}\right]}{\log\left[5.1 \times 10^{-1}\right]}$	pH = 6.10
[HA] = 0.27 M $[A^-]$ = 0.55 M Ka = 8.4×10^{-5}	$\text{pH} = 4.08 + \frac{\log\left[5.5 \times 10^{-1}\right]}{\log\left[2.7 \times 10^{-1}\right]}$	pH = 4.54
[HA] = 0.25 M $[A^-]$ = 0.37 M Ka = 6.2×10^{-7}	$\text{pH} = 6.21 + \frac{\log\left[3.7 \times 10^{-1}\right]}{\log\left[2.5 \times 10^{-1}\right]}$	pH = 6.93

10.15

Formula	Weak electrolyte	Strong electrolyte
H_2SO_4		X
NH_4OH	X	
NH_4Cl		X
MgI_2		X

10.16 a. $HCl + NaOH \rightarrow NaCl + H_2O$

$$\text{M HCl} = 21.9 \text{ mL NaOH} \times \frac{0.338 \text{ mole NaOH}}{1000 \text{ mL NaOH}} \times \frac{1 \text{ mole HCl}}{1 \text{ mole NaOH}} \times \frac{1000 \text{ mL}}{0.0416 \text{ L HCl}} = 0.178 \text{ M HCl}$$

b. $H_2SO_4 + 2NaOH \rightarrow Na_2SO_4 + 2H_2O$

$$\text{M } H_2SO_4 = 33.2 \text{ mL NaOH} \times \frac{0.225 \text{ mole NaOH}}{1000 \text{ mL NaOH}} \times \frac{1 \text{ mole } H_2SO_4}{2 \text{ moles NaOH}} \times \frac{1000 \text{ mL}}{0.0138 \text{ L } H_2SO_4} = 0.271 \text{ M } H_2SO_4$$

Answers to Self-Test

The numbers in parentheses refer to sections in your textbook:
1. F; 1.00×10^{-14} (10.8) **2**. T (10.1) **3**. F; base (10.2) **4**. F; more than (10.8)
5. T (10.3) **6**. T (10.9) **7**. F; water (10.7) **8**. T (10.11) **9**. F; 4 (10.9) **10**. T (10.12)
11. F; base (10.2) **12**. c (10.9) **13**. a (10.9) **14**. d (10.9) **15**. a (10.9) **16**. c (10.15)
17. c (10.2) **18**. a (10.5) **19**. b (10.7) **20**. d (10.14) **21**. b (10.12)

Chapter Overview

Chemical reactions involve the exchange or sharing of the electrons of the atoms as chemical bonds are broken or formed. In **nuclear reactions** (Sec. 11.1) an atom's nucleus changes, absorbing or emitting particles or rays.

In this chapter you will compare three kinds of nuclear radiation and write equations for radioactive decay. You will study the rate of radioactive decay, defining the concept of the half-life of a radionuclide and using it in calculations. You will compare nuclear fission and nuclear fusion, study the effects of ionizing radiation on the human body, and learn some of the ways that radiation and radionuclides are used in medicine.

Practice Exercises

11.1 The **radioactive decay** (Sec. 11.3) of naturally radioactive substances results in the emission of three types of radiation. They differ in mass and charge.

Complete the following table summarizing the three types of radiation:

Type of radiation	Mass number	Charge	Symbol
alpha (Sec. 11.2)			
beta (Sec. 11.2)			
gamma (Sec. 11.2)			

11.2 The emission of an alpha particle from a nucleus results in the formation of a **nuclide** (Sec. 11.1) of a different element. The **daughter nuclide** (Sec. 11.3) has an atomic number that is two less and a mass number that is four less than the **parent nuclide** (Sec. 11.3).

Write the **balanced nuclear equations** (Sec. 11.3) for the alpha particle decay of the following nuclides. Give the complete symbol of the new nuclide produced in each process.

a. ${}^{149}_{65}\text{Tb} \rightarrow$

b. ${}^{231}_{91}\text{Pa} \rightarrow$

11.3 Beta particle decay results in the formation of a nuclide of a different element. The daughter nuclide has a mass number that is the same and an atomic number that is one greater than the parent nuclide.

Write the balanced nuclear equations for the beta particle decay of the following nuclides. Give the complete symbol for the new nuclide produced in each process.

a. ${}^{31}_{14}\text{Si} \rightarrow$

b. ${}^{59}_{26}\text{Fe} \rightarrow$

11.4 All **radioactive nuclides** (Sec. 11.1) do not decay at the same rate; the more unstable the nucleus, the faster it decays. The **half-life** (Sec. 11.4) of a substance, the amount of time for 1/2 of a given quantity of nuclide to decay, is a measure of the nuclide's stability.

The amount of radioactive material remaining after radioactive decay can be calculated from the following formula, where n = number of half-lives.

$$\begin{pmatrix}\text{amount of radionuclide}\\ \text{undecayed after } n \text{ half-lives}\end{pmatrix} = \begin{pmatrix}\text{original amount}\\ \text{of radionuclide}\end{pmatrix} \times \left(\frac{1}{2^n}\right)$$

For the following problems, an original sample of 0.43 mg of plutonium-239 is used (half-life = 24,400 years).

a. How much plutonium-239 would remain after three half-lives?

b. How much plutonium-239 would remain after 122,000 years?

11.5 **The number of half-lives that have elapsed since the original measurement can be determined by the fraction of the original nuclide that remains.**

A sample of iron-59 after 135 days has 1/8 of the iron-59 of the original sample. What is the half-life of iron-59?

11.6 **Transmutation** (Sec. 11.5) **of some nuclides into nuclides of other elements can be attained by bombardment of the nuclei with small particles traveling at very high speeds.**

Complete these bombardment reaction (Sec. 11.5) **equations by supplying the missing parts:**

a. $^{10}_{5}B$ + ____ $\rightarrow$ $^{7}_{3}Li + ^{4}_{2}\alpha$

b. $^{7}_{3}Li + ^{1}_{1}p \rightarrow$ ____ $+ ^{4}_{2}\alpha$

c. $^{10}_{5}B + ^{4}_{2}\alpha \rightarrow ^{13}_{7}N +$ ____

d. ____ $+ ^{14}_{7}N \rightarrow ^{247}_{99}Es + 5\,^{1}_{0}n$

11.7 **Radionuclides with high atomic numbers decay through a series of steps to reach a stable nuclide of lower atomic number. A few of these steps in a radioactive decay series are shown below. Complete the equations by filling in the missing parts:**

step 1: $^{210}_{82}Pb \rightarrow$ ____ $+ ^{0}_{-1}\beta$

step 2: ____ $\rightarrow ^{210}_{84}Po + ^{0}_{-1}\beta$

step 3: $^{210}_{84}Po \rightarrow$ ____ $+ ^{4}_{2}\alpha$

11.8 The three types of naturally occurring radioactive emissions differ in their ability to penetrate matter and, therefore, in their biological effects. Complete the following table summarizing properties of the three types of radiation:

Type of radiation	Speed	Penetration	Biological damage
alpha			
beta			
gamma			

Self-Test

True-false: Indicate whether the following statements are true or false. If the statement is false, give the word or phrase that may be substituted for the underlined portion to make the statement true.

1. Elements retain their identity during <u>nuclear reactions</u>.
2. <u>Alpha particles</u> are the same type of radiation as X-rays.
3. When an atom loses an alpha particle, its atomic number is <u>decreased</u> by 2.
4. Beta particles are more penetrating than <u>alpha particles</u>.
5. The half-life of a radionuclide is the length of time needed for <u>all</u> of the nuclide to decay.
6. <u>Nuclear fission</u> is the reaction that provides the sun's energy.
7. The combination of two small nuclei to produce a larger nucleus is called <u>nuclear fusion</u>.
8. Ionizing radiation knocks <u>protons</u> off some atoms so that they become ions.
9. The energy involved in a nuclear reaction is <u>smaller</u> than the energy involved in a chemical reaction.
10. A <u>diagnostic application</u> of radionuclides is the use of ionizing radiation to kill cancer cells.
11. In a <u>Geiger counter</u> radiation is detected by the ionization of argon gas in a metal tube.
12. A free radical is an atom or molecule with <u>a positive charge</u> whose formation can be caused by radiation.
13. Different isotopes of an element have practically identical <u>chemical properties</u>.
14. An average American is exposed to <u>more</u> radiation from natural sources than from human-made sources.
15. A transmutation process is a nuclear reaction in which a nuclide of one element is changed into a nuclide of <u>the same</u> element.

Multiple choice:

16. The nuclide produced by the emission of an alpha particle from the platinum nuclide $^{186}_{78}Pt$ would be:
    a) $^{188}_{80}Hg$
    b) $^{182}_{76}Os$
    c) $^{184}_{75}Re$
    d) $^{186}_{79}Au$
    e) none of these
17. The nuclide produced by beta emission from the barium radionuclide $^{142}_{56}Ba$ would be:
    a) $^{142}_{57}La$
    b) $^{143}_{55}Cs$
    c) $^{138}_{54}Xe$
    d) $^{146}_{58}Ce$
    e) none of these

18. The half-life for tritium $^{3}_{1}H$ is 12.26 years. After 36.78 years, how much of an original 0.40 g sample of tritium would remain?
    a) 0.35 g
    b) 0.20 g
    c) 0.10 g
    d) 0.050 g
    e) none of these
19. Four radionuclides have the half-lives listed below. Which of the four has the most stable nucleus?
    a) 22 days
    b) 4000 years
    c) 16 minutes
    d) 56 seconds
    e) all are very unstable
20. An alpha particle is made up of:
    a) two protons and four neutrons
    b) two protons and two neutrons
    c) two protons, two neutrons and two electrons
    d) two protons, two electrons and four neutrons
    e) none of these
21. The nuclear chain reaction of uranium-235 provides the energy in:
    a) nuclear power plants
    b) nuclear fusion reactors
    c) the sun
    d) hydroelectric generators
    e) none of these
22. Radiation is harmful to the human body because:
    a) it causes the formation of free radicals.
    b) it causes molecules to fragment.
    c) it produces ions in the body.
    d) a and b are true.
    e) a, b, and c are true.
23. Fusion reactions are not generally used to provide energy on Earth because:
    a) they give off too little energy.
    b) a very high temperature is required to start the reactions.
    c) the reactions proceed only at very low pressures.
    d) the starting materials (reactants) are too expensive.
    e) none of these
24. Radiation therapy includes the following use of radiation:
    a) radiographs of bone tissue
    b) radiation of cancer cells
    c) dental X-rays
    d) use of radioactive tracers
    e) all of the above

Answers to Practice Exercises

11.1 Complete the following table summarizing the three types of radiation:

Type of radiation	Mass number	Charge	Symbol
alpha	4	+2	$^{4}_{2}\alpha$
beta	0	−1	$^{0}_{-1}\beta$
gamma	0	0	$^{0}_{0}\gamma$

11.2 a. $^{149}_{65}Tb \rightarrow ^{145}_{63}Eu + ^{4}_{2}\alpha$

b. $^{231}_{91}Pa \rightarrow ^{227}_{89}Ac + ^{4}_{2}\alpha$

11.3 a. $^{31}_{14}Si \rightarrow ^{31}_{15}P + ^{0}_{-1}\beta$

b. $^{59}_{26}Fe \rightarrow ^{59}_{27}Co + ^{0}_{-1}\beta$

11.4 a. $\begin{pmatrix}\text{amount of radionuclide} \\ \text{undecayed after 3 half - lives}\end{pmatrix} = (0.43 \text{ mg}) \times \left(\frac{1}{2^3}\right) = (0.43 \text{ mg}) \times \left(\frac{1}{8}\right) = 0.054 \text{ mg}$

b. $\text{number of half - lives} = \dfrac{122,000 \text{ years}}{24,400 \text{ years/half - live}} = 5 \text{ half - lives}$

$\begin{pmatrix}\text{amount of radionuclide} \\ \text{undecayed after 5 half - lives}\end{pmatrix} = (0.43 \text{ mg}) \times \left(\frac{1}{2^5}\right) = (0.43 \text{ mg}) \times \left(\frac{1}{32}\right) = 0.013 \text{ mg}$

11.5 $\left(\frac{1}{2}\right) \times \left(\frac{1}{2}\right) \times \left(\frac{1}{2}\right) = \frac{1}{8} = \frac{1}{2^3} = \frac{1}{2^n}$; $n = 3$ = number of half-lives

$\dfrac{135 \text{ days}}{3 \text{ half - lives}} = 45 \text{ days} = 1 \text{ half-life of iron-59}$

11.6 a. ${}^{10}_{5}\text{B} + {}^{1}_{0}\text{n} \rightarrow {}^{7}_{3}\text{Li} + {}^{4}_{2}\alpha$

b. ${}^{7}_{3}\text{Li} + {}^{1}_{1}\text{p} \rightarrow {}^{4}_{2}\text{He} + {}^{4}_{2}\alpha$

c. ${}^{10}_{5}\text{B} + {}^{4}_{2}\alpha \rightarrow {}^{13}_{7}\text{N} + {}^{1}_{0}\text{n}$

d. ${}^{238}_{92}\text{U} + {}^{14}_{7}\text{N} \rightarrow {}^{247}_{99}\text{Es} + 5\,{}^{1}_{0}\text{n}$

11.7 step 1: ${}^{210}_{82}\text{Pb} \rightarrow {}^{210}_{83}\text{Bi} + {}^{0}_{-1}\beta$

step 2: ${}^{210}_{83}\text{Bi} \rightarrow {}^{210}_{84}\text{Po} + {}^{0}_{-1}\beta$

step 3: ${}^{210}_{84}\text{Po} \rightarrow {}^{206}_{82}\text{Pb} + {}^{4}_{2}\alpha$

11.8

Type of radiation	Speed	Penetration	Biological damage
alpha	slow	very little, stopped by skin or paper	ingestion--damages internal organs
beta	faster	penetrating, stopped by wood or aluminum foil	skin burns, ingestion--damages internal organs
gamma	fastest (speed of light)	very penetrating, stopped by lead or concrete	ionizing radiation causes serious damage to all tissues

Answers to Self-Test

The numbers in parentheses refer to sections in your textbook:
1. F; chemical reactions (11.13) **2**. F; gamma rays (11.2) **3**. T (11.3) **4**. T (11.8)
5. F; one-half (11.4) **6**. F; nuclear fusion (11.12) **7**. T (11.12) **8**. F; electrons (11.7)
9. F; larger (11.13) **10**. F; therapeutic application (11.11) **11**. T (11.9)
12. F; an unpaired electron (11.7) **13**. T (11.13) **14**. T (11.10) **15**. F; another (11.5)
16. b (11.3) **17**. a (11.3) **18**. d (11.4) **19**. b (11.4) **20**. b (11.2) **21**. a (11.12)
22. e (11.8) **23**. b (11.12) **24**. b (11.11)

Saturated Hydrocarbons — Chapter 12

Chapter Overview

Carbon compounds are the basis of life on Earth; all organic materials are carbon-based. The hydrocarbons, which you will study in this chapter, make up petroleum and are thus an important part of the industrial world, as fuel and in the manufacture of synthetic materials.

In this chapter you will find out how carbon forms such a vast variety of compounds. You will write structural and condensed structural formulas for alkanes and cycloalkanes and name them according to the IUPAC rules. You will identify and draw structural isomers and *cis-trans* isomers. You will write equations for the two major reactions of hydrocarbons, and name halogenated hydrocarbons.

Practice Exercises

12.1 The structures of **alkanes** (Sec. 12.4) and other **organic** (Sec. 12.1) compounds are usually represented in two dimensions, rather than three. The common types of representations are: **expanded structural formula** (Sec. 12.5), which shows all atoms and all bonds, **condensed structural formula** (Sec. 12.5), which shows groupings of atoms, and **skeletal formula** (Sec. 12.5), which shows carbon atoms and bonds, but omits hydrogen atoms.

Complete the following table of structural representations. All carbons are connected in a straight chain.

Molecular formula	Condensed structural formula	Expanded structural formula	Skeletal formula
C_4H_{10}		H H H H │ │ │ │ H—C—C—C—C—H │ │ │ │ H H H H	
	CH_3—CH_2—CH_2—CH_2—CH_3		
			C—C—C—C—C—C

12.2 The molecular formulas of alkanes fit the general formula C_nH_{2n+2}, where n is the number of carbon atoms present. Using the general formula for alkanes, complete the information in the table below:

Number of carbon atoms	Molecular formula	Total number of atoms	Total number of bonds
7 carbon atoms			
	C_9H_{20}		
		38	

12.3 **Structural isomers** (Sec. 12.6) are compounds with the same molecular formula but different structural formulas. Draw and name three structural isomers of heptane, C_7H_{16}. Use condensed structural formulas. Rules for naming alkanes can be found in section 12.8 of your textbook. (There are nine possible structural isomers for this molecular formula.)

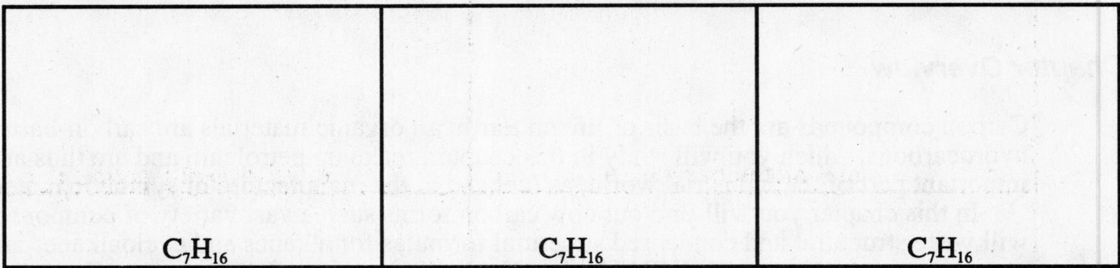

12.4 **Conformations** (Sec. 12.7) are differing orientations of a molecule made possible by rotation about a single bond. They are not isomers since one form can change to another without breaking or forming bonds.

Using condensed structural formulas, draw two conformations of the straight chain alkane, heptane, C_7H_{16}.

12.5 The IUPAC rules for naming organic compounds make it possible to give each compound a name that uniquely identifies it and also to draw its structural formula from that name.

Using the rules for nomenclature found in section 12.8 in your textbook, give the IUPAC name for each of the following structures:

a.
CH3 (on first CH2), CH3 (on CH)
$CH_2-CH_2-CH-CH_3$

b.
CH3 CH3
$CH_3-CH_2-CH_2-CH-CH-CH_3$

c.
CH3 (on first CH2); CH3–CH2 (on first CH); CH3 (on second CH)
$CH_2-CH_2-CH_2-CH-CH-CH_3$

d.
CH3–CH2–CH2 (on first CH); CH3 (on first CH2); CH3 CH3 (on second and third CH)
$CH_2-CH-CH_2-CH-CH-CH_2-CH_3$

12.6 Once you have learned the IUPAC rules for naming organic compounds, you can translate the name of an alkane into a structural formula.

Give the condensed structural formulas for the following alkanes:

a. 3-methylhexane	b. 2,2-dimethylpentane
c. 3,5-dimethylheptane	d. 2,2,4,4-tetramethylhexane

12.7 Each carbon atom in a **hydrocarbon** (Sec. 12.3) can be classified according to the number of other carbon atoms to which it forms bonds: a primary carbon atom is bonded to one other carbon atom, a secondary to two, a tertiary to three, and a quaternary to four other carbon atoms.

Using the structural formulas that you drew in Practice Exercise 12.6, determine the total number of **primary, secondary, tertiary, and quaternary carbons** (Sec. 12.9) in each compound.

Compound	Primary carbons	Secondary carbons	Tertiary carbons	Quaternary carbons
3-methylhexane				
2,2-dimethylpentane				
3,5-dimethylheptane				
2,2,4,4-tetramethylhexane				

12.8 There are four common **alkyl groups** (Sec. 12.8) containing **branched chains** (Sec. 12.6) that you should learn to name and draw.

Complete the following table by drawing the structural formulas for these four branched chain alkyl groups:

a. isopropyl	b. tert-butyl	c. isobutyl	d. sec-butyl

12.9 In a **cycloalkane** (Sec. 12.11), the carbon atoms are attached to one another in a ring-like arrangement. The general formula for cycloalkanes is C_nH_{2n}. IUPAC naming procedures for cycloalkanes are found in section 12.12 of your textbook. **Line-angle drawings** (Sec. 12.10) are often used to represent cycloalkane structures. A line-angle drawing can also denote a chain of carbon atoms using a sawtooth pattern of lines.

Give the IUPAC names for the following cycloalkanes:

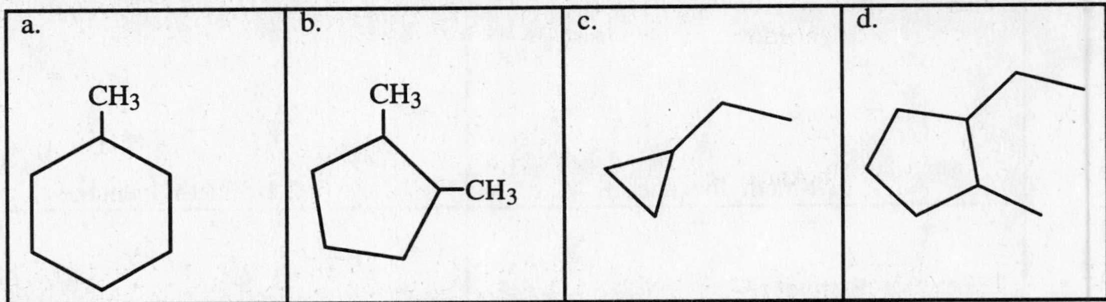

12.10 ***Cis-trans*** **isomers** (Sec. 12.13) are compounds that have the same molecular and structural formulas, but that have different spatial arrangements of their atoms because rotation is restricted around bonds.

Give the IUPAC name for the *cis-trans* isomers, a. and b.

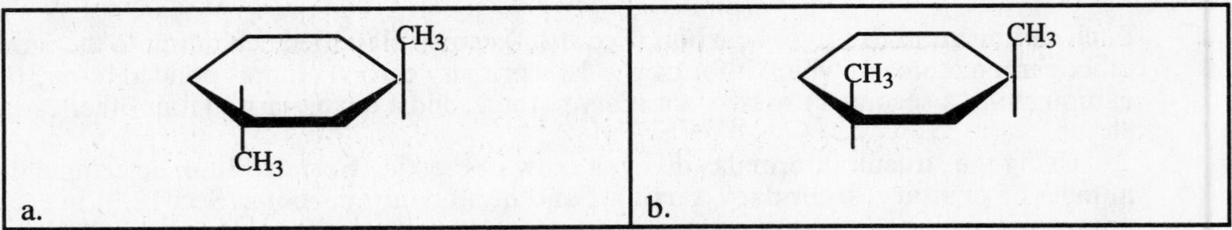

Draw a cyclohexane ring with 2 methyl groups, one on carbon-1 and one on carbon-2 (structure c.). Draw a second cyclohexane ring with 2 methyl groups on carbon-1 (structure d). Determine whether *cis-trans* isomerism is possible for structures c. and/or d. Give the IUPAC name for structures c. and d.

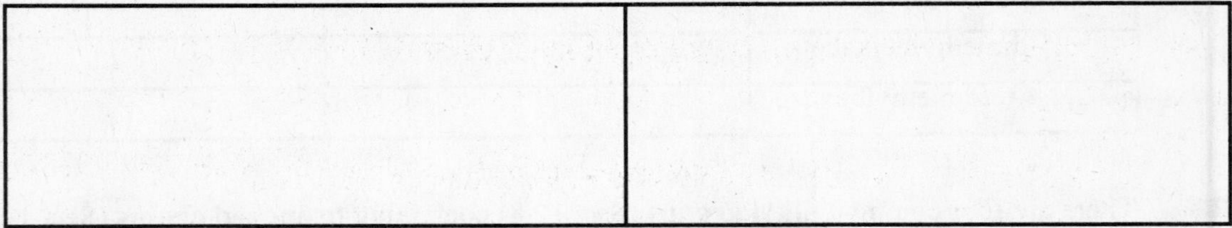

12.11 The problems in this exercise give a review of structural isomerism in alkanes. Compare the following pairs of molecules. Use line-angle notation to draw a structural formula for each molecule, and write its molecular formula. Tell whether the molecules in each pair are isomers of one another.

a. 1-methylcyclohexane and 1,3-dimethylcyclobutane

b. 2,3-dimethylpentane and 2,2,3-trimethylbutane

c. ethylcyclopropane and 2-methylbutane

12.12 One of the principal reactions of hydrocarbons is **combustion** (Sec. 12.16). Complete combustion is the reaction of a hydrocarbon with oxygen to produce carbon dioxide and water.

Write a balanced equation for the complete combustion of the following alkanes:

a. pentane

b. 2,3-dimethylhexane

12.13 The other important reaction of alkanes is **halogenation** (Sec. 12.16). The compounds shown below are products of halogenation of alkanes. Give the IUPAC name for each compound:

$CH_3-CH_2-CH_2-CH(Cl)-CH_3$ a.	$CH_3-CH_2-CH(Cl)-CH_2-Cl$ b.
$CH_3-CH_2-CH(Cl)-CH(CH_3)-CH_3$ c.	Br, Br (cyclopentane ring) d.

12.14 Draw and name the four structural isomers of dichloropropane, obtained by the substitution in propane of two atoms of chlorine for two atoms of hydrogen.

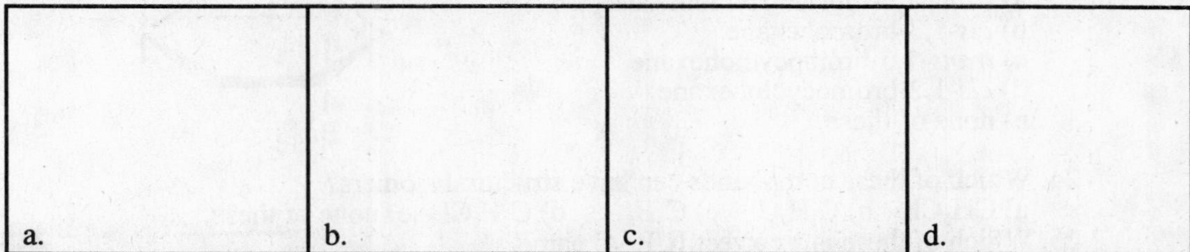

Self-Test

True-false: Indicate whether the following statements are true or false. If the statement is false, give the word or phrase that may be substituted for the underlined portion to make the statement true.

1. The smallest alkane is ethane.
2. Alkyl groups cannot rotate around the single bonds between the carbons in the ring of a cycloalkane.
3. Straight chain pentane would have a higher boiling point than its structural isomer, 2,2-dimethylpropane.
4. Carbon atoms can form compounds containing chains and rings because carbon has six valence electrons.
5. A hydrocarbon contains only carbon, hydrogen, and oxygen atoms.
6. A saturated hydrocarbon contains at least one carbon-carbon double bond or triple bond.
7. The alkane 2-methylpentane contains one tertiary carbon atom.
8. An acyclic hydrocarbon contains carbon atoms arranged in a ring structure.
9. The alkyl group $-CH_2-CH_2-CH_3$ is named propanyl.

10. It takes a minimum of three carbon atoms to form a cyclic arrangement of carbon atoms.
11. The formula for a cycloalkane is C_nH_{2n}.
12. Natural gas consists mainly of ethane.
13. Alkanes are good preservatives for metals because they have low boiling points.
14. The reaction of alkanes with oxygen to form carbon dioxide and water is a substitution reaction.
15. The IUPAC name for isopropyl chloride is 2-chloropropane.
16. The compounds called CFCs are chlorofluorocarbons.

Multiple choice:

17. The straight chain alkane having the molecular formula C_6H_{14} is called:
    a) hexane b) heptane c) pentane d) nonane e) none of these
18. Using IUPAC rules for naming, if the parent compound is pentane, an ethyl group could be attached to which carbon:
    a) 1 b) 2 c) 3 d) 4 e) 5
19. How many structural isomers can butane have?
    a) 2 b) 3 c) 4 d) 5 e) 6
20. Cyclopropane has the following molecular formula:
    a) CH_4 b) C_2H_6 c) C_3H_8 d) C_4H_{10} e) none of these
21. Which of the compounds below is an isomer of hexane?
    a) methylcyclopentane b) 2-methylbutane c) 3-ethylpentane d) 2-methylpentane e) none of these
22. Which of the following compounds forms *cis-trans* isomers?
    a) 2,3-dimethylpentane b) 2,2-dimethylpentane c) 1,1-dimethylcyclopentane d) 1,2-dimethylcyclopentane e) none of these
23. Choose the correct name for this compound:
    a) *trans*-1,3-dibromocyclohexane
    b) *cis*-1,3-bromohexane
    c) *trans*-1,5-bromocyclohexane
    d) *cis*-1,2-bromocyclohexane
    e) none of these

Br

Br

24. Which of these compounds can have structural isomers?
    a) CH_3Cl b) C_3H_7Cl c) C_3H_8 d) C_2H_5Cl e) none of these
25. Which of these is a correct IUPAC name?
    a) 2-methylcyclobutane b) *cis*-2,3-dimethylpentane c) 1-methylbutane d) 3-ethylhexane e) *cis*-1,2-methypropane
26. How many possible isomers can be written for dichloropropane?
    a) 2 b) 3 c) 4 d) 5 e) 6
27. How many possible isomers can be written for dimethylcyclobutane?
    a) 2 b) 3 c) 4 d) 5 e) 6

28. Draw structures for the following compounds:

2-bromo-3-methylbutane	*trans*-1,2-dichlorocyclopropane

Answers to Practice Exercises

12.1

Molecular formula	Condensed structural formula	Expanded structural formula	Skeletal formula
C_4H_{10}	$CH_3—CH_2—CH_2—CH_3$	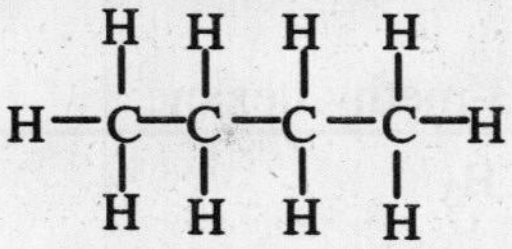	C—C—C—C
C_5H_{12}	$CH_3—CH_2—CH_2—CH_2—CH_3$	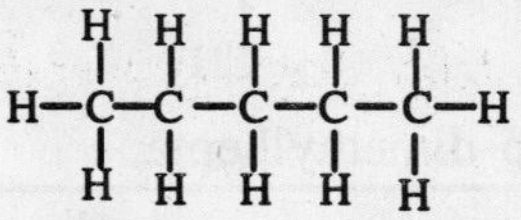	C—C—C—C—C
C_6H_{14}	$CH_3—CH_2—CH_2—CH_2—CH_2—CH_3$	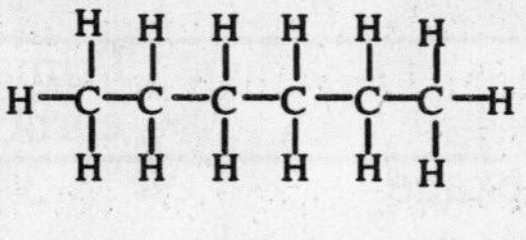	C—C—C—C—C—C

12.2

Number of carbons	Molecular formula	Total number of atoms	Total number of bonds
7 Carbons	C_7H_{16}	23	22
9 Carbons	C_9H_{20}	29	28
12 Carbons	$C_{12}H_{26}$	38	37

12.3

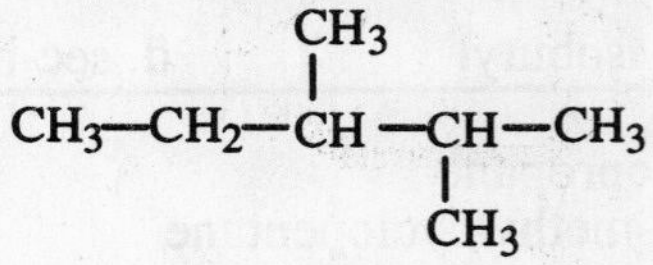	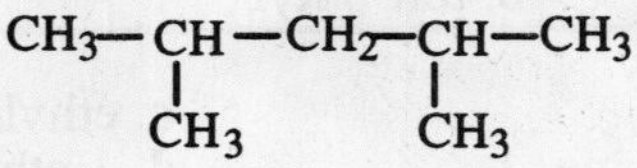	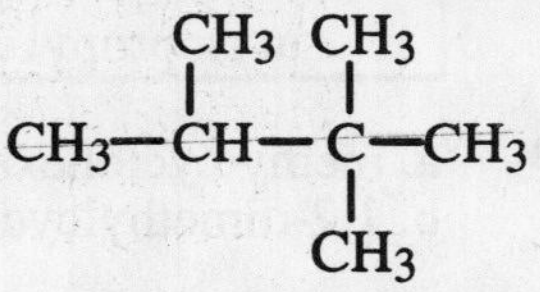
2,3-dimethylpentane	2,4-dimethylpentane	2,2,3-trimethylbutane

12.4

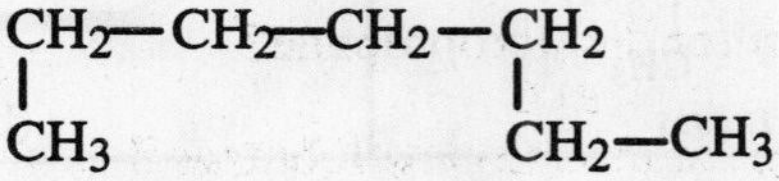	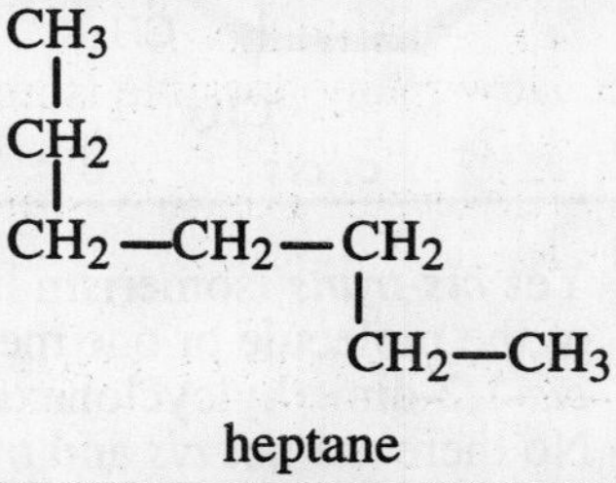
heptane	heptane

There are many more conformations of heptane; these are just two of the representations.

12.5 a. 2-methylpentane
b. 2,3-dimethylhexane
c. 3-ethyl-2-methylheptane
d. 6-ethyl-3,4-dimethylnonane

12.6

$CH_3-CH_2-CH(CH_3)-CH_2-CH_2-CH_3$ a. 3-methylhexane	$CH_3-C(CH_3)_2-CH_2-CH_2-CH_3$ b. 2,2-dimethylpentane
$CH_3-CH_2-CH(CH_3)-CH_2-CH(CH_3)-CH_2-CH_3$ c. 3,5-dimethylheptane	$CH_3-C(CH_3)_2-CH_2-C(CH_3)_2-CH_2-CH_3$ d. 2,2,4,4-tetramethylhexane

12.7

	Primary carbons	Secondary carbons	Tertiary carbons	Quaternary carbons
3-methylhexane	3	3	1	0
2,2-dimethylpentane	4	2	0	1
3,5-dimethylheptane	4	3	2	0
2,2,4,4-tetramethylhexane	6	2	0	2

12.8

$CH_3-CH(CH_3)-$ a. isopropyl	$CH_3-C(CH_3)_2-$ b. tert-butyl	$CH_3-CH(CH_3)-CH_2-$ c. isobutyl	$CH_3-CH_2-CH(CH_3)-$ d. sec-butyl

12.9 a. methylcyclohexane
b. 1,2-dimethylcyclopentane
c. ethylcyclopropane
d. 1-ethyl-2-methylcyclopentane

12.10 a. *trans*-1,3-dimethylcyclohexane
b. *cis*-1,3-dimethylcyclohexane

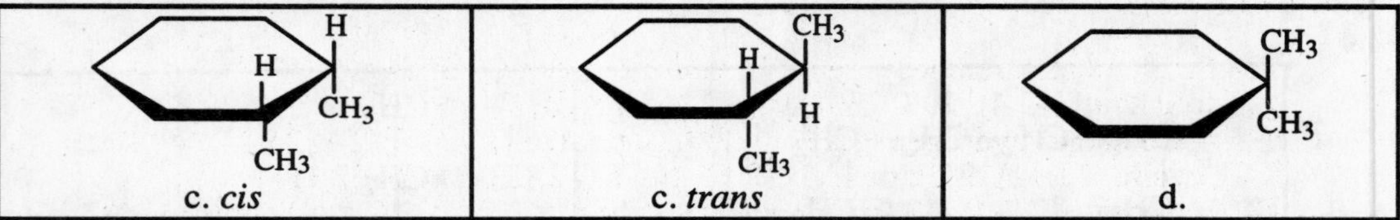

c. *cis* c. *trans* d.

c. Yes *cis-trans* isomerism is possible because the methyl groups can be on the same side of the molecule or one methyl group can be above and one below. The molecules are *cis*-1,2-dimethylcyclohexane and *trans*-1,2-dimethylcyclohexane.
d. No there are not *cis* and *trans* forms of this molecule since both methyl groups are attached to the same carbon.

12.11 a.

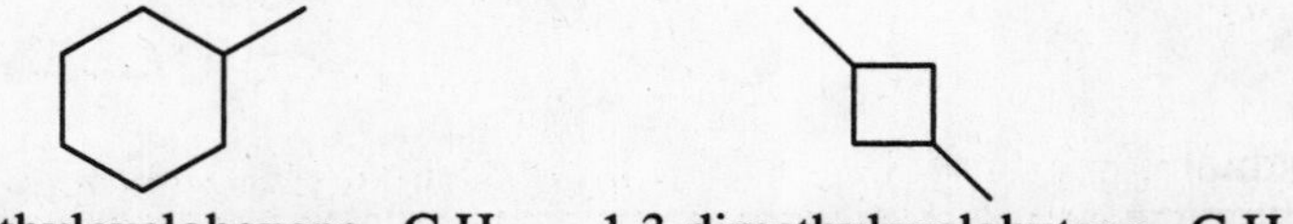
1-methylcyclohexane C_7H_{14} 1,3-dimethylcyclobutane C_6H_{12}

not isomers (different number of carbons and hydrogens)

b.

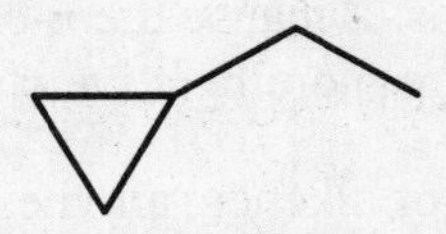

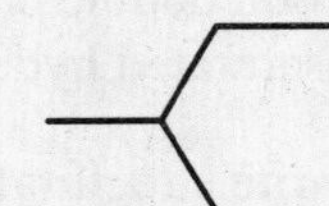

isomers (same molecular formula)

2,3-dimethylpentane C_7H_{16} 2,2,3-trimethylbutane C_7H_{16}

c.

not isomers (different numbers of hydrogens)

ethylcyclopropane C_5H_{10} 2-methylbutane C_5H_{12}

12.12 a. pentane

$$C_5H_{12} + 8O_2 \rightarrow 5CO_2 + 6H_2O$$

b. 2,3-dimethylhexane

$$2C_8H_{18} + 25O_2 \rightarrow 16CO_2 + 18H_2O$$

12.13 a. 2-chloropentane
b. 1,2-dichlorobutane
c. 3-chloro-2-methylpentane
d. *trans*-1,3-dibromocyclopentane

12.14

$CH_3-CH_2-CH(Cl)-Cl$	$CH_3-CH(Cl)-CH_2-Cl$	$CH_3-C(Cl)_2-CH_3$	$CH_2(Cl)-CH_2-CH_2(Cl)$
a.	b.	c.	d.

a. 1,1-dichloropropane
b. 1,2-dichloropropane
c. 2,2-dichloropropane
d. 1,3-dichloropropane

Answers to Self-Test

The numbers in parentheses refer to sections in your textbook:
1. F; methane (12.4) **2**. T (12.13) **3**. T (12.15) **4**. F; four (12.2)
5. F; carbon and hydrogen (12.3) **6**. F; unsaturated (12.3) **7**. T (12.9)
8. F; cyclic (12.11) **9**. F; propyl (12.8) **10**. T (12.11) **11**. T (12.11)
12. F; methane (12.4) **13**. F; low water solubility (12.15) **14**. F; combustion (12.16)
15. T (12.17) **16**. T (12.17) **17**. a (12.8) **18**. c (12.8) **19**. a (12.6) **20**. e; C_3H_6 (12.11)
21. d (12.6) **22**. d (12.13) **23**. a (12.13) **24**. b (12.6, 12.17) **25**. d (12.8, 12.12, 12.13)
26. c (12.6, 12.17) **27**. d (12.13)
28.

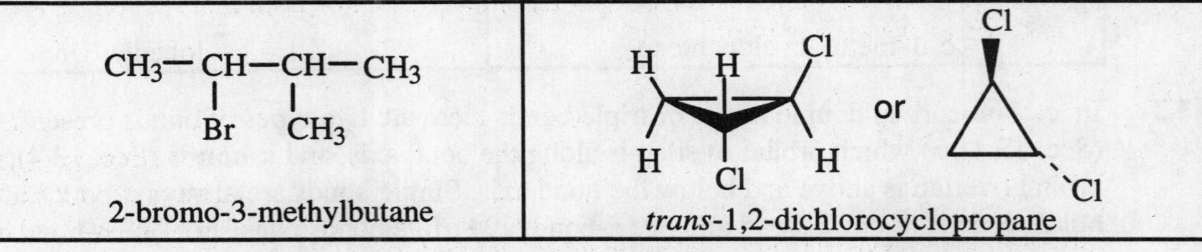

Unsaturated Hydrocarbons — Chapter 13

Chapter Overview

Unsaturated hydrocarbons contain fewer than the largest possible number of hydrogen atoms since they have one or more carbon-carbon double or triple bonds. Because the π-bond is more easily broken than the σ-bond, unsaturated hydrocarbons are more reactive chemically than saturated hydrocarbons.

In this chapter you will name and write structural formulas for alkenes, alkynes, and aromatic hydrocarbons. You will learn the characteristics of these compounds, their physical and chemical properties, and their most common reactions.

Practice Exercises

13.1 An **alkene** (Sec. 13.2) contains one or more carbon-carbon double bonds. The IUPAC names for alkenes are similar to those for alkanes, with the *-ene* ending replacing the *-ane* ending. The longest carbon chain must contain the double bond. The location of the double bond is indicated with a single number, that of the first carbon atom of the double bond.

Write the IUPAC names for the following alkenes and **cycloalkenes** (Sec. 13.2).

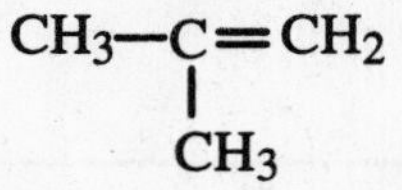 $CH_3-C(CH_3)=CH_2$ a.	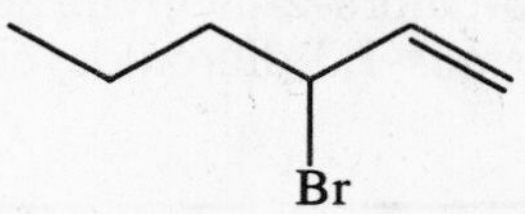 Br b.
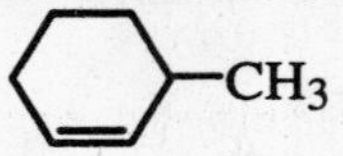 $-CH_3$ c.	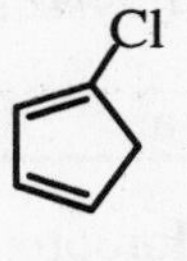 Cl d.

13.2 Draw structural formulas for the following alkenes and cycloalkenes.

a. 3-bromo-1-pentene	b. 2-methyl-1,3-pentadiene
c. 1-methylcyclopentene	d. 1,4-cycloheptadiene

13.3 In carbon-carbon double bonds or triple bonds there are two types of bonds present: **σ-bonds** (Sec. 13.4), in which orbital overlap is along the bond axis, and **π-bonds** (Sec. 13.4), in which orbital overlap is above and below the bond axis. Single bonds are always σ-bonds, double bonds consist of one σ-bond and one π-bond, and triple bonds consist of one σ-bond and two π-bonds.

Determine the number of carbon-carbon σ-bonds and carbon-carbon π-bonds in structural formulas a. through d. of Practice Exercise 13.1. Ignore carbon-hydrogen bonds.

a.
b.
c.
d.

13.4 *Cis-trans* isomerism is possible for some alkenes, since the π-bond prevents rotation around the double bond. For *cis-trans* isomers to exist, each of the two carbons of the double bond must have two different groups attached to it.

Determine whether each of these alkenes can exist as *cis-trans* isomers. If isomers do exist, draw and give the IUPAC name for each isomer.

$CH_3-\underset{\Large CH_3}{\underset{\vert}{C}}=CH_2$ a.	$CH_3-CH=CH-CH_2-CH_3$ b.
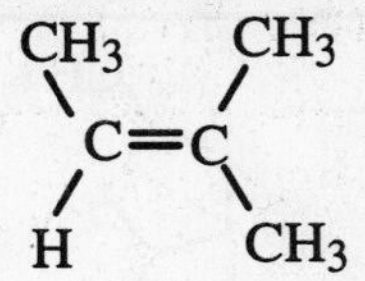 $(CH_3)(H)C=C(CH_3)(CH_3)$ [CH_3 and CH_3 above, H and CH_3 below] c.	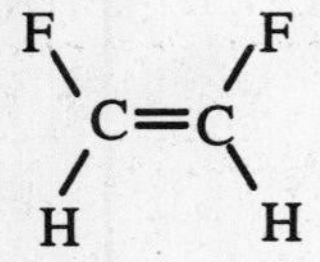 $(F)(H)C=C(F)(H)$ [F and F above, H and H below] d.

13.5 Draw the structural formulas for the following alkenes.

a. *trans*-3-hexene	b. *cis*-2-hexene	c. *trans*-1,2-dibromoethene

13.6 The most important reactions of alkenes are **addition reactions** (Sec. 13.7). Hydrogenation of an alkene involves the addition of a hydrogen atom to each carbon of the double bond. This is accomplished by heating the alkene and H_2 in the presence of a catalyst. Halogenation involves the use of Br_2 or Cl_2 to add a halogen atom to each carbon of the double bond.

Complete the following reactions. Give the structural formula for the product.

a. $CH_3-CH_2-CH=CH_2 + H_2 \xrightarrow{\text{catalyst}}$

b. $CH_3-CH_2-CH=CH_2 + Br_2 \longrightarrow$

13.7 Addition reactions may also be **unsymmetrical** (Sec. 13.7); different atoms or groups of atoms are added to the carbons of the double bond. Hydrohalogenation and hydration are important types of unsymmetrical addition. **Markovnikov's rule** (Sec. 13.7) states that, in an unsymmetrical addition, the hydrogen atom from the molecule being added becomes attached to the unsaturated carbon atom that already has the most hydrogen atoms.

Complete the following reactions. In each case give the structural formula for the major expected product.

a. $CH_3-CH_2-CH=CH_2 + HBr \longrightarrow$

b. $CH_3-CH_2-CH=CH_2 + H_2O \xrightarrow{H_2SO_4}$

c. 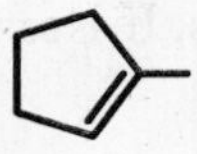 1-methylcyclopentene (ring)$-CH_3 + HBr \longrightarrow$

d. 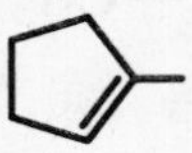 1-methylcyclopentene (ring)$-CH_3 + H_2O \xrightarrow{H_2SO_4}$

13.8 Write the name of the alkene that could be used to prepare each of the following compounds. Remember Markovnikov's rule.

a.	b.	c.
$CH_3-CH_2-CH_2-CH(Br)-CH_3$	$CH_3-CH(CH_3)-CH(OH)-CH_3$	cyclopentane ring with CH_3 and Br

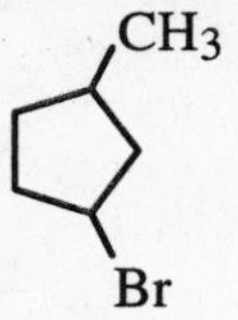

13.9 How many molecules of hydrogen gas, H_2, would react with one molecule of each of the following compounds? Give the structure and IUPAC name for each product.

a.	b.
$CH_2=CH-CH_2-CH=CH_2$	cycloheptadiene ring

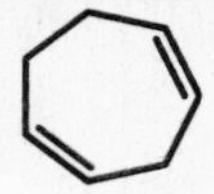

13.10 A **polymer** (Sec. 13.8) is a very large molecule composed of many identical repeating units. Alkenes can form **addition polymers** (Sec. 13.8) when the alkene monomers simply add together. The π-bond of the alkene is broken and those electrons become involved in carbon-carbon bonds between monomers.

Draw the "start" (first four repeating units) of the structural formula of the addition polymers prepared from the following monomers.

a. $CH_2=CH-CN$
b. $CH_2=CH-CH_2-CH_3$

13.11 An **alkyne** (Sec. 13.9) has one or more carbon-carbon triple bonds. The rules for naming alkynes are the same as those for naming alkenes, with the ending *-yne* instead of *-ene*.

Give the IUPAC names for the following alkynes:

a. $CH_3—CH_2—CH_2—C{\equiv}CH$	b. (skeletal structure with Br)	c. $CH_3—CH_2—C(CH_3)_2—C{\equiv}C—CH_3$

13.12 Write structural formulas for the following alkynes:

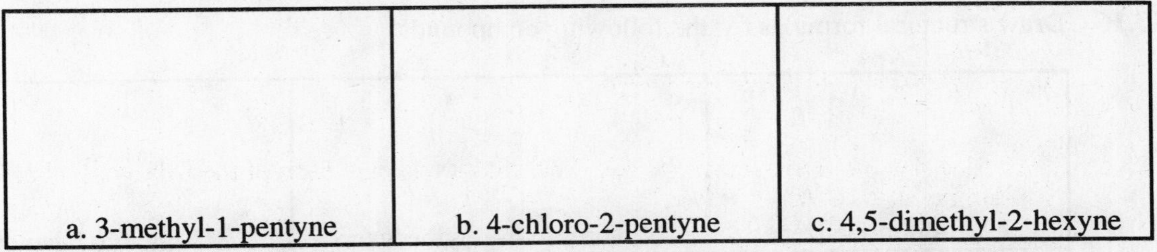

a. 3-methyl-1-pentyne
b. 4-chloro-2-pentyne
c. 4,5-dimethyl-2-hexyne

13.13 Addition reactions of alkynes are similar to those of alkenes. However, two molecules of a specific reactant can add to the triple bond.

Complete the following reactions. Give the structural formula for the major expected product.

a. $CH_3—CH_2—CH_2—C{\equiv}CH + 2\,Cl_2 \longrightarrow$

b. $CH_3—CH_2—CH_2—C{\equiv}CH + 2\,HBr \longrightarrow$

c. $CH_3—CH_2—CH_2—C{\equiv}CH + 1\,HCl \longrightarrow$

d. (skeletal alkyne) $+ \; 2\,H_2 \xrightarrow{Ni}$

13.14 **Aromatic hydrocarbons** (Sec. 13.10) are unsaturated cyclic compounds whose bonding can be represented by **resonance structures** (Sec. 13.10) indicating **delocalized bonding** (Sec. 13.10). The naming of the compounds below is based on the aromatic hydrocarbon benzene. The name of the substituent on the benzene ring is used as a prefix.

Give the IUPAC name for the following aromatic compounds:

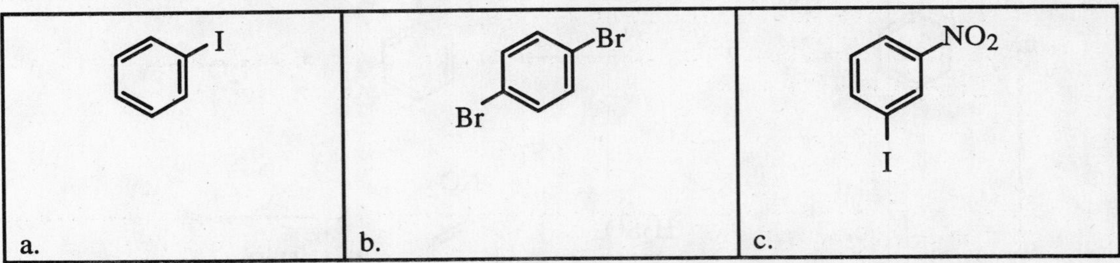

13.15 If the group attached to the benzene ring is not easily named as a substituent, the benzene ring is treated as the attached group and called a *phenyl* group. The compound is then named as an alkane, an alkene, or an alkyne.

Write the IUPAC name for the following phenyl-substituted compounds:

a. $CH_3-CH_2-CH_2-CH(C_6H_5)-CH_2Cl$	b. $CH_3-CH=CH-CH(C_6H_5)-CH_3$

13.16 Draw structural formulas for the following compounds:

a. 1,3-diiodobenzene	b. 1-bromo-4-ethylbenzene	c. 3-phenyl-1-hexene

13.17 Aromatic hydrocarbons do not readily undergo addition reactions since π-bonding stabilizes the aromatic ring. The most important reactions of aromatic compounds are substitution reactions. These include alkylation (using alkyl halides and the catalyst $AlCl_3$), halogenation (using Br_2 or Cl_2 in the presence of a catalyst), nitration (using nitric acid), and sulfonation (using sulfuric acid).

Complete the following equations by supplying the missing information.

a. ? $\xrightarrow{AlCl_3}$ $C_6H_5-CH_3$ + HCl

b. ? $\xrightarrow{FeBr_3}$ C_6H_5-Br + HBr

c. C_6H_6 + ? $\longrightarrow$ $C_6H_5-SO_2\text{-}OH$ + H_2O

d. ? $\xrightarrow[\text{heat}]{H_2SO_4}$ $C_6H_5-NO_2$ + H_2O

13.18 Use this identification exercise to review your knowledge of the structures introduced in this chapter. Using structures A through I, give the best choice for each of the terms below.

CH_3 H C=C H CH_3 A.	H H C=C CH_3 CH_3 B.	$CH_3—CH_2—CH_2—CH_3$ C.
CH_3—(cyclohexane ring) D.	CH_3 Cl (benzene ring) E.	$CH_3—CH_2—C≡CH$ F.

a. aromatic compound ________
b. isomer of compound B ________
c. alkyne ________
d. cyclic alkane ________
e. *cis*-isomer of an alkene ________
f. straight chain alkane ________

Give the IUPAC names for structures A through F above:
g. structure A ________
h. structure B ________
i. structure C ________
j. structure D ________
k. structure E ________
l. structure F ________

Self-Test

True-false: Indicate whether the following statements are true or false. If the statement is false, give the word or phrase that may be substituted for the underlined portion to make the statement true.

1. The specific part of a molecule that governs the molecule's chemical properties is called a <u>functional</u> group.
2. The general formula for an <u>alkene</u> is C_nH_{2n}.
3. When naming alkenes by the IUPAC system, select as the parent carbon chain the longest chain that contains <u>at least one carbon atom</u> of the double bond.
4. The name of a cycloalkene <u>does not include</u> the numbered location of the double bond.
5. The carbon atoms at each end of a double bond have a <u>trigonal planar</u> arrangement of bonds.
6. A carbon-carbon σ-bond is <u>weaker</u> than a carbon-carbon π-bond.
7. Benzene undergoes an <u>addition</u> reaction with bromine in the presence of a catalyst.
8. <u>Propene</u> is the simplest alkene that has *cis-trans* isomerism.
9. If 2-butene is halogenated with bromine gas, the most probable product is <u>2-bromobutane</u>.
10. Hydration of an alkene produces <u>an alcohol</u>.
11. Polyethylene contains <u>many</u> double bonds.

Multiple choice:

12. According to Markovnikov's rule, addition of HBr to the double bond of 1-hexene would produce:
    a) 1-bromohexane
    b) 1,2-dibromohexane
    c) 2-bromo-1-hexene
    d) 2-bromohexane
    e) none of these
13. The geometry of the carbon-carbon triple bond of an alkyne is:
    a) linear
    b) trigonal planar
    c) tetrahedral
    d) angular
    e) none of these
14. Which of these is a correct name according to IUPAC rules?
    a) 2-methylbenzene
    b) 1-chlorobenzene
    c) 1-bromo-2-chlorobenzene
    d) 2,4-dichlorobenzene
    e) none of these
15. Another name for *meta*-dichlorobenzene is:
    a) 1,2-dichlorobenzene
    b) 2,3-dichlorobenzene
    c) 1,3-dichlorobenzene
    d) 1,4-dichlorobenzene
    e) none of these
16. What is the maximum number of carbon-carbon *sigma* bonds in 1-butene?
    a) 4 b) 1 c) 5 d) 3 e) none of these
17. Aromatic compounds have structures based on what parent molecule?
    a) benzene
    b) cyclopropane
    c) cyclohexane
    d) hexane
    e) none of these
18. Which of the following compounds does **not** have the formula C_5H_8?
    a) 1-methylcyclobutene
    b) 3-methyl-1-butyne
    c) 2-methyl-1-butene
    d) 1,3-pentadiene
    e) none of these
19. If you wished to prepare 2-bromo-3-methylbutane by the addition of HBr to an alkene, which of these alkenes would you use?
    a) 2-methyl-2-butene
    b) 2-methyl-1-butene
    c) 3-methyl-1-butene
    d) 3-methyl-2-butene
    e) none of these
20. Which of the following compounds would be the most likely to undergo an addition reaction with HCl?
    a) toluene
    b) cyclohexene
    c) nitrobenzene
    d) heptane
    e) none of these

Answers to Practice Exercises

13.1 a. 2-methylpropene;
b. 3-bromo-1-hexene;
c. 3-methylcyclohexene;
d. 1-chloro-1,3-cyclopentadiene

13.2

$CH_3—CH_2—CH(Br)—CH{=}CH_2$ a. 3-bromo-1-pentene	$CH_3—CH{=}CH—C(CH_3){=}CH_2$ b. 2-methyl-1,3-pentadiene
—CH_3 c. 1-methylcyclopentene	d. 1,4-cycloheptadiene

13.3 a. three sigma bonds and one pi bond
b. four sigma bonds and two pi bonds
c. seven sigma bonds and one pi bond
d. five sigma bonds and two pi bonds

13.4 a. no
b. yes; *trans*-2-pentene and *cis*-2-pentene

c. no
d. yes; *cis*-1,2-difluoroethene and *trans*-1,2-difluoroethene

F, F / C=C / H, H	F, H / C=C / H, F
cis-1,2-difluoroethene	*trans*-1,2-difluoroethene

13.5

CH_3-CH_2, H / C=C / H, CH_2-CH_3	H, H / C=C / CH_3, $CH_2-CH_2-CH_3$	Br, H / C=C / H, Br
a. *trans*-3-hexene	b. *cis*-2-hexene	c. *trans*-1,2-dibromoethene

13.6 a. $CH_3-CH_2-CH=CH_2 + H_2 \xrightarrow{\text{catalyst}} CH_3-CH_2-CH_2-CH_3$

b. $CH_3-CH_2-CH=CH_2 + Br_2 \longrightarrow CH_3-CH_2-\underset{Br}{\underset{|}{CH}}-\underset{Br}{\underset{|}{CH_2}}$

13.7 a. $CH_3-CH_2-CH=CH_2 + HBr \longrightarrow CH_3-CH_2-\underset{Br}{\underset{|}{CH}}-CH_3$

b. $CH_3-CH_2-CH=CH_2 + H_2O \xrightarrow{H_2SO_4} CH_3-CH_2-\underset{OH}{\underset{|}{CH}}-CH_3$

c. 1-methylcyclopentene $-CH_3$ + HBr ⟶ 1-bromo-1-methylcyclopentane (CH_3, Br)

d. 1-methylcyclopentene $-CH_3$ + H_2O $\xrightarrow{H_2SO_4}$ 1-methylcyclopentanol (CH_3, OH)

13.8 a. 1-pentene
b. 3-methyl-1-butene
c. 4-methylcyclopentene

13.9 a. 2 moles of hydrogen gas.
b. 2 moles of hydrogen gas.

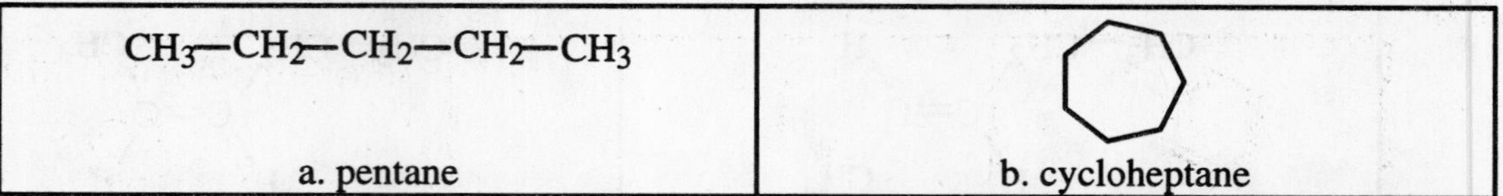

$CH_3-CH_2-CH_2-CH_2-CH_3$
a. pentane
b. cycloheptane

13.10 a.

$$\left(\underset{\text{H}}{\text{CH}}-\underset{\text{CN}}{\text{CH}}-\underset{\text{H}}{\text{CH}}-\underset{\text{CN}}{\text{CH}}-\underset{\text{H}}{\text{CH}}-\underset{\text{CN}}{\text{CH}}-\underset{\text{H}}{\text{CH}}-\underset{\text{CN}}{\text{CH}} \right)$$

b.

$$\left(\underset{\text{H}}{\text{CH}}-\underset{CH_2-CH_3}{\text{CH}}-\underset{\text{H}}{\text{CH}}-\underset{CH_2-CH_3}{\text{CH}}-\underset{\text{H}}{\text{CH}}-\underset{CH_2-CH_3}{\text{CH}}-\underset{\text{H}}{\text{CH}}-\underset{CH_2-CH_3}{\text{CH}} \right)$$

13.11 a. 1-pentyne
b. 4-bromo-1-butyne
c. 4,4-dimethyl-2-hexyne

13.12

$CH_3-CH_2-\underset{CH_3}{CH}-C\equiv CH$	$CH_3-\underset{Cl}{CH}-C\equiv C-CH_3$	$CH_3-\underset{CH_3}{CH}-\underset{CH_3}{CH}-C\equiv C-CH_3$
a. 3-methyl-1-pentyne	b. 4-chloro-2-pentyne	c. 4,5-dimethyl-2-hexyne

13.13

a. $CH_3-CH_2-CH_2-C\equiv CH + 2\,Cl_2 \longrightarrow CH_3-CH_2-CH_2-\overset{Cl}{\underset{Cl}{C}}-\overset{Cl}{\underset{Cl}{CH}}$

b. $CH_3-CH_2-CH_2-C\equiv CH + 2\,HBr \longrightarrow CH_3-CH_2-CH_2-\overset{Br}{\underset{Br}{C}}-CH_3$

c. $CH_3-CH_2-CH_2-C\equiv CH + 1\,HCl \longrightarrow CH_3-CH_2-CH_2-\underset{Cl}{C}=CH_2$

d. (1-pentyne, line structure) + $2\,H_2$ $\xrightarrow{Ni}$ (pentane, line structure)

13.14 a. iodobenzene
b. 1,4-dibromobenzene
c. 1-iodo-3-nitrobenzene

13.15 a. 1-chloro-2-phenylpentane
b. 4-phenyl-2-pentene

13.16

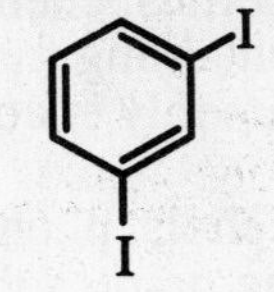 a. 1,3-diiodobenzene	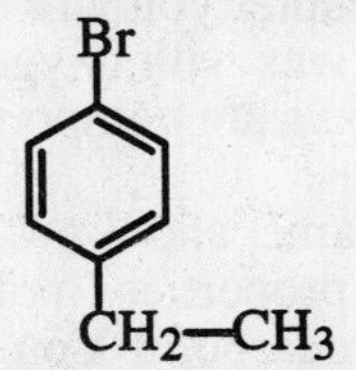 b. 1-bromo-4-ethylbenzene	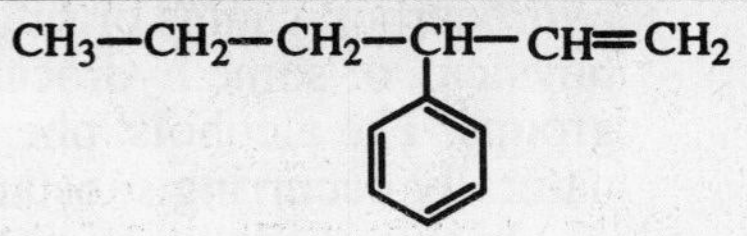 c. 3-phenyl-1-hexene
(benzene ring with I at positions 1 and 3)	Br–C_6H_4–CH_2–CH_3	CH_3—CH_2—CH_2—CH(C_6H_5)—CH=CH_2

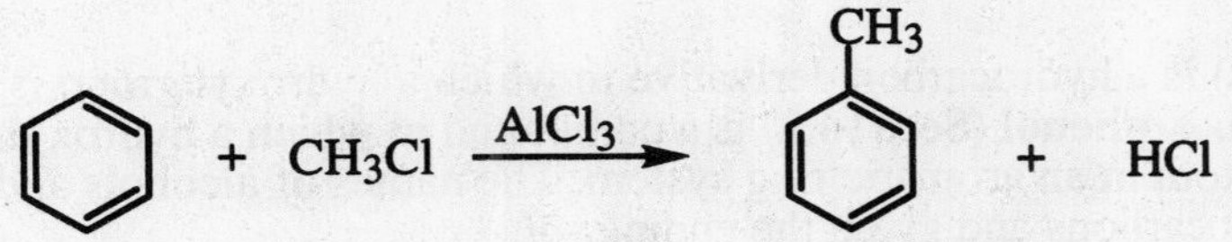

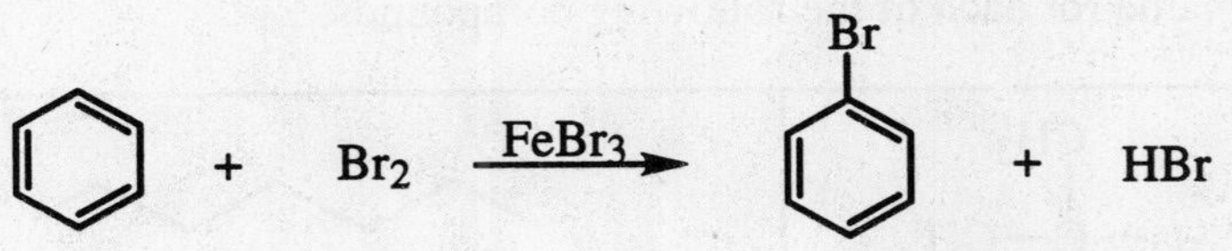

13.17 a. C_6H_6 + CH_3Cl $\xrightarrow{AlCl_3}$ $C_6H_5CH_3$ + HCl

b. C_6H_6 + Br_2 $\xrightarrow{FeBr_3}$ C_6H_5Br + HBr

c. C_6H_6 + H_2SO_4 $\longrightarrow$ $C_6H_5SO_2\text{-}OH$ + H_2O

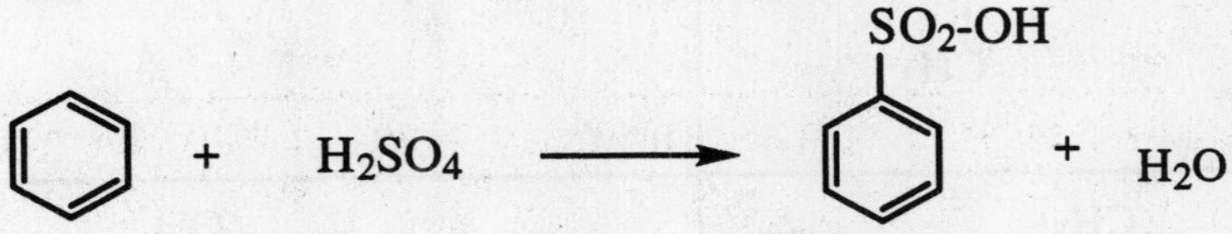

d. C_6H_6 + HO—NO_2 $\xrightarrow[\text{heat}]{H_2SO_4}$ $C_6H_5NO_2$ + H_2O

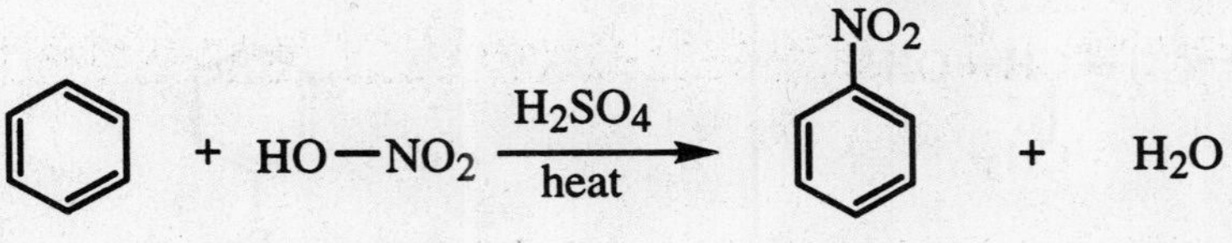

13.18 a. E; b. A; c. F; d. D; e. B; f. C;
g. *trans*-2-butene;
h. *cis*-2-butene;
i. butane;
j. methylcyclohexane;
k. 2-chlorotoluene or 1-chloro-2-methylbenzene
l. 1-butyne.

Answers to Self-Test

The numbers in parentheses refer to sections in your textbook:
1. T (13.1) **2**. T (13.2) **3**. F; both carbon atoms (13.3) **4**. T (13.3) **5**. T (13.2)
6. F; stronger (13.4) **7**. F; substitution (13.13) **8**. F; 2-butene (13.5)
9. F; 2,3-dibromobutane (13.7) **10**. T (13.7) **11**. F; no (13.8) **12**. d (13.7)
13. a (13.9) **14**. c (13.11) **15**. c (13.11) **16**. d (13.4) **17**. a (13.10) **18**. c (13.2, 13.3)
19. c (13.7) **20**. b (13.7, 13.13)

Alcohols, Phenols, and Ethers — Chapter 14

Chapter Overview

Most of the chemical reactions of organic molecules involve the functional groups on the hydrocarbon chains or rings. In this chapter you will consider the properties, chemical and physical, of some hydrocarbon derivatives with oxygen- and sulfur-containing functional groups. The alcohols, phenols, and ethers are compounds that are commonly encountered in naturally occurring substances.

In this chapter you will identify, name, and draw structures for alcohols, phenols, and ethers. You will compare the physical properties (melting point, boiling point, solubility in water) of these compounds to the hydrocarbons. You will write equations for the mild oxidation of alcohols, and for other reactions that alcohols undergo.

Practice Exercises

14.1 An **alcohol** (Sec. 14.2) is a hydrocarbon derivative in which a hydroxyl group is attached to a saturated carbon atom. A **phenol** (Sec. 14.2) is a compound in which a hydroxyl group is attached to a carbon atom in an aromatic ring system. The names of alcohols and phenols are derived from the hydrocarbons and given the ending *-ol*.

Give the IUPAC name for each of the following compounds:

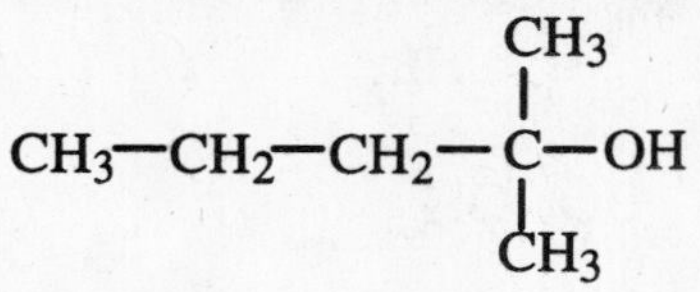 a. $CH_3-CH_2-CH_2-C(CH_3)_2-OH$	 b. (skeletal structure) OH
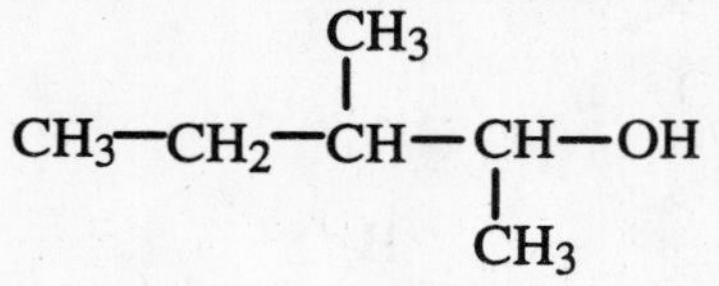 c. $CH_3-CH_2-CH(CH_3)-CH(CH_3)-OH$	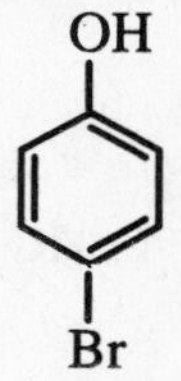 d. OH, Br (para-substituted benzene ring)

14.2 Draw the structural formula for the following compounds:

a. 3-hexanol	b. 2,3-dimethyl-1-butanol
c. 1-chloro-2-methyl-2-butanol	d. 3-propylphenol

14.3 One method for preparing alcohols is the hydration of alkenes in the presence of sulfuric acid. The predominant alcohol product can be determined from Markovnikov's rule.

Write an equation for the preparation of each of the following alcohols from the appropriate alkene.

a. 2-hexanol

b. cyclopentanol

14.4 Alcohols undergo **dehydration** (Sec. 14.7), the removal of a water molecule, when they are heated in the presence of a sulfuric acid catalyst. The product formed is dependent on the temperature of the reaction. At 180°C intramolecular dehydration, an **elimination reaction** (Sec. 14.7), produces an alkene. According to **Zaitsev's rule** (Sec. 14.7), the alkene with the greatest number of alkyl groups attached to the double bond will be formed. At a lower temperature (140°C) intermolecular dehydration, a **condensation reaction** (Sec. 14.7), produces an **ether** (Sec. 14.2).

Complete the following equations by supplying the missing information:

a. $CH_3{-}\underset{\displaystyle CH_3}{\underset{|}{CH}}{-}CH_2{-}OH \xrightarrow[H_2SO_4]{140°C}$?

b. $CH_3{-}CH_2{-}CH_2{-}\underset{\displaystyle OH}{\underset{|}{CH}}{-}CH_3 \xrightarrow[H_2SO_4]{180°C}$?

c. ? $\xrightarrow[H_2SO_4]{140°C} CH_3{-}\underset{\displaystyle CH_3}{\underset{|}{CH}}{-}O{-}\underset{\displaystyle CH_3}{\underset{|}{CH}}{-}CH_3$

d. ? $\xrightarrow[H_2SO_4]{180°C} CH_3{-}CH_2{-}CH_2{-}CH{=}\underset{\displaystyle CH_3}{\underset{|}{C}}{-}CH_3$

14.5 **Primary and secondary alcohols** (Sec. 14.7) undergo oxidation in the presence of a mild oxidizing agent to produce compounds that contain a carbon-oxygen double bond. **Tertiary alcohols** (Sec. 14.7) cannot be oxidized in this way since the carbon attached to the hydroxyl group is attached to three other carbons and cannot form a carbon-oxygen double bond.

Alcohols also undergo substitution reactions. An example is the replacement of the hydroxyl group by a halogen atom using PCl_3 or PBr_3 as a reagent.

Complete the following equations by supplying the missing information:

a. ? $\xrightarrow[\text{agent}]{\text{mild oxidizing}} CH_3{-}CH_2{-}CH_2{-}CH_2{-}\overset{\displaystyle H}{\overset{|}{C}}{=}O$

b. 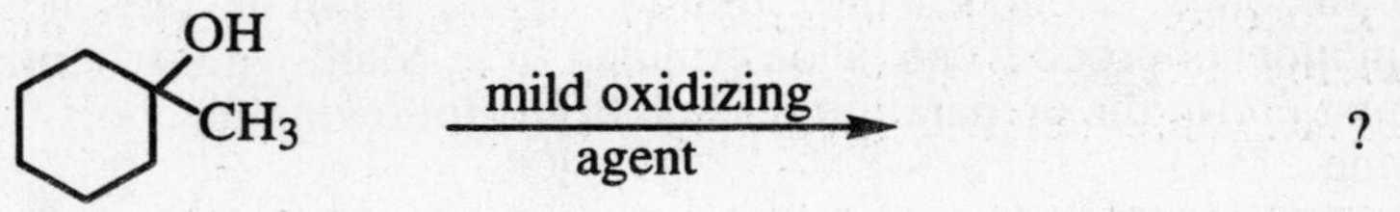 1-methylcyclohexanol (OH and CH_3 on the same ring carbon) $\xrightarrow[\text{agent}]{\text{mild oxidizing}}$?

c. 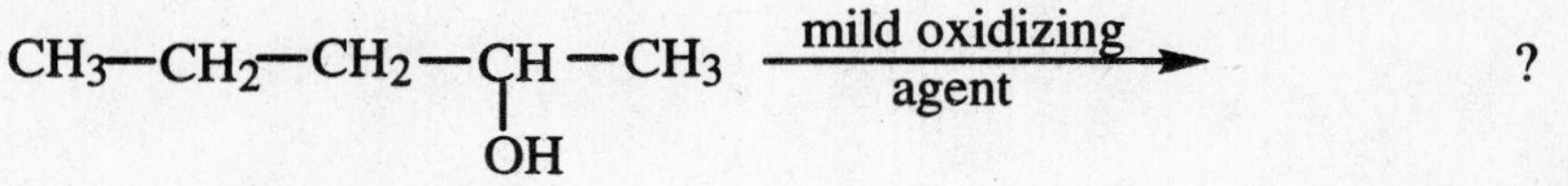 $CH_3-CH_2-CH_2-\underset{\displaystyle OH}{\underset{|}{CH}}-CH_3$ $\xrightarrow[\text{agent}]{\text{mild oxidizing}}$?

d.  ? $\xrightarrow{PCl_3}$ $CH_3-\underset{\displaystyle CH_3}{\underset{|}{CH}}-CH_2-\underset{\displaystyle Cl}{\underset{|}{CH}}-CH_3$

14.6 Ethers are compounds with two hydrocarbon groups attached to an oxygen atom. According to the IUPAC system, ethers are named as substituted hydrocarbons. The smaller hydrocarbon and the oxygen atom are called an **alkoxy group** (Sec. 14.10) and considered as a substituent on the larger hydrocarbon.

Write the IUPAC name for each of the following compounds:

 a. line-angle structure: five-carbon chain bonded to O, O bonded to CH_3	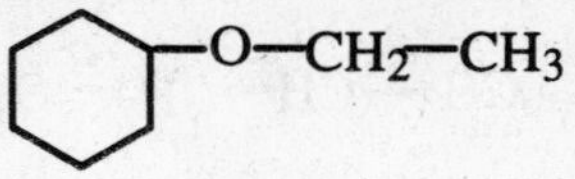 b. cyclohexane ring bonded to $-O-CH_2-CH_3$

14.7 Draw the structural formula and the line angle drawing for each of the following ethers:

a. 2-ethoxypentane	b. 1-chloro-2-ethoxycyclopentane

14.8 Many of the physical properties of an organic compound having an oxygen-containing functional group are determined by the molecule's ability to form hydrogen bonds with a like molecule or with water molecules.

Indicate which of the following pairs of compounds would have the higher boiling point and the greater solubility in water.

Compounds	Higher boiling point	Greater solubility in water
1-propanol and propane		
2-methyl-2-propanol and diethyl ether		
phenol and toluene		

14.9 **Sulfur analogs of alcohols are called *thiols*. These compounds contain a sulfhydryl group and are named by adding *-thiol* to the name of the parent alkane.**

Give the IUPAC name for the following thiols:

$CH_3-CH_2-CH_2-CH_2-CH_2-SH$	(cyclohexane ring)$-SH$
a.	b.

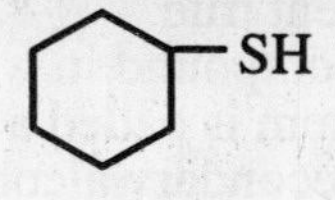

14.10 **Write the structural formulas for the following thiols:**

a. 2-methyl-1-butanethiol	b. 2-bromocyclobutanethiol

14.11 **Use this identification exercise to review your knowledge of the structures introduced in this chapter. Using structures A through I, give the best match for the terms below:**

$HO-CH_2-CH_2-OH$ A.	$CH_3-O-CH_2-CH_3$ B.	$CH_3-CH_2-CH_2-CH_3$ C.
$CH_3-CH_2-C(CH_3)_2-OH$ D.	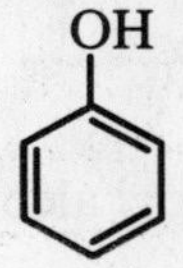 OH on benzene ring E.	$CH_3-CH_2-CH_2-OH$ F.
$CH_3-CH_2-CH=CH_2$ G.	CH_3-S-CH_3 H.	CH_3-CH_2-SH I.

a. 3° alcohol __________
b. glycol __________
c. alkane __________
d. ether __________
e. phenol __________
f. alkene __________
g. thiol __________
h. sulfide (thioether) __________
i. 1° alcohol __________

Give the IUPAC names for structures A through I.

j. structure A __________
k. structure B __________
l. structure C __________
m. structure D __________
n. structure E __________
o. structure F __________
p. structure G __________
q. structure H __________
r. structure I __________

Self-Test

True-false: Indicate whether the following statements are true or false. If the statement is false, give the word or phrase that may be substituted for the underlined portion to make the statement true.

1. A compound in which a hydroxyl group is attached to a carbon atom in an aromatic ring system is called a thiol.
2. A secondary alcohol has two hydrogen atoms attached to the hydroxyl carbon.
3. Ether molecules cannot form hydrogen bonds with other ether molecules.
4. Diethyl ether and 1-butanol are structural isomers.
5. Oxygen has six valence electrons, which include two nonbonding electron pairs.
6. 2-Butanol is an example of a tertiary alcohol.
7. Alcohols are more soluble in water than alkanes of similar molecular mass.
8. A symmetrical ether can be prepared by the intramolecular dehydration of an alcohol.
9. Phenols are used as antioxidants because they are low-melting solids.
10. Furan is an example of a heterocyclic organic compound.
11. Storage of ethers can be hazardous if unstable peroxides form.

Multiple choice:

12. An alcohol may be prepared by:
    a) hydrogenation of an alkene
    b) hydration of an alkene
    c) dehydration of a carbonyl group
    d) dehydrogenation of an alkane
    e) none of these
13. Mercaptans is an older term for:
    a) thiols
    b) phenols
    c) glycols
    d) ethers
    e) none of these
14. The mild oxidation of a secondary alcohol can result in the production of a(n):
    a) acid
    b) aldehyde
    c) ketone
    d) alkene
    e) none of these
15. The IUPAC name for isopropyl alcohol is:
    a) 1-propyl alcohol
    b) 2-propanol
    c) 2-propyl alcohol
    d) 1-propanol
    e) none of these
16. The common name for 1,2-ethanediol is:
    a) ethylene glycol
    b) glycerol
    c) diethyl ether
    d) 1,2-ethyl alcohol
    e) cresol
17. According to Zaitsev's rule, the main product of the dehydration of 2-methyl-2-butanol would be:
    a) 2-methyl-2-butene
    b) 2-methyl-1-butene
    c) 1-methyl-2-butene
    d) 1-methyl-1-butene
    e) none of these
18. Which of the following names is correct according to IUPAC rules?
    a) 2-ethyl-1-butene-2-ol
    b) 3-propylene-2-ol
    c) 3,4-butanediol
    d) 1,2-dimethylphenol
    e) none of these
19. Rank butane, butanol, and diethyl ether by boiling points, lowest boiling point first and highest boiling point last:
    a) butane, butanol, diethyl ether
    b) butanol, butane, diethyl ether
    c) butane, diethyl ether, butanol
    d) butanol, diethyl ether, butane
    e) diethyl ether, butane, butanol
20. The intramolecular dehydration of 2-pentanol produces:
    a) 2-methoxypentane
    b) 2-pentene
    c) 1-pentene
    d) 2-methylpentene
    e) none of these

21. An example of a cyclic ether is:
    a) phenol
    b) methoxybenzene
    c) catechol
    d) pyran
    e) methyl *tert*-butyl ether
22. Mild reduction of a disulfide produces:
    a) a thiol
    b) a peroxide
    c) a thioether
    d) a cyclic ether
    e) none of these

Answers to Practice Exercises

14.1 a. 2-methyl-2-pentanol;
b. 1-hexanol;
c. 3-methyl-2-pentanol;
d. 4-bromophenol

14.2

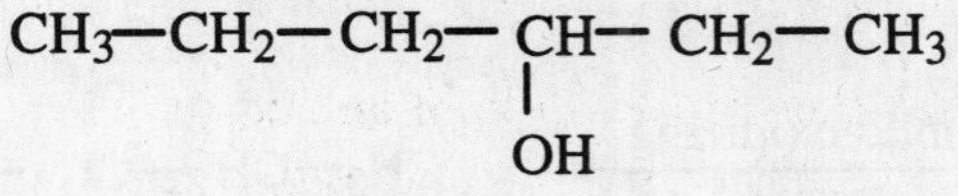

$CH_3-CH_2-CH_2-CH(OH)-CH_2-CH_3$ a. 3-hexanol	$CH_3-CH(CH_3)-CH(CH_3)-CH_2-OH$ b. 2,3-dimethyl-1-butanol
$CH_2(Cl)-C(CH_3)(OH)-CH_2-CH_3$ c. 1-chloro-2-methyl-2-butanol	benzene ring with OH and $CH_2-CH_2-CH_3$ in meta positions d. 3-propylphenol

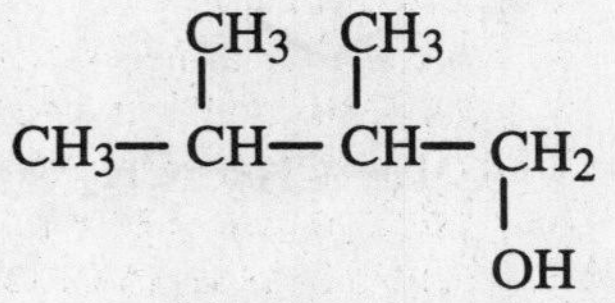

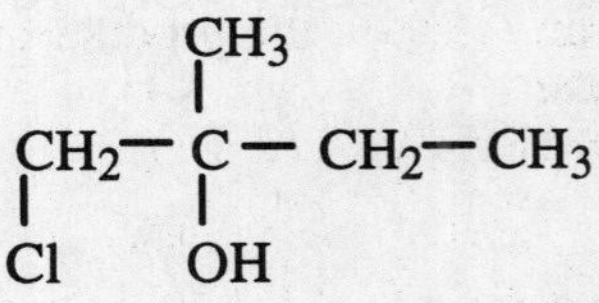

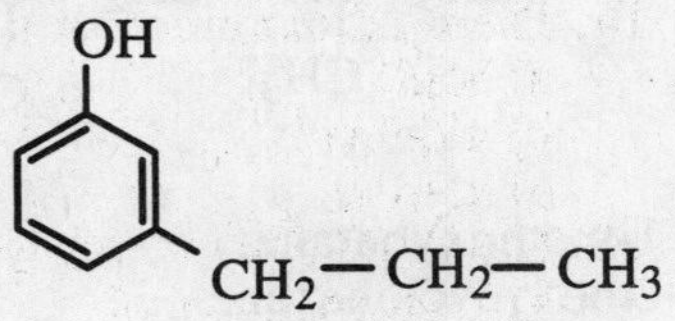

14.3 a. $CH_3-CH_2-CH_2-CH_2-CH=CH_2 + H_2O \xrightarrow{H_2SO_4} CH_3-CH_2-CH_2-CH_2-CH(OH)-CH_3$

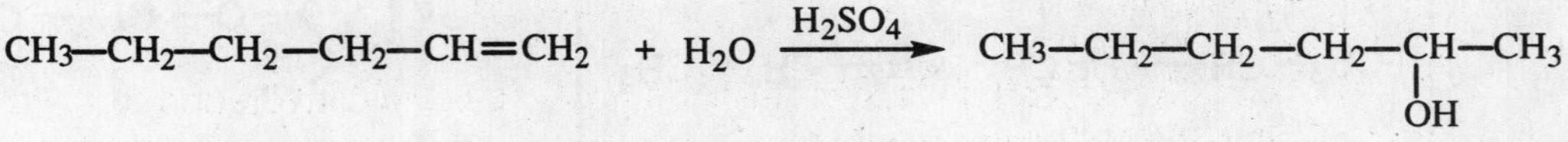

b. cyclopentene $+ H_2O \xrightarrow{H_2SO_4}$ cyclopentanol (cyclopentane–OH)

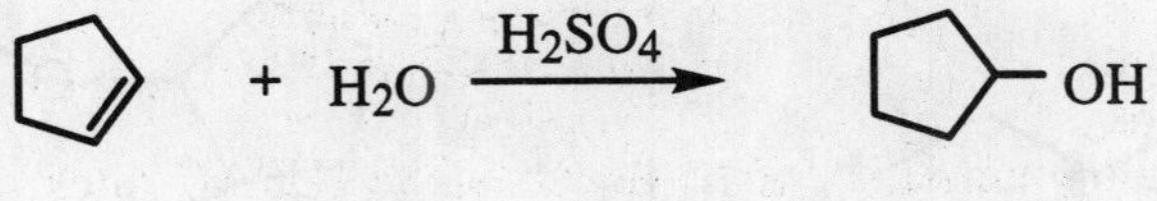

14.4 a. $CH_3-CH(CH_3)-CH_2-OH \xrightarrow[H_2SO_4]{140^\circ C} CH_3-CH(CH_3)-CH_2-O-CH_2-CH(CH_3)-CH_3$

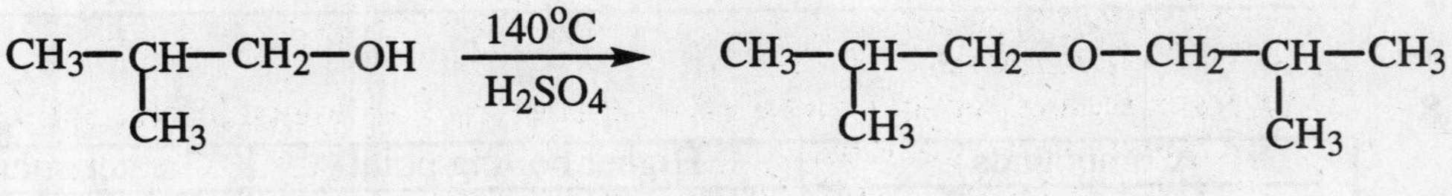

b. $CH_3-CH_2-CH_2-CH(OH)-CH_3 \xrightarrow[H_2SO_4]{180^\circ C} CH_3-CH_2-CH=CH-CH_3$

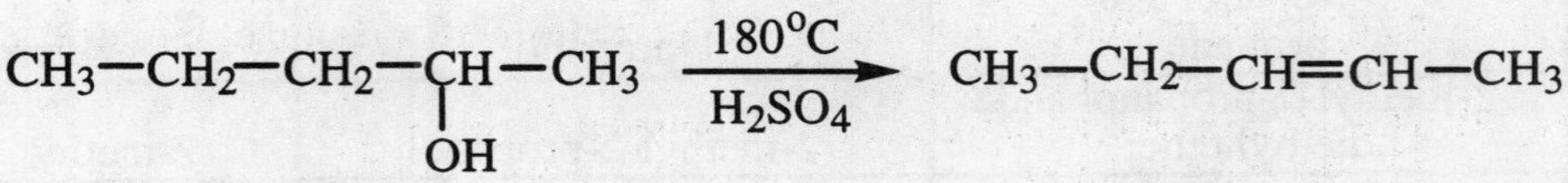

c. 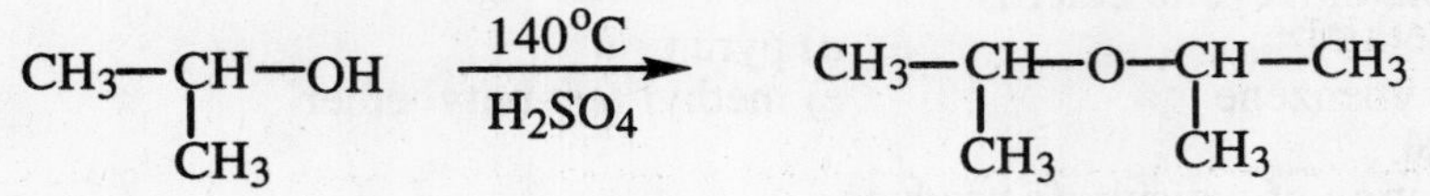 $CH_3-CH(CH_3)-OH \xrightarrow[H_2SO_4]{140^oC} CH_3-CH(CH_3)-O-CH(CH_3)-CH_3$

d. 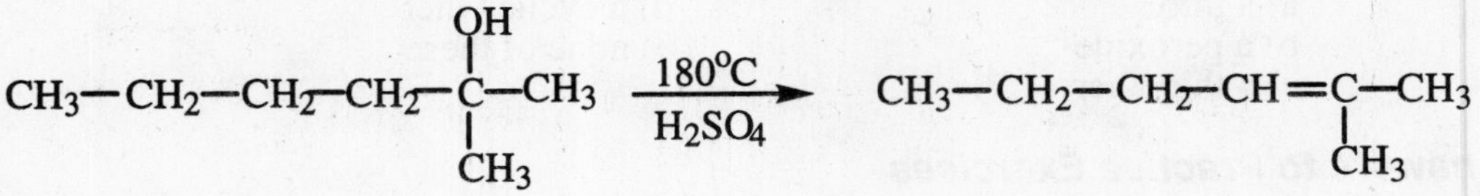 $CH_3-CH_2-CH_2-CH_2-C(OH)(CH_3)-CH_3 \xrightarrow[H_2SO_4]{180^oC} CH_3-CH_2-CH_2-CH=C(CH_3)-CH_3$

14.5 a. 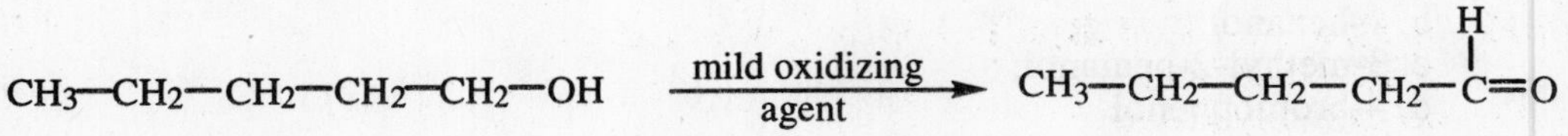 $CH_3-CH_2-CH_2-CH_2-CH_2-OH \xrightarrow[\text{agent}]{\text{mild oxidizing}} CH_3-CH_2-CH_2-CH_2-C(H)=O$

b. 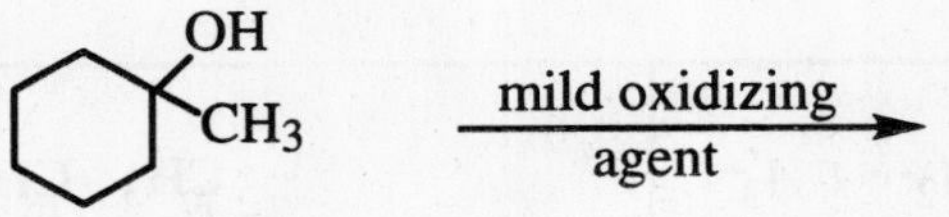 1-methylcyclohexanol (cyclohexane ring bearing OH and CH_3) $\xrightarrow[\text{agent}]{\text{mild oxidizing}}$ No reaction, tertiary alcohol

c. 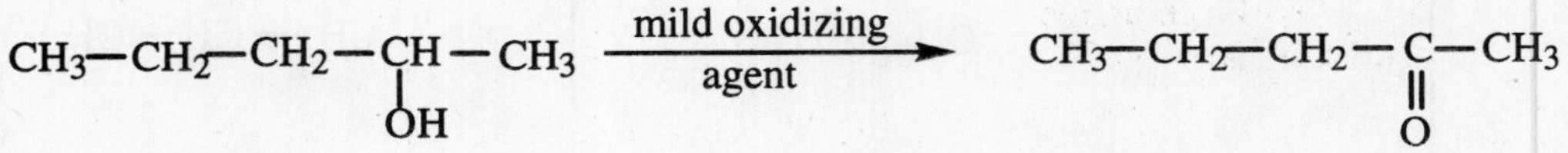 $CH_3-CH_2-CH_2-CH(OH)-CH_3 \xrightarrow[\text{agent}]{\text{mild oxidizing}} CH_3-CH_2-CH_2-C(=O)-CH_3$

d. 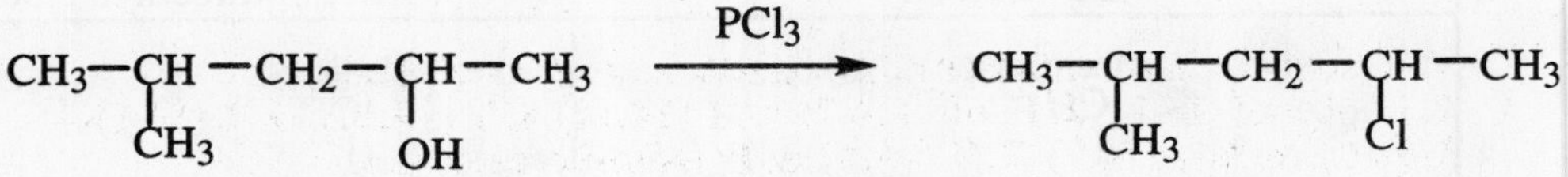 $CH_3-CH(CH_3)-CH_2-CH(OH)-CH_3 \xrightarrow{PCl_3} CH_3-CH(CH_3)-CH_2-CH(Cl)-CH_3$

14.6 a. 1-methoxybutane;
b. ethoxycyclohexane

14.7

a. 2-ethoxypentane	b. 1-chloro-2-ethoxycyclopentane
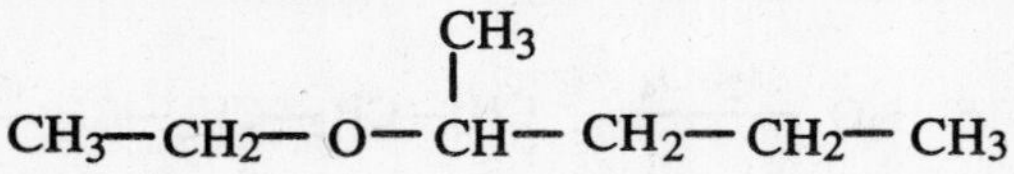 $CH_3-CH_2-O-CH(CH_3)-CH_2-CH_2-CH_3$	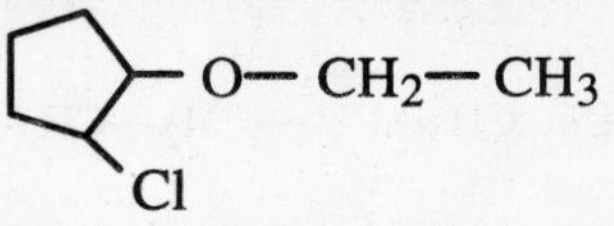 cyclopentane ring with $-O-CH_2-CH_3$ and adjacent Cl
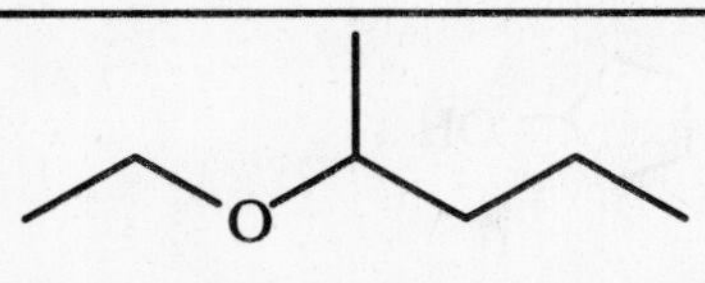 line-angle structure: O	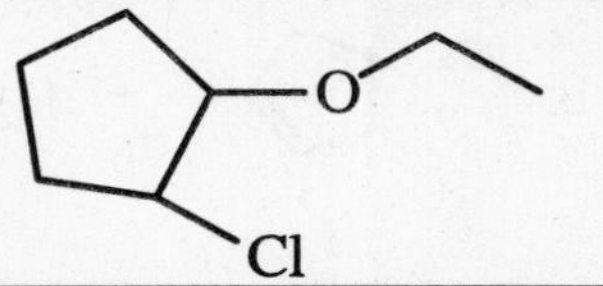 line-angle structure: O, Cl

14.8

Compounds	Higher boiling point	Greater solubility in water
1-propanol and propane	1-propanol	1-propanol
2-methyl-2-propanol and diethyl ether	2-methyl-2-propanol	2-methyl-2-propanol
phenol and toluene	phenol	phenol

14.9 a. 1-pentanethiol;
b. cyclohexanethiol

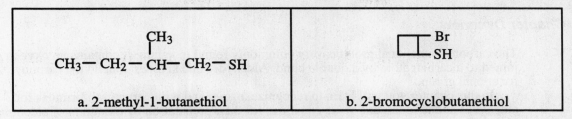

14.10

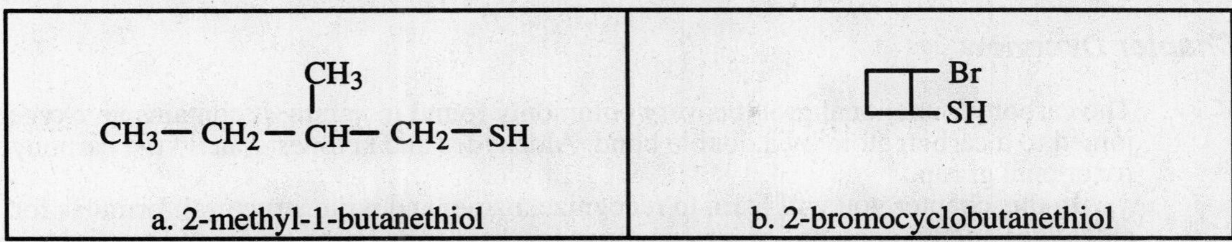

14.11 a. D; b. A; c. C; d. B; e. E; f. G; g. I; h. H; i. F
j. 1,2-ethanediol
k. methoxyethane
l. butane
m. 2-methyl-2-butanol
n. phenol
o. 1-propanol
p. 1-butene
q. methylthiomethane
r. ethanethiol

Answers to Self-Test

The numbers in parentheses refer to sections in your textbook:
1. F; phenol (14.2) **2**. F; one (14.7) **3**. T (14.11) **4**. T (14.10) **5**. T (14.1)
6. F; secondary (14.7) **7**. T (14.5) **8**. F; intermolecular (14.7)
9. F; easily oxidized (14.9) **10**. T (14.12) **11**. T (14.11) **12**. b (14.6) **13**. a (14.13)
14. c (14.7) **15**. b (14.4) **16**. a (14.4) **17**. a (14.7) **18**. e (14.3) **19**. c (14.5, 14.11)
20. b (14.7) **21**. d (14.3, 14.10) **22**. a (14.13)

Aldehydes and Ketones — Chapter 15

Chapter Overview

The carbonyl functional group is very commonly found in nature. It contains an oxygen atom joined to a carbon atom by a double bond. Aldehydes and ketones contain the carbonyl functional group.

In this chapter you will learn to recognize, name, and write structural formulas for aldehydes and ketones, and write equations for their preparation by oxidation of alcohols. You will compare the physical properties of aldehydes and ketones with those of other organic compounds. You will learn some tests used to distinguish aldehydes from ketones, and write equations for reactions involving addition to the carbonyl group of aldehydes and ketones.

Practice Exercises

15.1 The functional group that identifies **aldehydes and ketones** (Sec. 15.2) is the **carbonyl group** (Sec. 15.1), a carbon atom and an oxygen atom joined by a double bond. Aldehydes are compounds in which the carbonyl carbon is bonded to at least one hydrogen. Ketones are compounds in which the carbonyl carbon is bonded to two other carbons.

Classify each of the following structural formulas as an aldehyde, a ketone, or neither.

$CH_3-CH_2-CH_2-O-CH_3$ a.	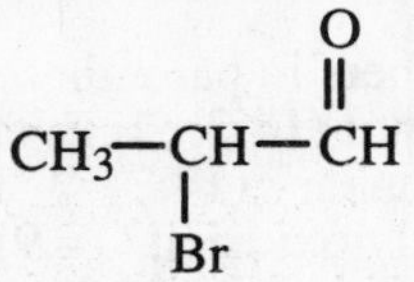 O ‖ $CH_3-CH-CH$ | Br b.
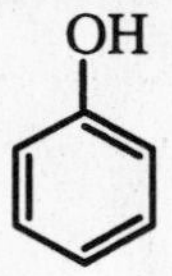 OH (benzene ring) c.	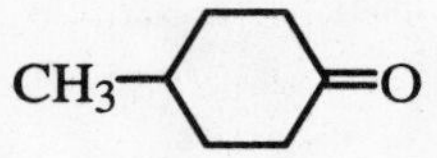 CH_3-(cyclohexane ring)$=O$ d.
Cl | $CH_3-C-CH-CH-CH_3$ ‖ | O CH_3 e.	CH_3 | CH_3-C-CH_2-CHO | Cl f.
O ‖ $CH_3-CH-C-CH_3$ | (benzene ring) g.	O Cl ‖ | $HC-CH_2-C-Cl$ | Cl h.

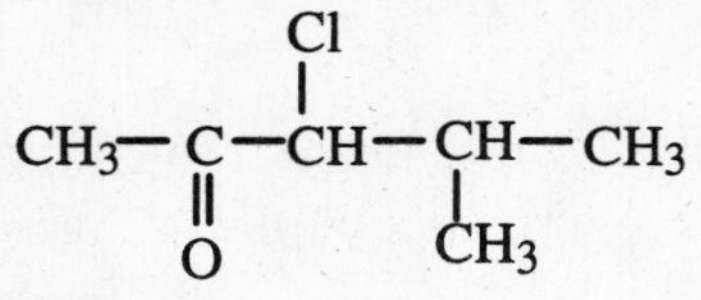

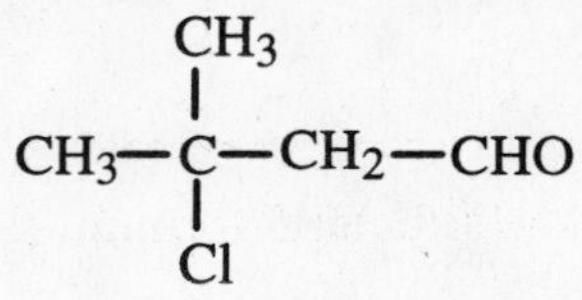

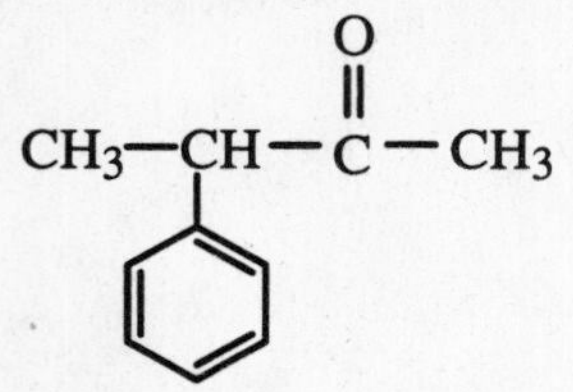

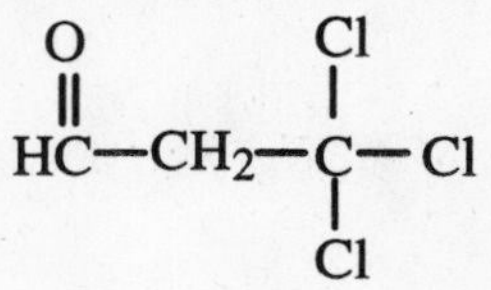

15.2 In naming aldehydes, change the *-e* ending of the hydrocarbon to *-al*, and name any substituents, starting the counting from the carbonyl carbon. No number is specified for the carbonyl group.

Give the IUPAC name for the following aldehydes:

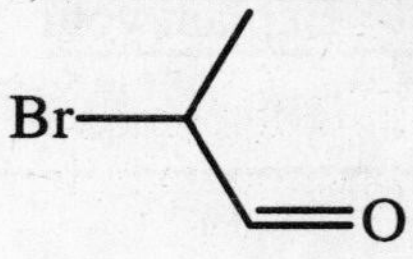 a. Br (line-angle structure, =O)	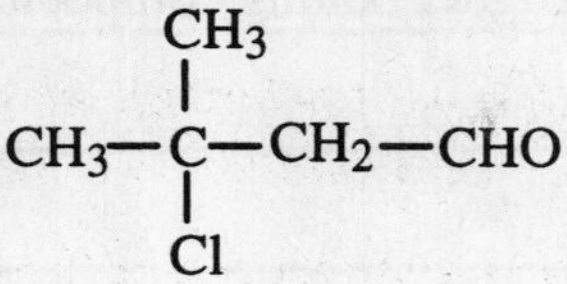 b. $CH_3-C(CH_3)(Cl)-CH_2-CHO$	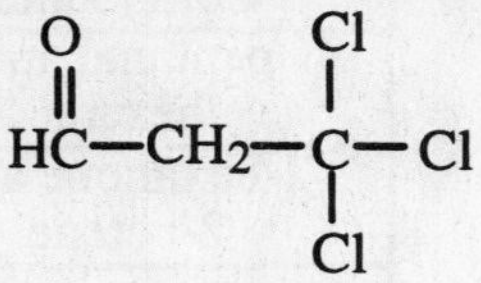 c. $HC(=O)-CH_2-CCl_3$

15.3 Draw the structural formula and the line angle drawing for each of the following aldehydes.

a. 4-chloro-3-methyl-3-nitrobutanal	b. 4-bromo-5-chloro-3-methylpentanal

15.4 In naming ketones, change the *-e* ending to *-one*. Give the carbonyl group the lowest possible number on the chain, and then name the substituents. The numbered position of the carbonyl group is included in the name.

Give the IUPAC names for the following ketones.

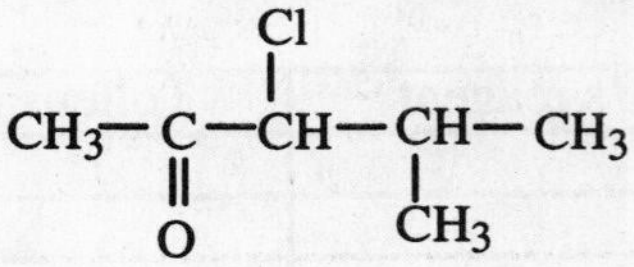 a. $CH_3-C(=O)-CH(Cl)-CH(CH_3)-CH_3$	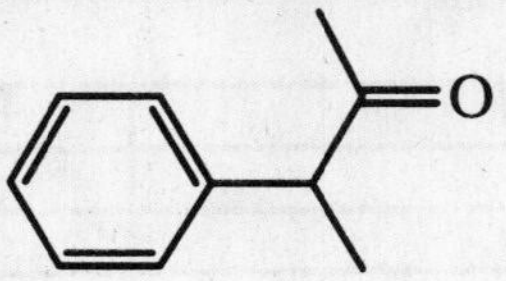 b. (line-angle structure, =O)	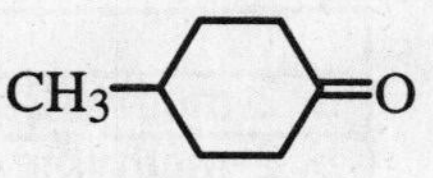 c. CH_3–(cyclohexane ring)=O

15.5 Draw the structural formula and the line angle drawing for each of the following ketones.

a. 5-bromo-6-chloro-4-methyl-3-hexanone	b. 3-ethyl-2-methylcyclohexanone

15.6 Hydrogen bonding cannot occur between the molecules of an aldehyde or between the molecules of a ketone. However, dipole-dipole attractions occur between molecules.

Determine which one of each pair of compounds would have a higher boiling point and explain your choice.

Compounds	Higher boiling compound	Explain your choice
pentanal and hexane		
2-hexanone and 2-octene		
pentanal and 1-pentanol		

15.7 Aldehydes and ketones can be produced by the oxidation of primary and secondary alcohols, respectively. Write an equation for the preparation of each of the following compounds by oxidation of an alcohol. Assume no further oxidation of aldehydes to carboxylic acids occurs.

a. 3-methyl-2-phenylpentanal

b. 3-ethyl-2-methylcyclohexanone

15.8 Aldehydes are easily oxidized to carboxylic acids; ketones are resistant to oxidation. Tests used to distinguish between aldehydes and ketones use metal ions (Ag^+ and Cu^{2+}) as oxidizing agents. The appearance of the reduced metal (Ag) or metal oxide (Cu_2O) indicates that an aldehyde is present.

Indicate whether each of the following compounds will give a positive test with Benedict's or Tollens reagent.

Compound	Benedict's reagent	Tollens reagent
a. 2-methyl-2-propanol		
b. 2-methylhexanal		
c. 4-chloro-3-methyl-2-pentanone		
d. 3,3-dimethylpentane		
e. 4-chlorocyclohexanone		

15.9 Aldehydes and ketones are easily reduced by hydrogen gas (H_2), in the presence of a catalyst (Ni, Pt, or Cu), to form alcohols.

Write an equation for the preparation of each of the following alcohols by the reduction of the appropriate aldehyde or ketone.

a. 3-chloro-1-butanol

b. 1-bromo-3-methyl-2-pentanol

15.10 Aldehydes and ketones easily undergo addition to the double bond of the carbonyl group. Addition of an alcohol molecule to an aldehyde molecule forms a **hemiacetal** (Sec. 15.9); addition of an alcohol molecule to a ketone molecule forms a **hemiketal** (Sec. 15.9).

Draw the structural formula of the hemiacetal or hemiketal formed when one molecule of methanol, CH_3OH, reacts with one molecule of each of the following carbonyl compounds.

a. 2-bromopropanal + methanol	b. 4-methylcyclohexanone + methanol	c. 3-chloro-2-pentanone + methanol

15.11 If a hemiacetal molecule reacts with a second molecule of alcohol, an **acetal** (Sec. 15.9) is produced. In the same way, addition of another molecule of alcohol to a hemiketal produces a **ketal** (Sec. 15.9).

Draw the structural formula of the acetal or ketal formed when one additional molecule of methanol is added to each of the hemiacetal or hemiketal molecules in Practice Exercise 15.10.

a.	b.	c.

15.12 The hydrolysis of an acetal or a ketal (or a hemiacetal or hemiketal) produces the aldehyde or ketone and the alcohol that originally reacted to form the acetal or ketal. Complete the following acid hydrolysis reactions.

a. $CH_3-CH_2-C(OH)(O-CH_2-CH_3)-CH_2-CH_3 + H_2O \xrightleftharpoons{\text{acid catalyst}}$? + ?

b. $CH_3-CH(CH_3)-CH(O-CH_3)(O-CH_3) + H_2O \xrightleftharpoons{\text{acid catalyst}}$? + ?

c. $C_5H_9-C(CH_3)(O-CH_2-CH_3)(O-CH_2-CH_3) + H_2O \xrightleftharpoons{\text{acid catalyst}}$? + ? (cyclopentyl ring attached to C)

15.13 Use this identification exercise to review your knowledge of the structures introduced in this chapter. Using structures A through I, give the best choice for each of the terms below.

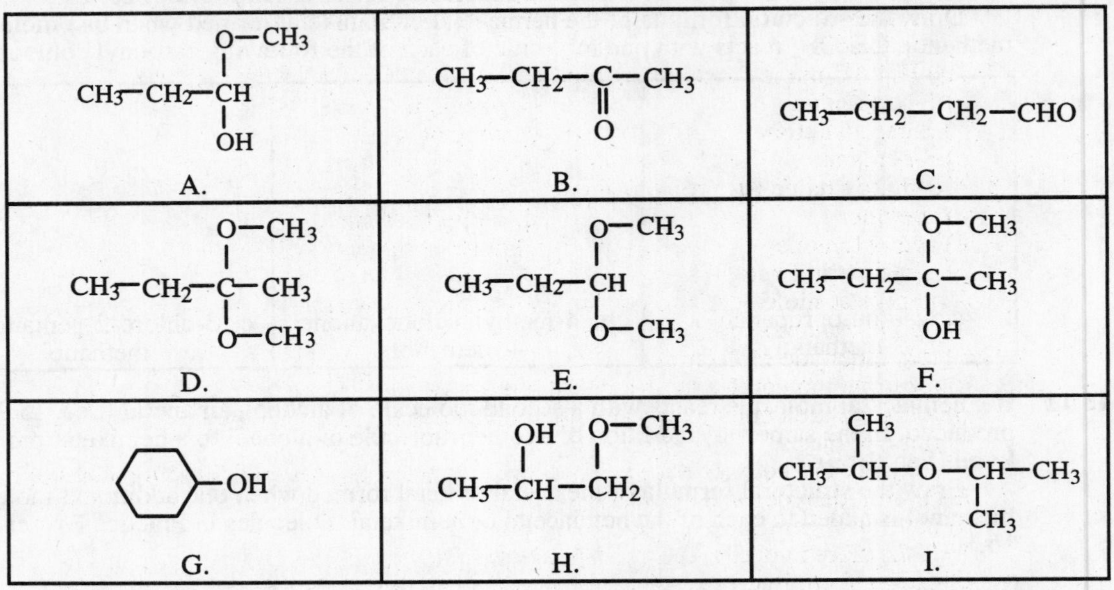

$O-CH_3$ CH_3-CH_2-CH OH A.	$CH_3-CH_2-C-CH_3$ $\Vert$ O B.	$CH_3-CH_2-CH_2-CHO$ C.
$O-CH_3$ $CH_3-CH_2-C-CH_3$ $O-CH_3$ D.	$O-CH_3$ CH_3-CH_2-CH $O-CH_3$ E.	$O-CH_3$ $CH_3-CH_2-C-CH_3$ OH F.
(cyclohexane ring)$-OH$ G.	$OH \quad O-CH_3$ $CH_3-CH-CH_2$ H.	CH_3 $CH_3-CH-O-CH-CH_3$ CH_3 I.

a. aldehyde ____________
b. ketone ____________
c. symmetrical ether ____________
d. hemiketal ____________
e. acetal ____________
f. ketal ____________
g. cyclic alcohol ____________
h. hemiacetal ____________

Self-Test

True-false: Indicate whether the following statements are true or false. If the statement is false, give the word or phrase that may be substituted for the underlined portion to make the statement true.

1. The simplest ketone contains two carbon atoms.
2. The simplest aldehyde has the common name formaldehyde.
3. The oxidation of 2-butanol produces a ketone.
4. Many important steroid hormones are aldehydes.
5. Aldehydes and ketones have higher boiling points than the corresponding alcohols.
6. Low-molecular-mass aldehydes and ketones are water-soluble because water forms hydrogen bonds with them.
7. Ketones are often produced by oxidation of the corresponding tertiary alcohol.
8. If an aldehyde is produced by oxidation of an alcohol, further oxidation to a ketone may take place.
9. A positive Tollens test indicates that an aldehyde is present.
10. Benedict's test uses Ag^+ ion as an oxidizing agent for aldehydes.
11. A ketone can be reduced by hydrogen gas in the presence of a Ni catalyst.
12. Addition to the carbon-oxygen double bond of a carbonyl group takes place less easily than addition to a carbon-carbon double bond.
13. Hemiacetals and hemiketals are much more stable than acetals and ketals.
14. An acetal molecule is the product of the addition of two molecules of an alcohol to one molecule of an aldehyde.
15. Formaldehyde-based polymers are made up of monomers connected in a chain.

Multiple choice:

16. Oxidation of ethanol cannot yield:
    a) acetaldehyde
    b) acetone
    c) acetic acid
    d) ethanal
    e) carbon dioxide
17. The IUPAC name for the compound $CH_3{-}CH_2{-}CH_2{-}CH_2{-}CHO$ is:
    a) 1-pentanone
    b) 1-pentyl ketone
    c) 1-pentanal
    d) pentanal
    e) none of these
18. A common name for propanone is:
    a) acetone
    b) acetophenone
    c) ethyl methyl ketone
    d) diethyl ketone
    e) none of these
19. A hemiketal molecule is formed by the reaction between:
    a) two ketone molecules
    b) two aldehyde molecules
    c) a ketone molecule and an alcohol molecule
    d) a ketone molecule and an aldehyde molecule
    e) none of these
20. Aldehydes and ketones have boiling points lower than the corresponding alcohols because aldehydes and ketones:
    a) have stronger dipole-dipole attractions
    b) form more hydrogen bonds than alcohols do
    c) cannot form hydrogen bonds between molecules
    d) form bonds with water molecules
    e) none of these
21. How many ketones are isomeric with 2-methylbutanal?
    a) one b) two c) three d) four e) none of these
22. Acetaldehyde can be produced by the oxidation of:
    a) acetone
    b) ethanol
    c) dimethyl ether
    d) ethane
    e) none of these

Answers to Practice Exercises

15.1 a. neither (an ether); b. aldehyde; c. neither (a phenol); d. ketone; e. ketone; f. aldehyde; g. ketone; h. aldehyde

15.2

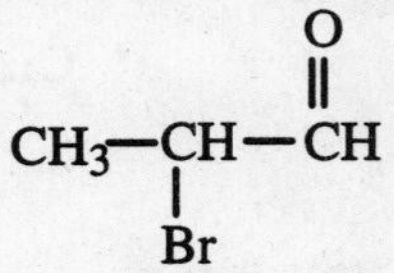 O ‖ CH_3—CH—CH | Br	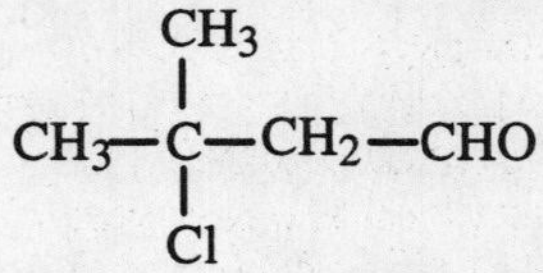 CH_3 | CH_3—C—CH_2—CHO | Cl	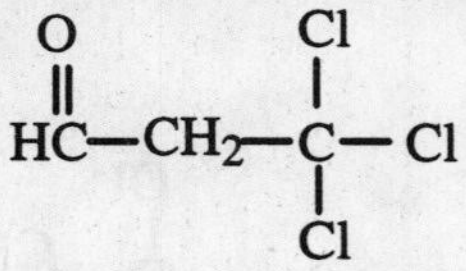 O Cl ‖ | HC—CH_2—C—Cl | Cl
a. 2-bromopropanal	b. 3-chloro-3-methylbutanal	c. 3,3,3-trichloropropanal

15.3

a. 4-chloro-3-methyl-3-nitrobutanal	b. 4-bromo-5-chloro-3-methylpentanal
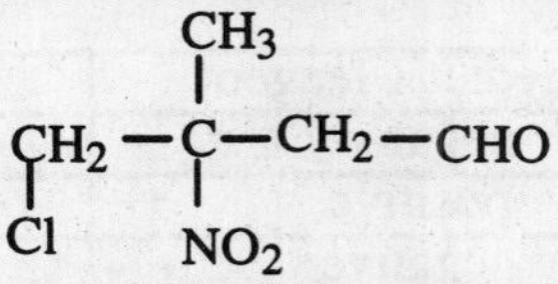 CH_3 | CH_2—C—CH_2—CHO | | Cl NO_2	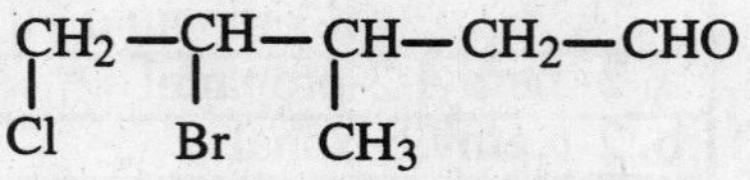 CH_2—CH—CH—CH_2—CHO | | | Cl Br CH_3
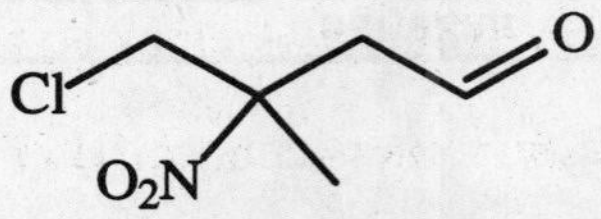 Cl O O_2N	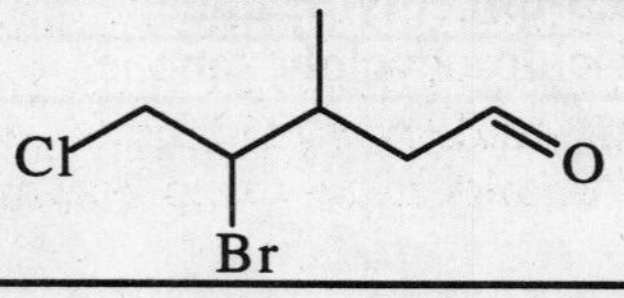 Cl O Br

15.4 a. 3-chloro-4-methyl-2-pentanone;
b. 3-phenyl-2-butanone;
c. 4-methylcyclohexanone

15.5

a. 5-bromo-6-chloro-4-methyl-3-hexanone	b. 3-ethyl-2-methylcyclohexanone
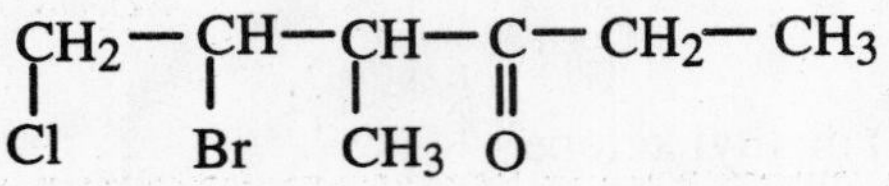 $CH_2-CH-CH-C-CH_2-CH_3$ (Cl on C_1; Br on C_2; CH_3 on C_3; O on C_4)	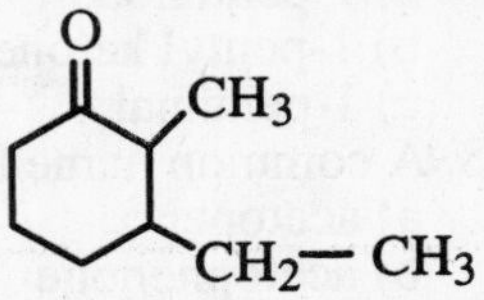 O, CH_3, CH_2-CH_3
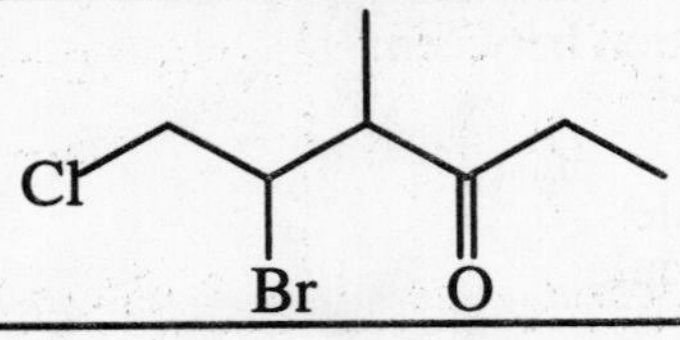 Cl, Br, O	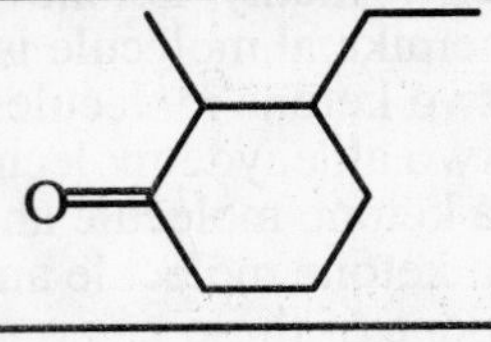 O

15.6

Compounds	Higher boiling point	Explain your choice
pentanal and hexane	pentanal	Pentanal has a higher boiling point because of dipole-dipole attractions.
2-hexanone and octane	2-hexanone	2-Hexanone has a higher boiling point because of dipole-dipole attractions.
pentanal and 1-pentanol	1-pentanol	1-Pentanol has hydrogen bonding between molecules.

15.7 a. $CH_3-CH_2-CH(CH_3)-CH(C_6H_5)-CH_2-OH \xrightarrow[\text{oxidation}]{\text{mild}} CH_3-CH_2-CH(CH_3)-CH(C_6H_5)-CHO$

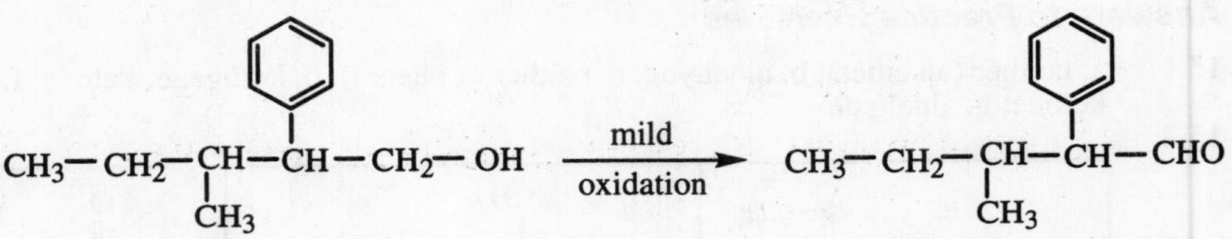

b. OH, CH_3, CH_2-CH_3 $\xrightarrow[\text{oxidation}]{\text{mild}}$ O, CH_3, CH_2-CH_3

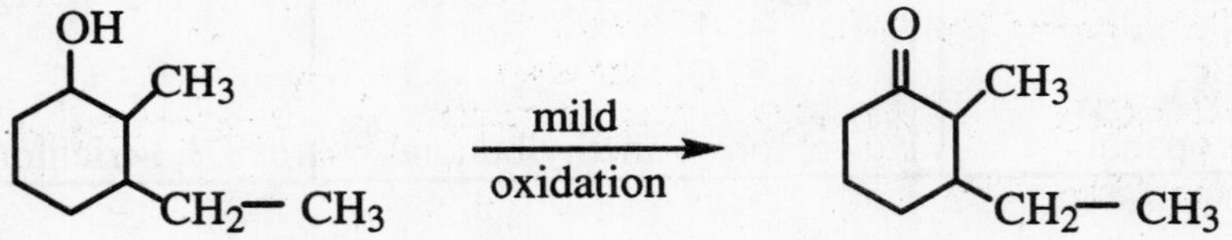

15.8

Compound	Benedict's reagent	Tollens reagent
a. 2-methyl-2-propanol	negative	negative
b. 2-methylhexanal	positive	positive
c. 4-chloro-3-methyl-2-pentanone	negative	negative
d. 3,3-dimethylpentane	negative	negative
e. 4-chlorocyclohexanone	negative	negative

15.9 a. $CH_3-CH(Cl)-CH_2-CHO + H_2 \xrightarrow{\text{Catalyst}} CH_3-CH(Cl)-CH_2-CH_2-OH$

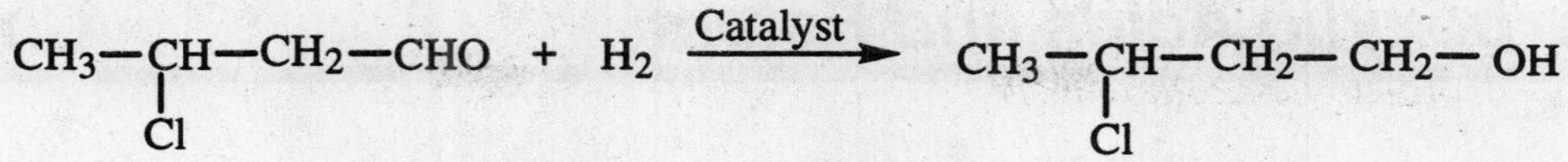

b. $CH_3-CH_2-CH(CH_3)-C(=O)-CH_2-Br + H_2 \xrightarrow{\text{Catalyst}} CH_3-CH_2-CH(CH_3)-CH(OH)-CH_2-Br$

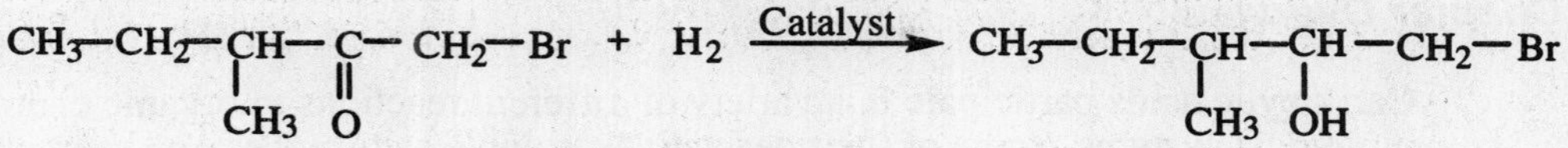

15.10

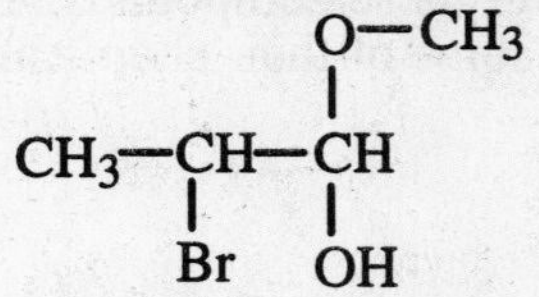

a. 2-bromopropanal	b. 4-methylcyclohexanone	c. 3-chloro-2-pentanone
$CH_3-CH(Br)-CH(OH)(O-CH_3)$	CH_3–cyclohexane–C(O–CH_3)(OH)	$CH_3-C(OH)(O-CH_3)-CH(Cl)-CH_2-CH_3$

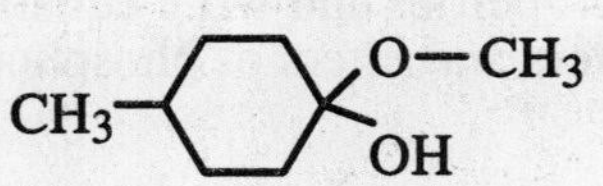

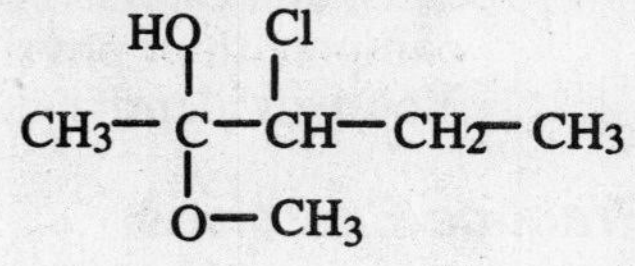

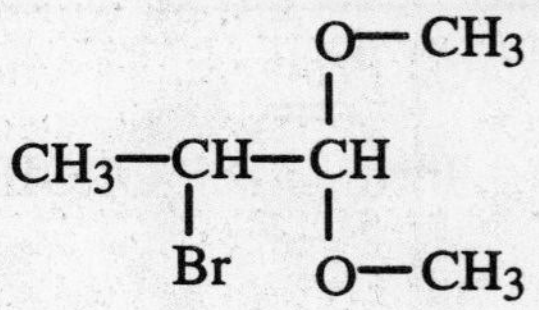

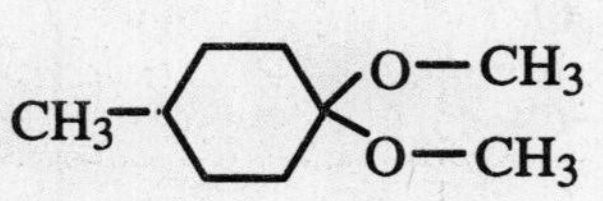

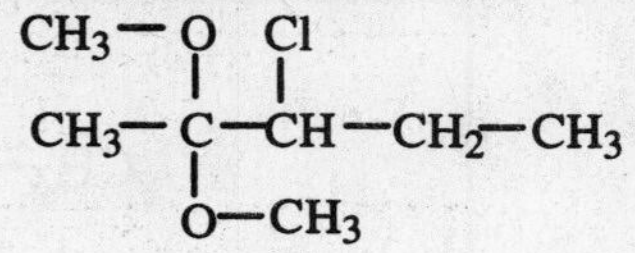

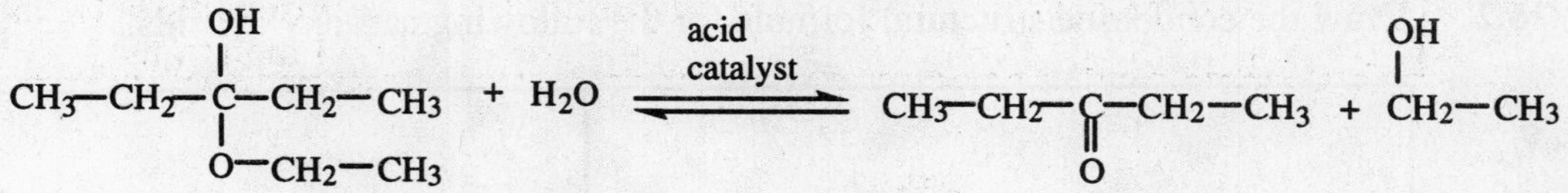

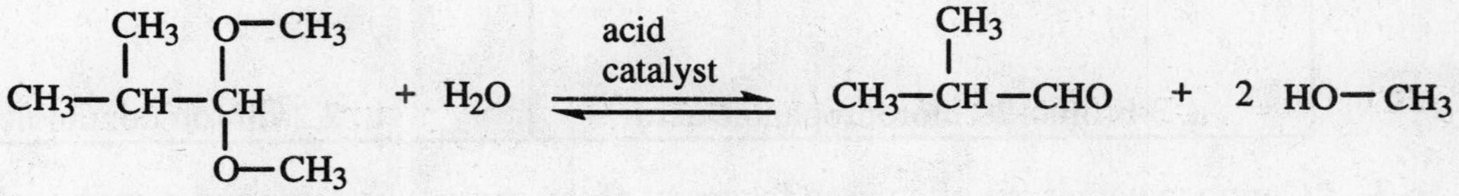

15.11

a.	b.	c.
$CH_3-CH(Br)-CH(O-CH_3)_2$	CH_3–cyclohexane–C(O–CH_3)(O–CH_3)	$CH_3-C(O-CH_3)_2-CH(Cl)-CH_2-CH_3$

15.12 a. $CH_3-CH_2-C(OH)(O-CH_2-CH_3)-CH_2-CH_3 + H_2O \underset{}{\overset{\text{acid catalyst}}{\rightleftharpoons}} CH_3-CH_2-C(=O)-CH_2-CH_3 + CH_2(OH)-CH_3$

b. $CH_3-CH(CH_3)-CH(O-CH_3)(O-CH_3) + H_2O \overset{\text{acid catalyst}}{\rightleftharpoons} CH_3-CH(CH_3)-CHO + 2\ HO-CH_3$

c. cyclopentyl$-C(O-CH_2-CH_3)_2-CH_3 + H_2O \overset{\text{acid catalyst}}{\rightleftharpoons}$ cyclopentyl$-C(=O)-CH_3 + 2\ CH_2(OH)-CH_3$

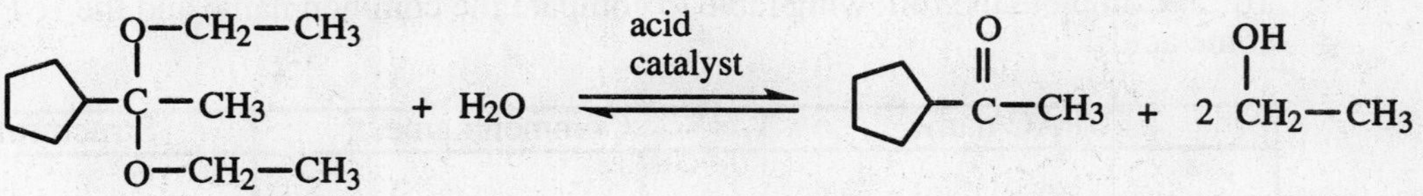

15.13 a. C; b. B; c. I; d. F; e. E; f. D; g. G; h. A.

Answers to Self-Test

The numbers in parentheses refer to sections in your textbook:
1. F; three (15.2) **2**. T (15.3) **3**. T (15.2) **4**. F; ketones (15.5) **5**. F; lower (15.6) **6**. T (15.6) **7**. F; secondary (15.7) **8**. F; a carboxylic acid (15.7) **9**. T (15.8) **10**. F; Cu^{2+} (15.8) **11**. T (15.8) **12**. F; more easily (15.9) **13**. F; much less stable (15.9) **14**. T (15.9) **15**. F; network (15.10) **16**. b (15.7, 15.8) **17**. d (15.3) **18**. a (15.4) **19**. c (15.9) **20**. c (15.6) **21**. c (15.2, 15.3) **22**. b (15.3, 15.7)

Chapter Overview

Carboxylic acids participate in a variety of different reactions in organic chemistry and in biological systems. Some of their derivatives, soluble salts and esters, keep our world clean and sweet smelling.

In this chapter you will learn to recognize and name carboxylic acid, carboxylic acid salts, and esters. You will write equations for preparation of these compounds. You will compare their physical properties and write equations for some of their chemical reactions. You will identify thioesters and esters of phosphoric acid.

Practice Exercises

16.1 **Carboxylic acids** (Sec. 16.1) contain the carboxyl functional group. Their naming is similar to that of aldehydes, with the ending *-oic acid*.

Give the IUPAC name for each of the following carboxylic acids.

a. $CH_3-C(Br)(CH_3)-C(=O)-OH$	b. I–C_6H_4–COOH (para)

16.2 Draw the condensed structural formula for the following carboxylic acids.

a. 3-bromo-2-chloropropanoic acid	b. 2,2-dibromobutanoic acid

16.3 Common names are often used for many of the carboxylic acids and **dicarboxylic acids** (Sec. 16.2). Complete the following table to compare the common name and the IUPAC name for some acids.

IUPAC name	Common name	Structural formula
butanoic acid		
	formic acid	
		HOOC—COOH
	adipic acid	

16.4 A number of carboxylic acids are polyfunctional, and some of these are important in reactions that occur in the human body. Complete the following table giving structures and names for some common polyfunctional carboxylic acids.

Common name	Structure of carboxylic acid
pyruvic acid	
	$CH_3-CH(OH)-COOH$
fumaric acid	

16.5 Carboxylic acids have very high boiling points because two molecules can form two hydrogen bonds with one another, from each of the double-bonded oxygens to the hydrogens of the -OH groups. Carboxylic acids also form hydrogen bonds with water.

Draw a diagram of the hydrogen bonding between the following molecules.

a. two molecules of propanoic acid	b. one molecule of propanoic acid and water

16.6 Carboxylic acids can be prepared by oxidation of a primary alcohol or an aldehyde. Aromatic acids can be prepared by oxidation of an alkyl side chain on a benzene derivative.

Write an equation for the preparation of a carboxylic acid from the indicated reactants. Show the necessary reagents used in each reaction.

a. 1-butanol

b. butanal

c. 4-bromo-1-ethylbenzene

16.7 Carboxylic acids react with strong bases to produce water and a **carboxylic acid salt** (Sec. 16.8). The negative ion of the salt is called a **carboxylate ion** (Sec. 16.7).

Complete the following table of carboxylic acids and their salts. Use sodium salts for the carboxylate ions.

Name of acid	Structural formula of carboxylic acid salt	IUPAC name of carboxylic acid salt
propanoic acid		
	$CH_3-CH_2-CH_2-\overset{\overset{\displaystyle O}{\|}}{C}-O^{-}\,Na^{+}$	
benzoic acid		sodium benzoate

16.8 A carboxylic acid salt can be converted to the carboxylic acid by reacting the salt with a strong acid such as HCl or H_2SO_4.

Write an equation for the reaction of potassium ethanoate with hydrochloric acid to form the carboxylic acid.

16.9 The reaction of a carboxylic acid with an alcohol, in the presence of a strong-acid catalyst, produces an **ester** (Sec. 16.9), whose functional group is –COOR. The –OH from the carboxyl group combines with –H from the alcohol to form a molecule of water as a by-product.

Complete the following table showing the ester structures and names and the reactants that form the ester. In ester names, the "alcohol part" comes first followed by the "acid part" with an *-ate* ending.

Acid and alcohol	Structure of ester formed	IUPAC name
ethanoic acid and methanol		
	$CH_3-CH_2-CH_2-\overset{\overset{\displaystyle O}{\|}}{C}-O-CH_3$	
		ethyl benzoate

16.10 Esters are often known by common names, which are based on the common names of the acid parts of the esters. Complete the following table comparing the IUPAC and common names of some esters.

IUPAC name	Common name
ethyl methanoate	
	methyl valerate
	propyl caproate

16.11 Ester molecules do not have a hydrogen atom bonded to an oxygen atom so they cannot form hydrogen bonds to each other. They can, however, form hydrogen bonds with water.

Indicate which compound in each of the following pairs would have a higher boiling point and explain your answer.

Compounds	Higher boiling point	Explain your answer
butanoic acid and methyl propanoate		
methyl propanoate and 1-butanol		
methyl hexanoate and decane		

16.12 Esters undergo hydrolysis with an acid catalyst to produce the acid and alcohol from which they were formed. They also undergo base-catalyzed hydrolysis, which is called **ester saponification** (Sec. 16.13). In this case the products are the alcohol and the salt of the carboxylic acid.

Give the names of the products formed in the following reactions.

a. Ethyl butanoate undergoes basic hydrolysis with sodium hydroxide (saponification).

b. Ethyl butanoate undergoes acidic hydrolysis.

16.13 Thiols react with carboxylic acids to form **thioesters** (Sec. 16.14). A molecule of water is formed as a by-product. Complete the following reactions showing thioester formation.

a. $HCOOH + CH_3-CH_2-SH \longrightarrow ?$

b. $? + ? \longrightarrow$ (cyclopentyl)$-\overset{O}{\overset{\|}{C}}-S-CH_3$

16.14 Inorganic acids react with alcohols to form esters in a manner similar to that for carboxylic acid esters. Phosphoric acid has three hydroxyl groups and so can form **phosphate esters** (Sec. 16.16) with one, two, or three molecules of an alcohol. Draw the structural formula of the ester formed from each set of reactants named below.

a. One molecule of ethanol and one molecule of phosphoric acid.

b. Two molecules of ethanol and one molecule of phosphoric acid.

16.15 Use this identification exercise to review your knowledge of the structures introduced in this chapter. Using structures A through I, give the best choice for each of the terms below.

$CH_3-C(Cl)(CH_3)-C(=O)-OH$ A.	$CH_3-CH_2-C(=O)-O^- K^+$ B.	$CH_3CH_2C(=O)OCH_3$ C.
$CH_3-CH_2-CH_2-SH$ D.	$CH_3-CH_2-CH_2-OH$ E.	$HOOC-(CH_2)_2-COOH$ F.
$CH_3-CH_2-C(=O)-C(=O)-OH$ G.	$CH_3-CH_2-CH(OH)-C(=O)-OH$ H.	$CH_3-C(=O)-CH_2-C(=O)-OH$ I.

a. dicarboxylic acid ___________
b. carboxylic acid ___________
c. β-keto carboxylic acid ___________
d. alcohol ___________
e. potassium salt ___________
f. ester ___________
g. α-hydroxy carboxylic acid ___________
h. thiol ___________

Give the IUPAC names for compounds A through F.

A. ___________

B. ___________

C. ___________

D. ___________

E. ___________

F. ___________

Self-Test

True-false: Indicate whether the following statements are true or false. If the statement is false, give the word or phrase that may be substituted for the underlined portion to make the statement true.

1. Formic acid is the simplest carboxylic acid.
2. A carboxylic acid with six carbons in a straight chain is named 1-hexanoic acid.
3. Vinegar is made of glacial acetic acid.
4. Another name for 2-methylbutanoic acid is β-methylbutyric acid.
5. Glycolic acid is a dicarboxylic acid.
6. Oxidation of 2-butanol produces butanoic acid.
7. Because of their extensive hydrogen bonding, carboxylic acids have low boiling points.
8. Benzoic acid cannot be prepared by the oxidation of ethylbenzene.
9. A carboxylate ion is formed by the loss of an acidic hydrogen atom.

10. The reaction between sodium hydroxide and benzoic acid would produce sodium benzoate.
11. According to Le Chatelier's principle, adding an excess of alcohol to a carboxylic acid will decrease the amount of ester that is formed.
12. The pleasant fragrances of many flowers and fruits are produced by mixtures of carboxylic acids.
13. Aspirin has an acid functional group and an ester functional group.
14. The boiling points of esters are lower than those of alcohols and acids with comparable molecular mass.
15. Ester saponification is the base-catalyzed hydrolysis of an ester.
16. A polyester is an addition polymer with ester linkages.

Multiple choice:

17. The reaction between methanol and ethanoic acid will produce:
    a) methyl acetate
    b) ethyl formate
    c) methyl butyrate
    d) ethyl ethanoate
    e) none of these
18. Rank these three types of compounds -- alcohol, carboxylic acid, ester of carboxylic acid (of comparable molecular mass) -- in order of boiling point, highest to lowest:
    a) alcohol, acid, ester
    b) ester, acid, alcohol
    c) acid, alcohol, ester
    d) ester, alcohol, acid
    e) acid, ester, alcohol
19. The products of the acid-catalyzed hydrolysis of an ester include:
    a) a carboxylate ion
    b) a ketone
    c) an alcohol
    d) an ether
    e) none of these
20. A thioester is formed by the reaction between:
    a) sulfuric acid and a carboxylate
    b) an alcohol and sulfuric acid
    c) a carboxylic acid and a thiol
    d) a carboxylic acid and a sulfate
    e) none of these
21. When a carboxylic acid and an alcohol react to form an ester
    a) a molecule of water is incorporated in the ester.
    b) a molecule of water is produced.
    c) a molecule of oxygen is incorporated in the ester.
    d) a molecule of hydrogen is produced.
    e) none of these
22. A carboxylate ion is produced in which of these reactions?
    a) esterification
    b) saponification
    c) acid hydrolysis of an ester
    d) reaction of an alcohol with an inorganic acid
    e) none of these
23. An example of a dicarboxylic acid is:
    a) succinic acid
    b) butyric acid
    c) caproic acid
    d) lactic acid
    e) none of these
24. The hydrolysis of methyl butanoate in the presence of strong acid would yield:
    a) methanoic acid and butyric acid
    b) methanol and butanoic acid
    c) methanoic acid and 1-butanol
    d) sodium butanoate and methanol
    e) none of these

Answers to Practice Exercises

16.1

a. 2-bromo-2-methylpropanoic acid
b. 4-iodobenzoic acid

16.2

$CH_2(Br)-CH(Cl)-C(=O)-OH$	$CH_3-CH_2-CBr_2-COOH$
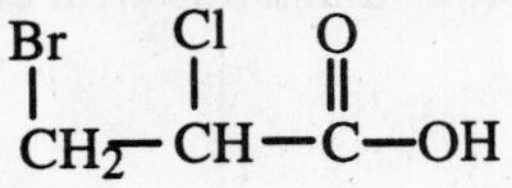 a. 3-bromo-2-chloropropanoic acid	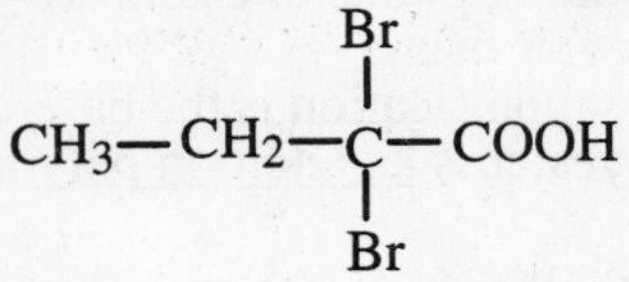 b. 2,2-dibromobutanoic acid

16.3

IUPAC name	Common name	Structural formula
butanoic acid	butyric acid	$CH_3-(CH_2)_2-COOH$
methanoic acid	formic acid	$HCOOH$
ethanedioic acid	oxalic acid	$HOOC-COOH$
hexanedioic acid	adipic acid	$HOOC-(CH_2)_4-COOH$

16.4

Common name	Structure of carboxylic acid
pyruvic acid	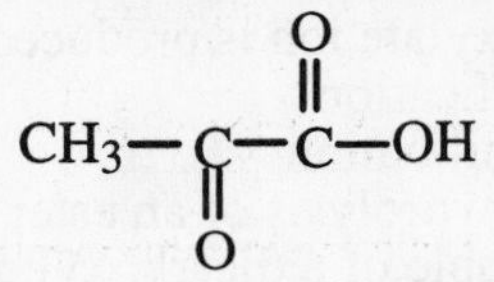 $CH_3-C(=O)-C(=O)-OH$
lactic acid	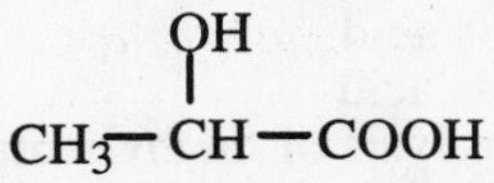 $CH_3-CH(OH)-COOH$
fumaric acid	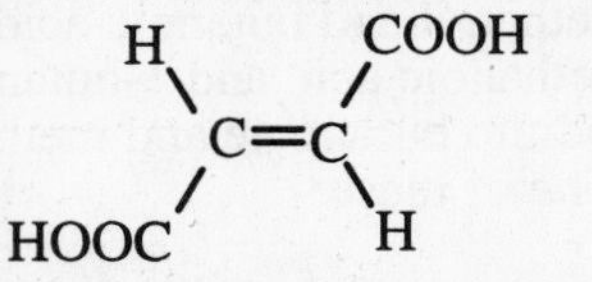 $H(HOOC)C=C(COOH)H$ (trans)

16.5 a. two molecules of propanoic acid

$$\begin{array}{l} \quad\quad\quad\quad\ddot{O}: \cdots H-\ddot{O}-C(=\ddot{O}:)-CH_2-CH_3 \\ CH_3-CH_2-C(=\ddot{O})-\ddot{O}-H \cdots :O \end{array}$$

b. one molecule of propanoic acid and water

$$CH_3-CH_2-C(=\ddot{O}\cdots H-O-H)-\ddot{O}(\cdots H-O-H)-H\cdots :\ddot{O}(H)-H$$

16.6 a. $CH_3-CH_2-CH_2-CH_2-OH \xrightarrow[\text{oxidizing agent}]{CrO_3} CH_3-CH_2-CH_2-COOH$

b. $CH_3-CH_2-CH_2-\overset{O}{\overset{\|}{C}}H \xrightarrow[\text{oxidizing agent}]{K_2Cr_2O_7} CH_3-CH_2-CH_2-COOH$

c. $BrC_6H_4-CH_2-CH_3 \text{ (para)} \xrightarrow[H_2SO_4]{K_2Cr_2O_7} BrC_6H_4-COOH \text{ (para)}$

16.7

Name of acid	Structural formula of carboxylic acid salt	IUPAC name of carboxylic acid salt
propanoic acid	$CH_3-CH_2-\overset{O}{\overset{\|}{C}}-\overset{-}{O}\ \overset{+}{Na}$	sodium propanoate
butanoic acid	$CH_3-CH_2-CH_2-\overset{O}{\overset{\|}{C}}-\overset{-}{O}\ \overset{+}{Na}$	sodium butanoate
benzoic acid	$C_6H_5-\overset{O}{\overset{\|}{C}}-\overset{-}{O}\ \overset{+}{Na}$	sodium benzoate

16.8 $CH_3-\overset{\overset{\displaystyle O}{\|}}{C}-O^{-}\,K^{+} + HCl \longrightarrow CH_3-\overset{\overset{\displaystyle O}{\|}}{C}-OH + K^{+}Cl^{-}$

16.9

Acid and alcohol	Structure of ester formed	IUPAC name
ethanoic acid and methanol	$CH_3-\overset{\overset{\displaystyle O}{\|}}{C}-O-CH_3$	methyl ethanoate
butanoic acid and methanol	$CH_3-CH_2-CH_2-\overset{\overset{\displaystyle O}{\|}}{C}-O-CH_3$	methyl butanoate
benzoic acid and ethanol	$C_6H_5-\overset{\overset{\displaystyle O}{\|}}{C}-O-CH_2-CH_3$	ethyl benzoate

16.10

IUPAC name	Common name
ethyl methanoate	ethyl formate
methyl pentanoate	methyl valerate
propyl hexanoate	propyl caproate

16.11

Compounds	Higher boiling points	Explain your answer
butanoic acid and methyl propanoate	butanoic acid	hydrogen bonding between two molecules of the carboxylic acid (produces a dimer)
methyl propanoate and 1-butanol	1-butanol	hydrogen bonding between the alcohol molecules
methyl hexanoate and decane	methyl hexanoate	dipole-dipole attraction between ester molecules

16.12 a. ethanol and sodium butanoate
b. ethanol and butanoic acid

16.13 a. HCOOH + $CH_3-CH_2-SH \longrightarrow HC(=O)-S-CH_2-CH_3 + H_2O$

b. cyclopentyl–COOH + $CH_3-SH \longrightarrow$ cyclopentyl–$C(=O)-S-CH_3 + H_2O$

16.14 a. $HO-P(=O)(OH)-O-CH_2-CH_3$

b. $HO-P(=O)(O-CH_2-CH_3)-O-CH_2-CH_3$

16.15 a. F; b. A; c. I; d. E; e. B; f. C; g. H; h. D.

A. 2-chloro-2-methylpropanoic acid
B. potassium propanoate
C. methyl propanoate
D. 1-propanethiol
E. 1-propanol
F. butanedioic acid

Answers to Self-Test

The numbers in parentheses refer to sections in your textbook:
1. T (16.1) **2**. F; hexanoic acid (16.2) **3**. F; dilute aqueous acetic acid solution (16.3)
4. F; α-methylbutyric acid (16.3) **5**. F; an α-hydroxy acid (16.4) **6**. F; butanone (16.6)
7. F; high boiling points (16.5) **8**. F; can be prepared (16.6) **9**. T (16.7) **10**. T (16.8)
11. F; increase (16.9) **12**. F; esters (16.11) **13**. T (16.11) **14**. T (16.12) **15**. T (16.13)
16. F; a condensation polymer **17**. a (16.10) **18**. c (16.5, 16.12) **19**. c (16.13)
20. c (16.14) **21**. b (16.9) **22**. b (16.9, 16.13) **23**. a (16.3, 16.4) **24**. b (16.13)

Amines and Amides

Chapter 17

Chapter Overview

Amines and amides are nitrogen-containing organic compounds that include many substances of importance in the human body as well as many natural and synthetic drugs.

In this chapter you will learn to recognize, name, and draw structural formulas for amines and amides. You will write equations for the preparation of amines and amides and for the hydrolysis of amides. You will learn the names and functions of some biologically important amines and amides.

Practice Exercises

17.1 **Amines** (Sec. 17.2) are organic derivatives of ammonia in which one or more hydrogens have been replaced by an alkyl group. Both common and IUPAC names are used for amines.

Complete the following table. Follow the rules for naming given in section 17.3 of your textbook.

Structure	IUPAC name	Common name
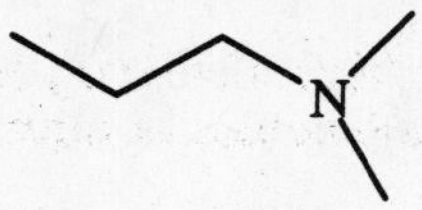 (skeletal structure: propyl chain attached to N bearing two methyl groups)	a.	
	b. N-ethyl-1-propanamine	
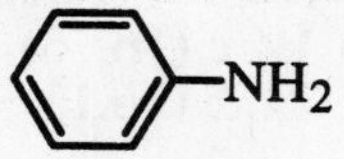 (benzene ring)$-NH_2$	c.	
	d.	*sec*-butylethylmethylamine
$CH_3-\underset{\underset{NH_2}{\vert}}{CH}-CH_2-\overset{\overset{O}{\Vert}}{C}-OH$	e.	

17.2 Amines are classified as primary, secondary, or tertiary on the basis of the number of alkyl groups attached to the nitrogen atom.

Classify each of the amines in Practice Exercise 17.1 as a primary, secondary, or tertiary amine.

a.

b.

c.

d.

e.

17.3 Simple aromatic amines are named as derivatives of aniline. Additional alkyl groups attached to the nitrogen atom are located using a capital N-. Name each of the following aromatic amines as derivatives of aniline.

O_2N (ring) $-NH_2$	(ring) $-NH$ (propyl chain)	$CH_2-CH_2-CH_3$ / (ring) $-N-CH_2-CH_3$
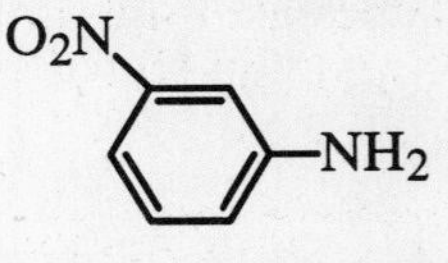 a.	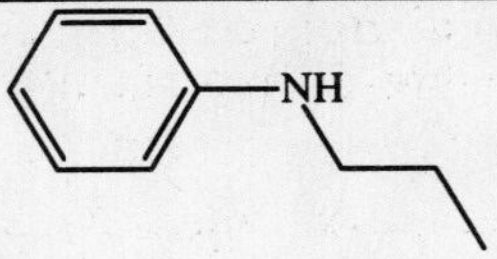 b.	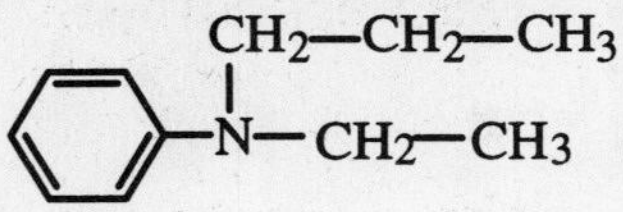 c.

17.4 Amines are capable of forming hydrogen bonds between molecules; however, since nitrogen is less electronegative than oxygen, hydrogen bonds between molecules of amines are weaker than those between alcohol molecules.

Determine which compound of each pair would have a higher boiling point and explain your choice.

Compounds	Higher boiling compound	Explain your choice
1-pentanamine and 1-pentanol		
1-pentanamine and hexane		

17.5 Since amines are bases, their reaction with an acid produces an **amine salt** (Sec. 17.6). The amine may be regenerated by reaction of the salt with a strong base. Complete the following equations.

a. $CH_3-CH_2-CH_2-CH_2(NH_2) + HCl \longrightarrow$?

b. ? + ? $\longrightarrow$ (ring)$-NH_2 + NaBr + H_2O$

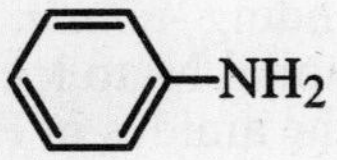

17.6 Amine salts are named in the same way as other ionic compounds are: the positive ion, the **substituted ammonium ion** (Sec. 17.5), is named first followed by the name of the negative ion. Give the name of each of the amine salts in Practice Exercise 17.5.

a.

b.

17.7 Amines can be prepared by the reaction of ammonia with an alkyl halide in the presence of a strong base. Primary amines formed by this method react further with more molecules of the alkyl halide to form secondary and tertiary amines.

Draw structural formulas and write names for the primary, secondary, and tertiary amines, and the **quaternary ammonium salt** (Sec. 17.7) formed when ammonia reacts with iodoethane in the presence of sodium hydroxide, a strong base. There are four reactions.

a.

b.

c.

d.

17.8 An **amide** (Sec. 17.11) is a carboxylic acid derivative in which the carboxyl –OH group is replaced by an amino or substituted amino group. Amides are prepared by the reaction between a carboxylic acid and ammonia or a primary or secondary amine.

Give the structure of the amide formed by the reaction of the following acids and amines.

a. 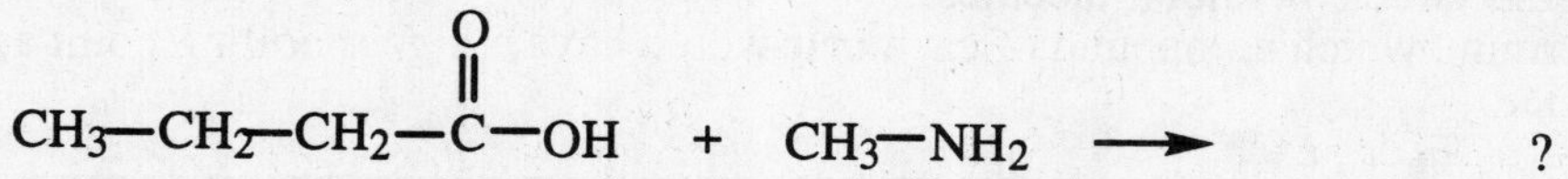 $CH_3-CH_2-CH_2-\overset{\overset{\displaystyle O}{\|}}{C}-OH \quad + \quad CH_3-NH_2 \longrightarrow \quad ?$

b. $CH_3-\underset{\underset{\displaystyle Cl}{|}}{CH}-\overset{\overset{\displaystyle O}{\|}}{C}-OH \quad + \quad CH_3-CH_2-\overset{\overset{\displaystyle CH_3}{|}}{NH} \longrightarrow \quad ?$

c. (3-bromophenyl)–COOH + $CH_3-\overset{\overset{\displaystyle CH_3}{|}}{NH} \longrightarrow \quad ?$

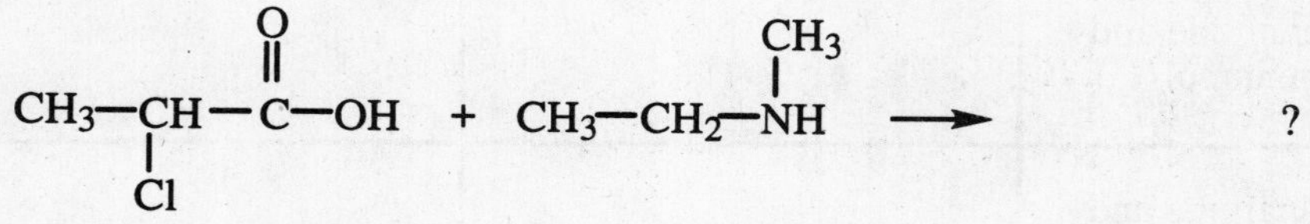

Br

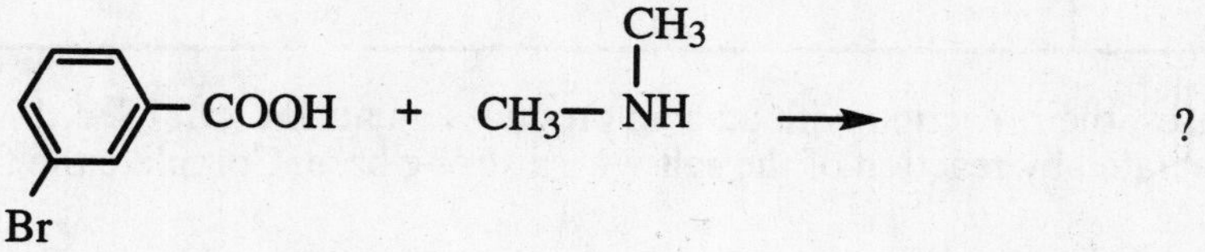

17.9 Amides can be classified as primary, secondary, or tertiary based on the number of carbon atoms attached to the nitrogen atom. Amide names in the IUPAC system use the name of the parent acid with the ending *-amide*. Alkyl groups attached to the nitrogen atom are included as prefixes, using a capital N- to locate them.

Classify each of the amides in Practice Exercise 17.8 as a primary, secondary, or tertiary amide and give the IUPAC name for the amide produced.

a.

b.

c.

17.10 Common names for amides are similar to IUPAC names, but the common name of the parent acid is used. Give the common names for the amides in Practice Exercise 17.8.

a.

b.

c.

17.11 Draw structural formulas for the following substituted amides.

a. N-ethyl-N-propylpentanamide	b. N,N-dimethylbenzamide

17.12 Amides undergo hydrolysis in a manner similar to that of esters. The products depend on the catalyst that is used: an acid catalyst produces a carboxylic acid and an amine salt, a basic catalyst produces a carboxylic acid salt and an amine.

Complete the following equations for the hydrolysis of each of these amides:

a. $CH_3-CH_2-CH_2-\overset{O}{\overset{\|}{C}}-\overset{CH_2-CH_3}{\overset{|}{NH}}$ $\xrightarrow[HCl]{Heat}$? + ?

b. $C_6H_5-\overset{O}{\overset{\|}{C}}-\overset{CH_3}{\overset{|}{N}}-CH_2-CH_3$ $\xrightarrow[HCl]{Heat}$? + ?

c. $CH_3CH_2CH_2-C(=O)-NH-CH_2CH_3$ (line structure) $\xrightarrow[NaOH]{Heat}$? + ?

17.13 Give the IUPAC names for the organic products formed in Practice Exercise 17.12.

a.

b.

c.

17.14 Use this identification exercise to review your knowledge of common structures. Using structures A through I, give the best choice for each of the terms below.

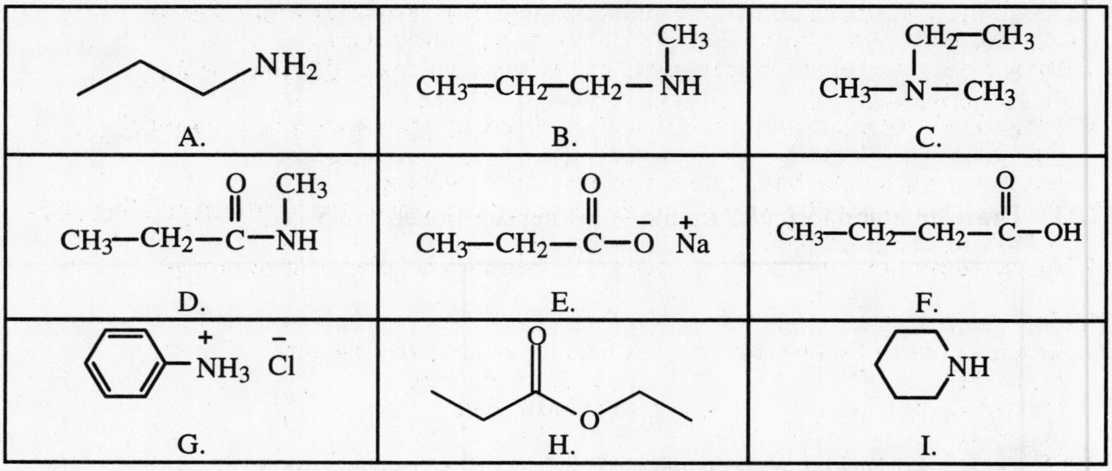

NH_2 A.	CH_3 $CH_3-CH_2-CH_2-NH$ B.	CH_2-CH_3 CH_3-N-CH_3 C.
O CH_3 CH_3-CH_2-C-NH D.	O $CH_3-CH_2-C-\bar{O}\ \overset{+}{Na}$ E.	O $CH_3-CH_2-CH_2-C-OH$ F.
$\overset{+}{N}H_3\ \bar{Cl}$ G.	O O H.	NH I.

a. cyclic amine ___________
b. salt of a carboxylic acid ___________
c. carboxylic acid ___________
d. amine salt ___________
e. primary amine ___________
f. amide ___________
g. secondary amine ___________
h. tertiary amine ___________
i. ester ___________

Give the IUPAC names of compounds A through H.

A. ______________________________
B. ______________________________
C. ______________________________
D. ______________________________
E. ______________________________
F. ______________________________
G. ______________________________
H. ______________________________

Self-Test

True-false: Indicate whether the following statements are true or false. If the statement is false, give the word or phrase that may be substituted for the underlined portion to make the statement true.

1. Nitrogen forms <u>two covalent bonds</u> to complete its octet of electrons.
2. *tert*-Butyl amine is a <u>primary amine</u>.
3. The IUPAC name for an amine having two methyl groups and one ethyl group attached to a nitrogen atom is <u>ethyldimethylamine</u>.
4. A benzene ring with an attached amino group is called <u>aniline</u>.
5. Amines are noted for their <u>pleasant odors</u>.

6. Hydrogen bonding between the molecules of an amine is stronger than hydrogen bonding between alcohol molecules.
7. Reaction of an alkyl halide with ammonia produces an amine salt.
8. The nitrogen atom of a heterocyclic amine cannot be part of an aromatic system.
9. Urea is a naturally occurring diamine with only one carbon atom.
10. The various types of nylon are synthesized by the reactions of diamines with dicarboxylic acids.
11. Kevlar is a very tough polyamide that owes its strength to the presence of aromatic rings in its "backbone."
12. Amides do not exhibit basic properties in water as amines do.
13. Disubstituted amides have no hydrogens attached to nitrogen for hydrogen bonding and so have low melting points.
14. The symptoms of Parkinson's disease are caused by a deficiency of serotonin.

Multiple choice:

15. An amine can be formed from its amine salt by treating the salt with:
    a) NaOH
    b) an alcohol
    c) H_2SO_4
    d) a carboxylic acid
    e) none of these
16. Which of the following is **not** a heterocyclic amine derivative?
    a) caffeine
    b) nicotine
    c) heme
    d) choline
    e) none of these
17. Alkaloids, a group of nitrogen-containing compounds obtained from plants, include all of the following compounds except:
    a) quinine
    b) nicotine
    c) atropine
    d) heroin
    e) morphine
18. Which of the following compounds is a secondary amine?
    a) 2-butanamine
    b) ethylmethylamine
    c) N,N-dimethylaniline
    d) 2-methylaniline
    e) none of these
19. N-methylpropanamide could be prepared as a product of the reaction between:
    a) N-propanamine and methanol
    b) propylamine and acetic acid
    c) methylamine and propanoic acid
    d) acetic acid and methylamine
    e) none of these
20. Basic hydrolysis of an amide produces:
    a) an amine salt and a carboxylic acid
    b) an amine salt and a carboxylic acid salt
    c) an amine and a carboxylic acid
    d) an amine and a carboxylic acid salt
    e) none of these
21. Mental depression may be caused by a deficiency of:
    a) histamine
    b) serotonin
    c) atropine
    d) porphyrin
    e) none of these
22. A correct name for the compound containing nitrogen bonded to two ethyl groups and one hydrogen is:
    a) diethylamine
    b) 2-ethylamine
    c) 2-ethyl amine
    d) diethyl amine
    e) none of these

Answers to Practice Exercises

17.1

Structure	IUPAC name	Common name
$CH_3-CH_2-CH_2-N(CH_3)-CH_3$	a. N,N-dimethyl-1-propanamine	dimethylpropylamine
$CH_3-CH_2-CH_2-NH-CH_2-CH_3$	b. N-ethyl-1-propanamine	ethylpropylamine
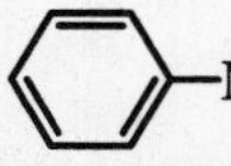 $C_6H_5-NH_2$	c. aniline	aniline
$CH_3-CH_2-CH(CH_3)-N(CH_3)-CH_2-CH_3$	d. N-ethyl-N-methyl-2-butanamine	*sec*-butylethylmethylamine
$CH_3-CH(NH_2)-CH_2-C(=O)-OH$	e. 3-aminobutanoic acid	β-aminobutyric acid

17.2
a. tertiary
b. secondary
c. primary
d. tertiary
e. primary

17.3
a. 3-nitroaniline
b. N-propylaniline
c. N-ethyl-N-propylaniline

17.4

Compounds	Higher boiling compound	Explain your choice
1-pentanamine and 1-pentanol	1-pentanol	-OH hydrogen bonds are stronger than -NH hydrogen bonds because oxygen is more electronegative than nitrogen
1-pentanamine and hexane	1-pentanamine	hydrogen bonding between amine molecules

17.5 a. $CH_3-CH_2-CH_2-CH_2(NH_2) + HCl \longrightarrow CH_3-CH_2-CH_2-CH_2-\overset{+}{N}H_3\ \overset{-}{Cl}$

b. 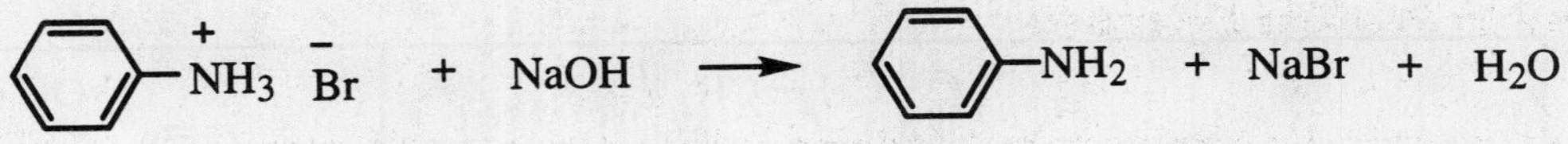 $C_6H_5-\overset{+}{N}H_3\ \overset{-}{Br} + NaOH \longrightarrow C_6H_5-NH_2 + NaBr + H_2O$

17.6 a. butylammonium chloride
b. anilinium bromide

17.7 a. primary
ethylamine

$$NH_3 + CH_3-CH_2-I \xrightarrow{OH^-} CH_3-CH_2-NH_2$$

b. secondary
diethylamine

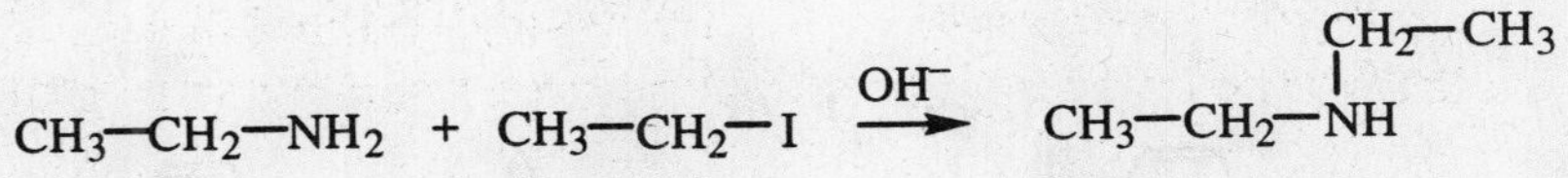

$$CH_3-CH_2-NH_2 + CH_3-CH_2-I \xrightarrow{OH^-} CH_3-CH_2-NH(CH_2-CH_3)$$

c. tertiary
triethylamine

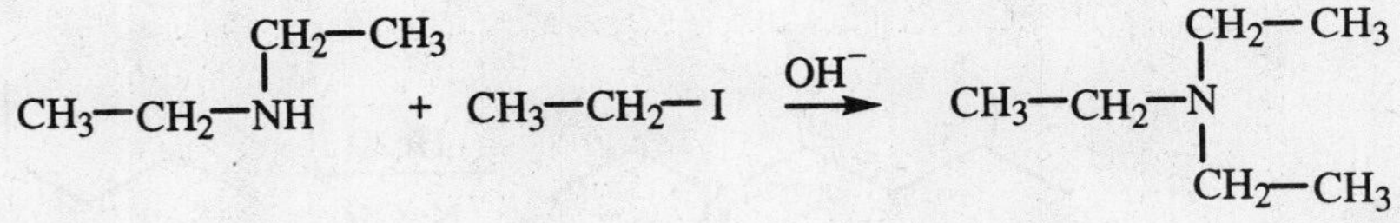

$$CH_3-CH_2-NH(CH_2-CH_3) + CH_3-CH_2-I \xrightarrow{OH^-} CH_3-CH_2-N(CH_2-CH_3)_2$$

d. quaternary
tetraethylammonium
iodide

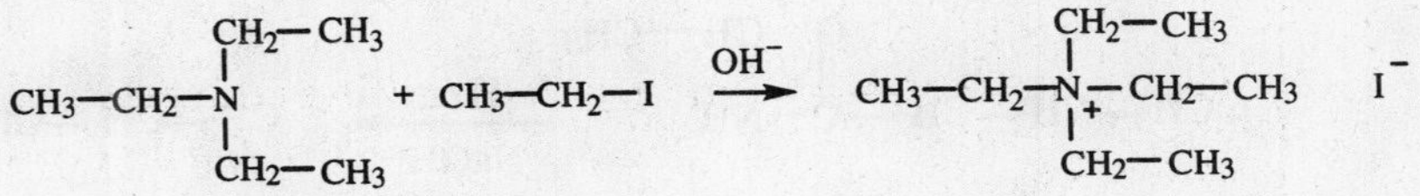

$$CH_3-CH_2-N(CH_2-CH_3)_2 + CH_3-CH_2-I \xrightarrow{OH^-} CH_3-CH_2-\overset{+}{N}(CH_2-CH_3)_3\ I^-$$

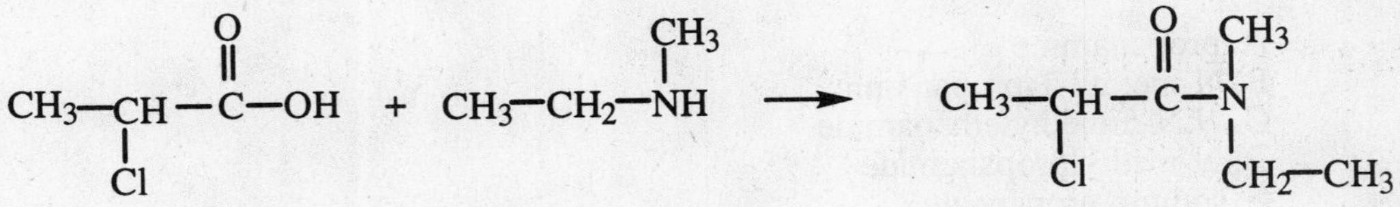

17.8 a. 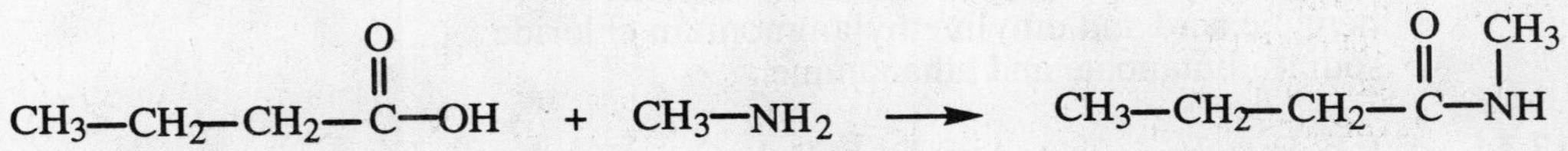 $CH_3-CH_2-CH_2-\overset{O}{\overset{\|}{C}}-OH + CH_3-NH_2 \longrightarrow CH_3-CH_2-CH_2-\overset{O}{\overset{\|}{C}}-NH-CH_3$

b. $CH_3-CH(Cl)-\overset{O}{\overset{\|}{C}}-OH + CH_3-CH_2-NH-CH_3 \longrightarrow CH_3-CH(Cl)-\overset{O}{\overset{\|}{C}}-N(CH_3)(CH_2-CH_3)$

c. m-$Br-C_6H_4-COOH + CH_3-NH-CH_3 \longrightarrow$ m-$Br-C_6H_4-\overset{O}{\overset{\|}{C}}-N(CH_3)_2$

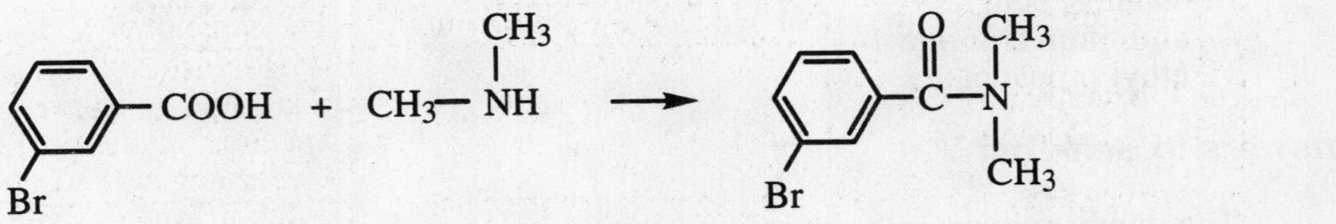

17.9 a. secondary amide, N-methylbutanamide
b. tertiary amide, N-ethyl-N-methyl-2-chloropropanamide
c. tertiary amide, N,N-dimethyl-m-bromobenzamide

17.10 a. N-methylbutyramide
b. N-ethyl-N-methyl-2-chloropropionamide
c. N,N-dimethyl-m-bromobenzamide

17.11

a.	b.
$CH_3-CH_2-CH_2-CH_2-\overset{O}{\overset{\|\|}{C}}-\underset{}{\overset{CH_2-CH_3}{\overset{\|}{N}}}-CH_2-CH_2-CH_3$	$C_6H_5-\overset{O}{\overset{\|\|}{C}}-N(CH_3)_2$
a. N-ethyl-N-propylpentanamide	b. N,N-dimethylbenzamide

17.12 a. $CH_3-CH_2-CH_2-\overset{O}{\overset{\|\|}{C}}-NH(CH_2-CH_3) \xrightarrow[HCl]{Heat} CH_3-CH_2-CH_2-\overset{O}{\overset{\|\|}{C}}-OH + CH_3-CH_2-\overset{+}{N}H_3\ Cl^-$

b. $C_6H_5-\overset{O}{\overset{\|\|}{C}}-N(CH_3)-CH_2-CH_3 \xrightarrow[HCl]{Heat} C_6H_5-COOH + CH_3-CH_2-\overset{+}{N}H_2(CH_3)\ Cl^-$

c. (line structures) butanamide, N-ethyl $\xrightarrow[NaOH]{Heat}$ butanoate O^-Na^+ + H_2N–ethyl

$CH_3-CH_2-CH_2-\overset{O}{\overset{\|\|}{C}}-NH(CH_2-CH_3) \xrightarrow[NaOH]{Heat} CH_3-CH_2-CH_2-\overset{O}{\overset{\|\|}{C}}-O^-\ Na^+ + CH_3-CH_2-NH_2$

17.13 a. butanoic acid and ethylammonium chloride
b. benzoic acid and ethylmethylammonium chloride
c. sodium butanoate and ethanamine

17.14 a. I; b. E; c. F; d. G; e. A; f. D; g. B; h. C; i. H.

A. propanamine
B. N-methyl-1-propanamine
C. N,N-dimethylethanamine
D. N-methylpropanamide
E. sodium propanoate
F. butanoic acid
G. anilinium chloride
H. ethyl propanoate

Answers to Self-Test

The numbers in parentheses refer to sections in your textbook:
1. F; three covalent bonds (17.1) **2**. T (17.2) **3**. F; N,N-dimethylethanamine (17.3)
4. T (17.3) **5**. F; unpleasant or fishlike odors (17.4) **6**. F; weaker (17.4) **7**. T (17.6)
8. F; can (17.8) **9**. F; diamide (17.12) **10**. T (17.16) **11**. T (17.16) **12**. T (17.13)
13. T (17.13) **14**. F; dopamine (17.9) **15**. a (17.6) **16**. d (17.8) **17**. d (17.10)
18. b (17.3) **19**. c (17.14) **20**. d (17.15) **21**. b (17.9) **22**. a (17.3)

Carbohydrates — Chapter 18

Chapter Overview

The remaining chapters in the book will be concerned with compounds that are important in biological systems. Carbohydrates are important energy-storage compounds. Their oxidation provides energy for the activities of living organisms.

In this chapter you will learn to identify various types of carbohydrates and write structural diagrams for some of the most common ones. You will be able to explain stereoisomerism in terms of chiral carbons and draw figures that represent the three-dimensional structures of compounds containing chiral carbons. You will identify the structural features of some important disaccharides and polysaccharides, and learn where in nature they are found and what functions they have.

Practice Exercises

18.1 **Monosaccharides** (Sec. 18.3) are **carbohydrates** (Sec. 18.3) that contain a single polyhydroxy aldehyde or polyhydroxy ketone unit. Carbohydrates can be classified by the number of monosaccharide units they contain. Give the names for carbohydrates having the following numbers of monosaccharide units.

a) 2 monosaccharide units ____________________

b) 2 to 10 monosaccharide units ____________________

c) 3 monosaccharide units ____________________

d) many (greater than 10) units ____________________

18.2 An object that cannot be superimposed on its **mirror image** (Sec. 18.4) is a **chiral object** (Sec. 18.4). An organic molecule is chiral if it contains a **chiral center** (Sec. 18.4), a single atom with four different atoms or groups of atoms attached to it. Indicate whether the circled carbon in each structure below is chiral or **achiral** (Sec. 18.4):

a. CH_3 (top), CH_3—C—Cl, Br (bottom)

b. CH_3 (top), H—C—Br, I (bottom)

c. CH_3 (top), CH_3—CH_2—C—CH_2—CH_2—CH_3, H (bottom)

18.3 Organic molecules may contain more than one chiral center. Circle the chiral centers in the following condensed structures.

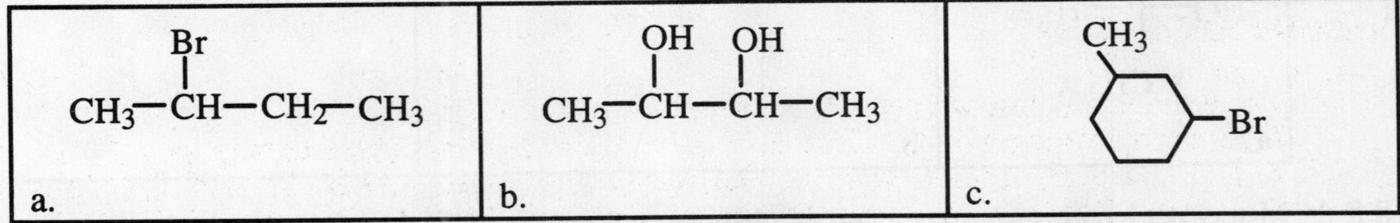

18.4 The three-dimensional nature of chiral carbons can be represented by using **Fischer projections** (18.6). A Fischer projection shows the chiral carbon as intersecting horizontal and vertical lines, with vertical lines representing bonds directed into the page and horizontal lines representing bonds directed out of the page.

Convert the following three-dimensional representations into Fischer projection formulas.

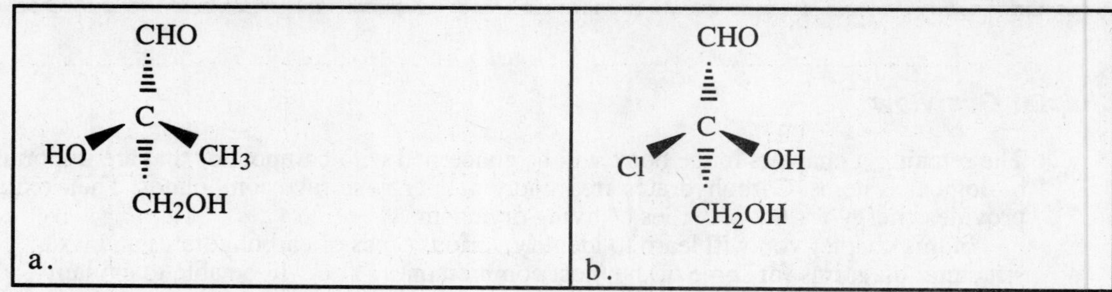

18.5 **Stereoisomers** (Sec. 18.5) are isomers whose atoms are connected in the same way but differ in their arrangement in space. **Enantiomers** (Sec. 18.5) are stereoisomers whose molecules are mirror images of one another.

Using Fischer projections, draw the enantiomers of the Fischer projection formulas in Practice Exercise 18.4.

a.

b.

18.6 By definition, the enantiomer with the -OH on the right in the Fischer projection of a monosaccharide is the right-handed enantiomer and has the designation D. The enantiomer with the –OH on the left is the left-handed enantiomer and has the designation L. When more than one chiral center is present, the D and L designations refer to the highest numbered chiral carbon.

Classify the Fischer projection formulas in Practice Exercise 18.4 as D or L.

a. ____________

b. ____________

18.7 **Diastereomers** (Sec. 18.5) are stereoisomers whose molecules are not mirror images of each other. For the following structure (a.), draw the enantiomer (b.) and draw one diasteromer (c.)

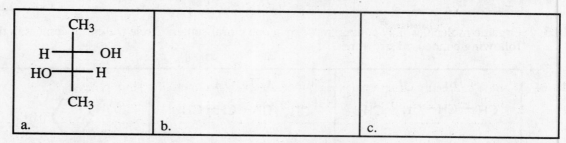

18.8 Monosaccharides are classified as **aldoses** or **ketoses** (Sec. 18.8) on the basis of the carbonyl group: an aldose contains an aldehyde group and a ketose contains a ketone group. They can be further classified by the number of carbons in the molecule (for example, aldopentose). Monosaccharides and **disaccharides** (Sec. 18.3) are often called **sugars** (Sec. 18.8).

A monosaccharide that has n chiral centers may exist in a maximum of 2^n steroisomeric forms.

Fill in information for each of the following carbohydrates: a. classify using carbon number and type of carbonyl group present; b. calculate the total number of stereoisomers; c. name the isomer shown.

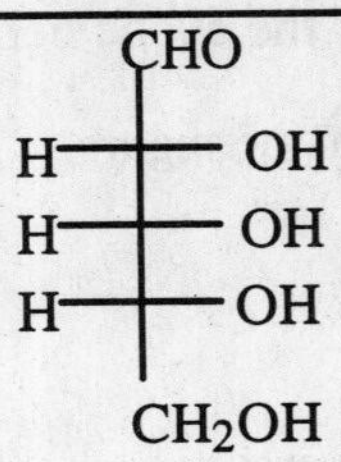

CHO H—OH H—OH H—OH CH_2OH	CH_2OH =O HO—H H—OH H—OH CH_2OH	CHO H—OH HO—H HO—H H—OH CH_2OH
a. b. c.	a. b. c.	a. b. c.

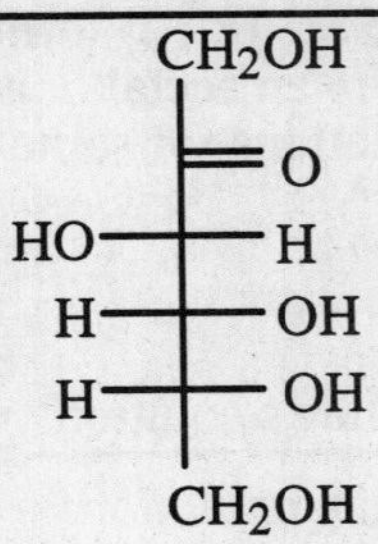

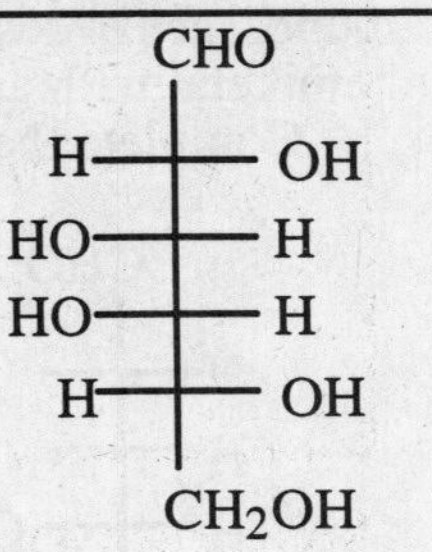

18.9 For monosaccharides containing five or more carbons, open-chain structures are in equilibrium with cyclic structures formed by the intramolecular reaction of the carbonyl group with a hydroxyl group to form a cyclic hemiacetal or cyclic hemiketal. The structural representations of these cyclic forms are called **Haworth projections** (Sec. 18.11). See section 18.11 in your textbook for conventions observed in drawing Haworth projections.

Complete the following equations by giving the structural formula for the cyclic or open-chain form.

a.

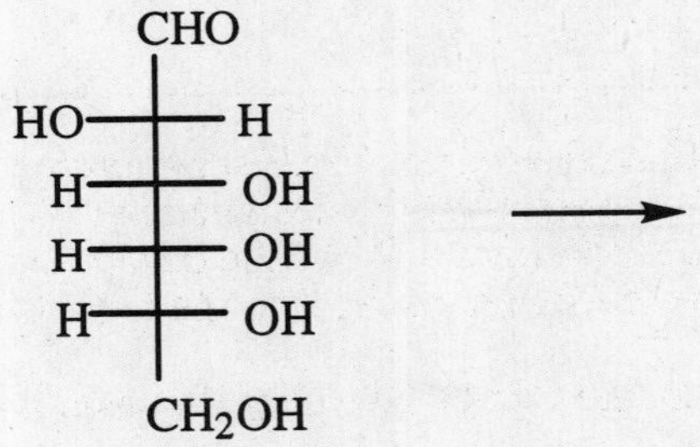

CHO
HO—H
H—OH
H—OH
H—OH
CH_2OH
→

b.

CH_2OH
O OH
HO
OH OH
→

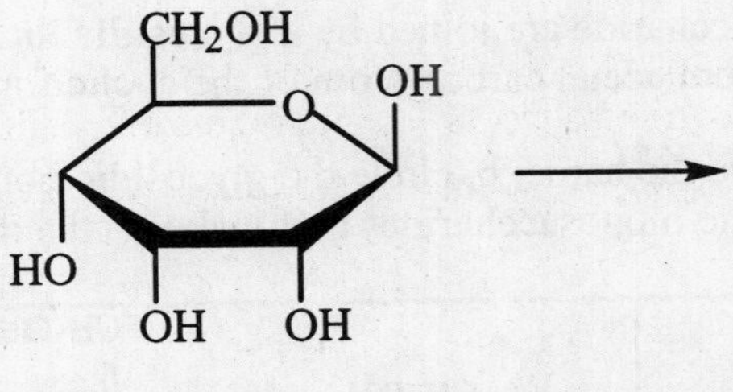

18.10 For the following structures, a. identify each of the following structures as an α or β monosaccharide, b. identify each structure as a hemiacetal or hemiketal.

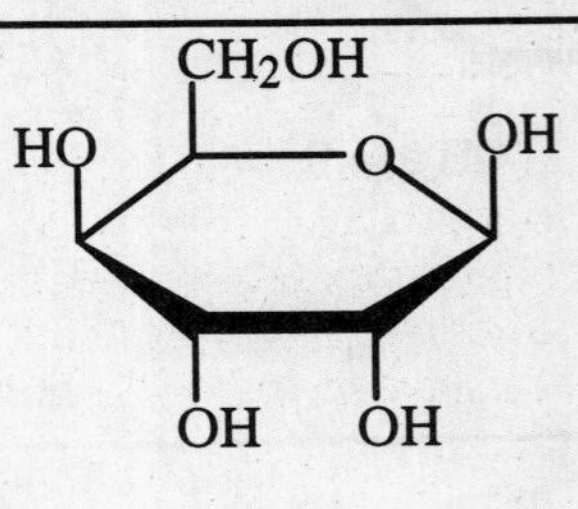

CH_2OH HO O OH OH OH	CH_2OH HO O OH OH OH	CH_2OH O CH_2OH OH HO HO
a. b.	a. b.	a. b.

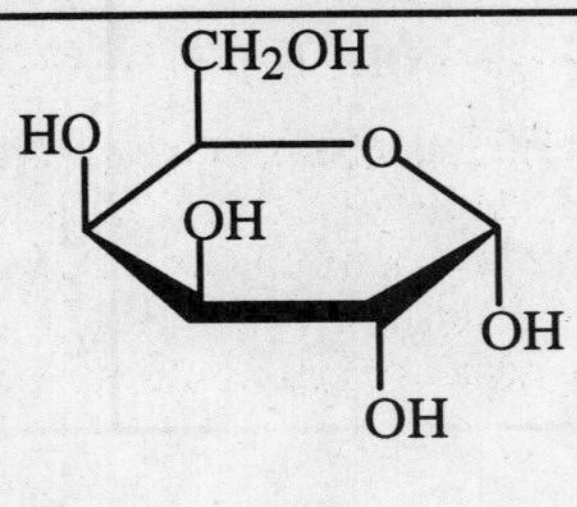

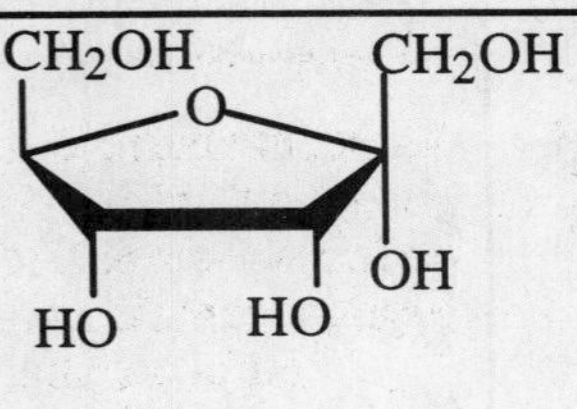

18.11 Weak oxidizing agents, such as Fehling's solution, oxidize the carbonyl group end of a monsaccharide to give an acid. Sugars that can be oxidized by these agents are called **reducing sugars** (Sec. 18.12). The carbonyl group of sugars can also be reduced to a hydroxyl group using H_2 as the reducing agent. Another important reaction of monosaccharides is **glycoside** (Sec. 18.12) formation, reaction of the cyclic hemiacetal or hemiketal with an alcohol to form an acetal or a ketal.

Complete the following equations for some common reactions of sugars.

a. 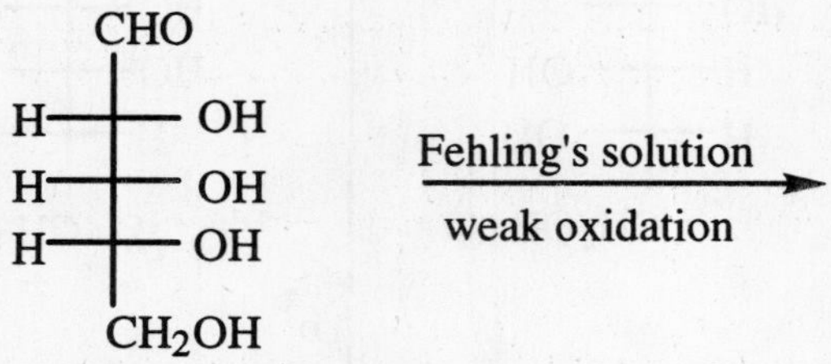 CHO / H—OH / H—OH / H—OH / CH_2OH —(Fehling's solution, weak oxidation)→

b. 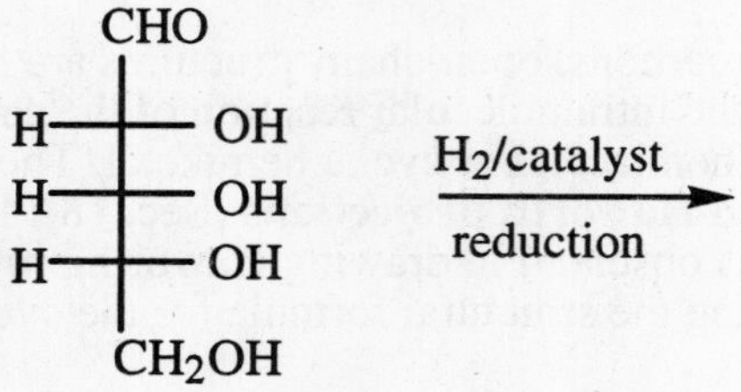 CHO / H—OH / H—OH / H—OH / CH_2OH —(H_2/catalyst, reduction)→

c. 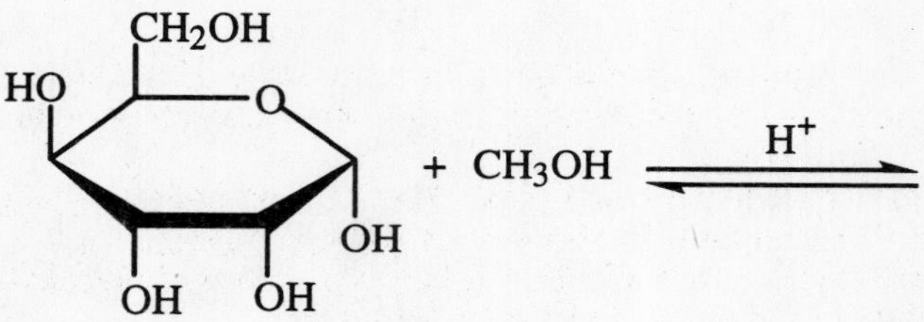 (cyclic sugar: CH_2OH, HO, O, OH, OH, OH) + CH_3OH $\xrightleftharpoons{H^+}$

18.12 The two monosaccharides that form a disaccharide are joined by a **glycosidic linkage** (Sec. 18.13). The configuration (α or β) of the hemiacetal carbon atom of the cyclic form is often very important.

For the following disaccharides: a. give the name; b. circle the glycosidic bond and name the linkage (α or β); c. give the names of the monosaccharides that make up the disaccharide.

(disaccharide: CH_2OH, CH_2OH, O, OH, OH, HO, OH, OH, OH)	(disaccharide: CH_2OH, CH_2OH, HO, O, OH, OH, OH, OH, OH)
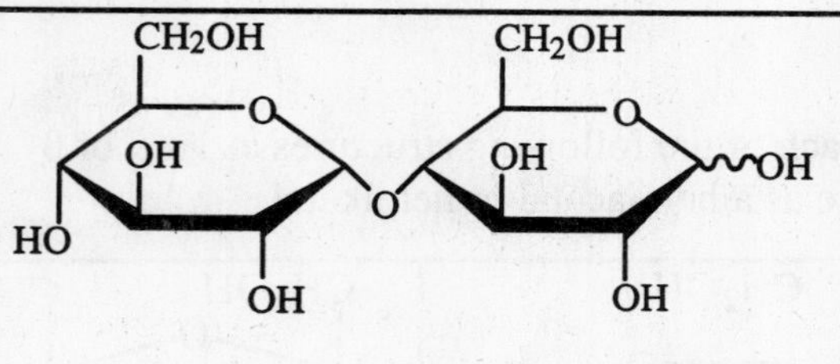 a. b. c.	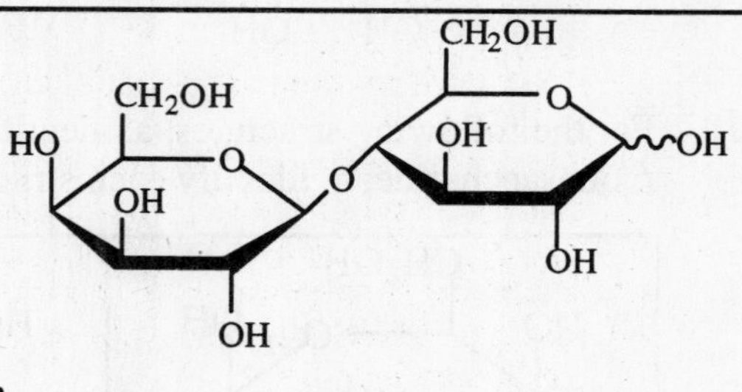 a. b. c.

18.13 Are the disaccharides in Practice Exercise 18.12 reducing or nonreducing sugars with Benedict's solution? ____________________ Place a box around the functional group in the structures above that is responsible for giving a positive test with Benedict's solution.

18.14 There are a number of important **polysaccharides** (Sec. 18.14) that are made up of D-glucose units. They differ in the types of linkages between monomers, the size, and the function. Complete the following table that summarizes these properties. Use information found in section 18.14 of your textbook.

Polysaccharide	Linkage	Size	Branching	Function
glycogen				
amylose				
amylopectin				
cellulose				

Self-Test

True-false: Indicate whether the following statements are true or false. If the statement is false, give the word or phrase that may be substituted for the underlined portion to make the statement true.

1. Bioorganic substances include carbohydrates, lipids, proteins, and <u>nucleic acids</u>.
2. Carbohydrates form part of the structural framework of <u>DNA and RNA</u>.
3. An oligosaccharide is a carbohydrate that contains <u>at least 20</u> monosaccharide units.
4. <u>A chiral object</u> is an object that is identical to its mirror image.
5. The compound 2-methyl-2-butanol contains <u>one</u> chiral center.
6. In a Fischer projection, <u>horizontal lines</u> represent bonds to groups directed into the printed page.
7. Enantiomers are stereoisomers whose molecules <u>are not</u> mirror images of each other.
8. <u>Diastereomers</u> have identical boiling and freezing points.
9. A dextrorotatory compound rotates plane-polarized light in a <u>clockwise</u> direction.
10. The direction of rotation of plane-polarized light by sugar molecules is designated in the compound's name as <u>D or L</u>.
11. Response of the human body to the two enantiomeric forms of a chiral molecule is <u>identical</u>.
12. A six-carbon monosaccharide with a ketone functional group is called <u>an aldopentose</u>.
13. Fructose can form a 5-membered ring that is <u>a hemiketal</u>.
14. The carbonyl group of a monosaccharide can be reduced to a hydroxyl group using <u>Tollens solution</u> as a reducing agent.
15. The bond between two monosaccharide units in a disaccharide is a <u>glycosidic</u> linkage.
16. Humans cannot digest cellulose because they lack an enzyme to catalyze hydrolysis of the <u>$\alpha(1\rightarrow 4)$ linkage</u>.
17. Amylopectin molecules are <u>more highly branched</u> than amylose molecules are.
18. A <u>homopolysaccharide</u> is a polysaccharide in which more than one type of monosaccharide unit is present.

Multiple choice:

19. Which of the following substances is not required for the production of carbohydrates by photosynthesis?
    a) carbon dioxide
    b) oxygen
    c) water
    d) sunlight
    e) chlorophyl
20. Which of these compounds contains three chiral centers?
    a) glyceraldehyde
    b) glucose
    c) fructose
    d) galactose
    e) none of these
21. Ribose is an example of a(n):
    a) ketohexose
    b) aldohexose
    c) aldotetrose
    d) aldopentose
    e) none of these
22. Dextrose, or blood sugar, is:
    a) D-glucose
    b) D-galactose
    c) D-fructose
    d) D-ribose
    e) none of these
23. The sugar sometimes known as levulose is:
    a) D-ribose
    b) L-glucose
    c) L-galactose
    d) D-fructose
    e) none of these
24. Which of the following compounds is **not** made up solely of D-glucose units?
    a) maltose
    b) lactose
    c) cellulose
    d) starch
    e) glycogen
25. Which of the following sugars is **not** a reducing sugar?
    a) lactose
    b) maltose
    c) sucrose
    d) galactose
    e) all are reducing sugars
26. The main storage form of D-glucose in animal cells is:
    a) glycogen
    b) amylopectin
    c) amylose
    d) sucrose
    e) none of these
27. Which of the following molecules has a chiral center?
    a) ethanol
    b) 1-chloro-1-bromoethane
    c) 1-chloro-2-bromoethane
    d) 1,2-dichloroethane
    e) none of these
28. Hydrolysis of sucrose yields:
    a) glucose and fructose
    b) glucose and galactose
    c) ribose and fructose
    d) ribose and glucose
    e) none of these

Answers to Practice Exercises

18.1
a. disaccharides
b. oligosaccharides
c. trisaccharides
d. polysaccharides

18.2
a. achiral
b. chiral
c. chiral

18.3

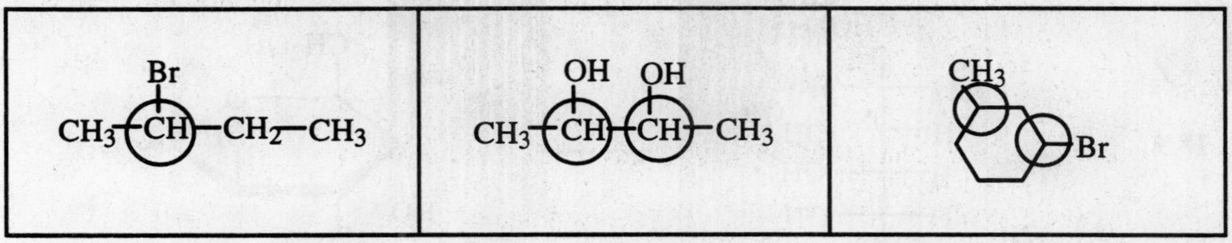

18.4

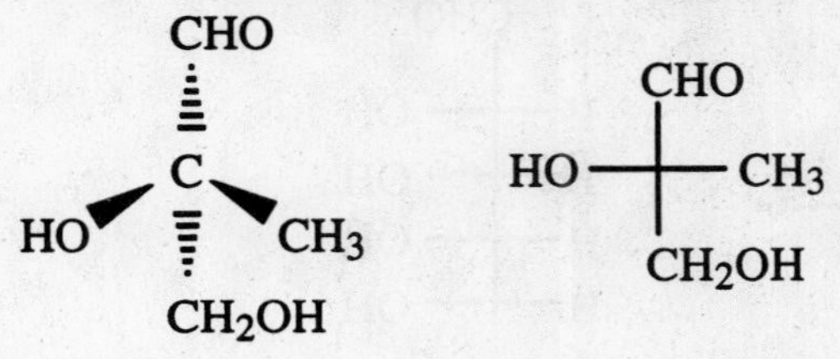

a.

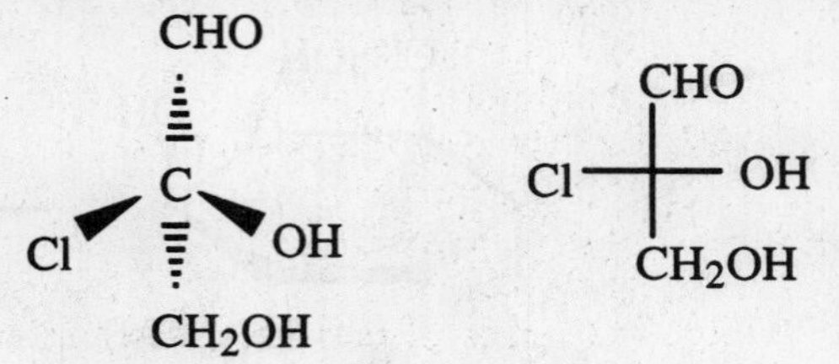

b.

18.5

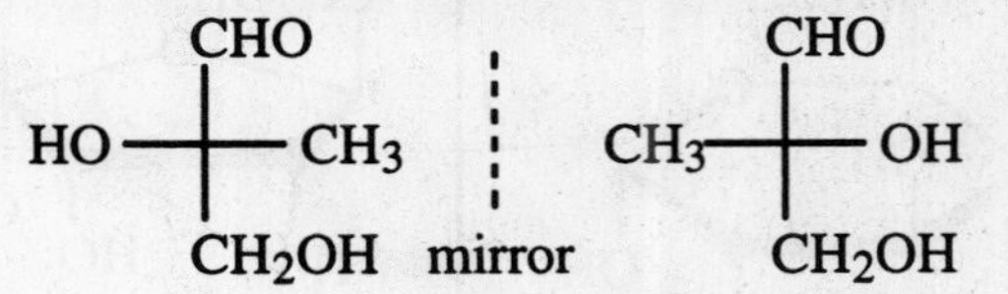

a.

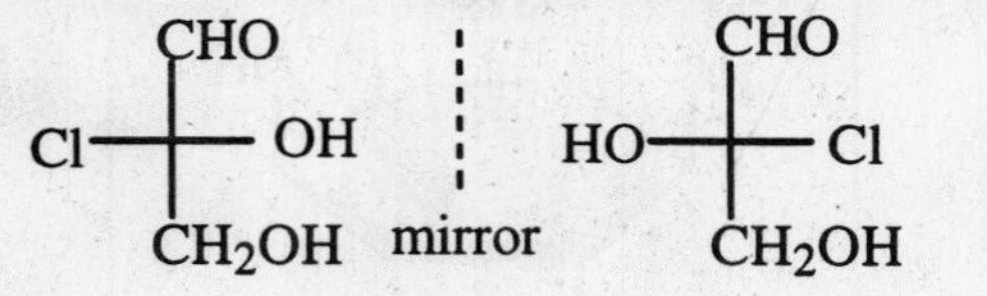

b.

18.6 a. L; b. D.

18.7

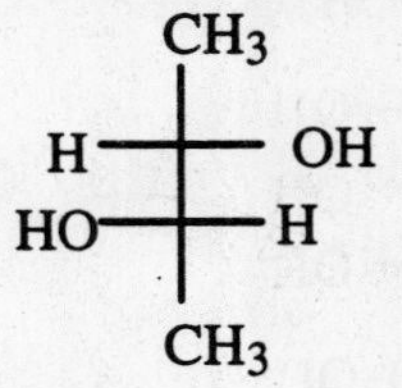

a.

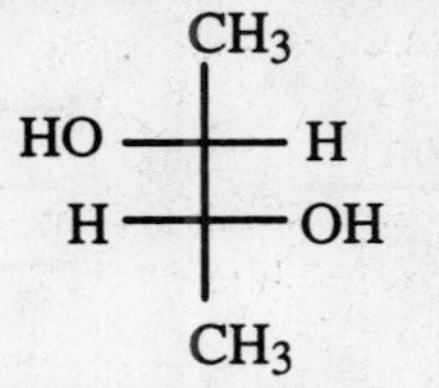

b. enantiomer

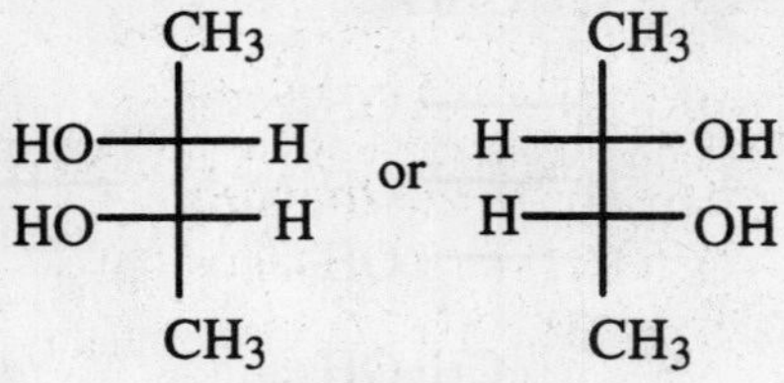

c. diastereomer

18.8

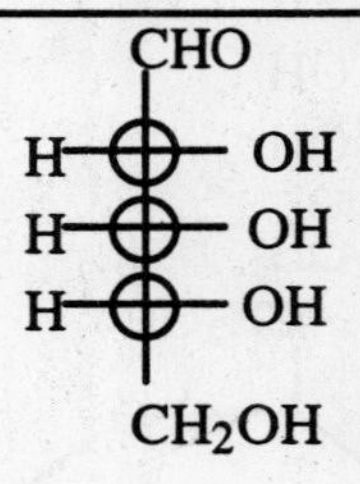

a. aldopentose
b. $2^n = 2^3 = 8$
c. D-ribose

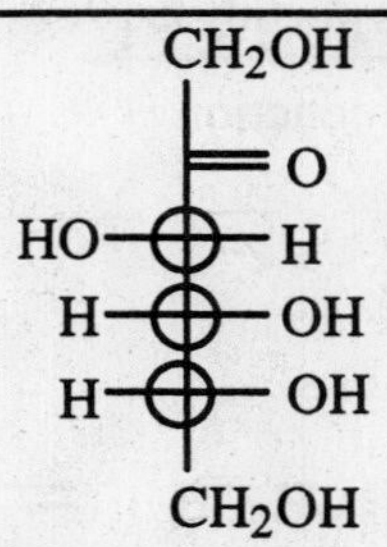

a. ketohexose
b. $2^n = 2^3 = 8$
c. D-fructose

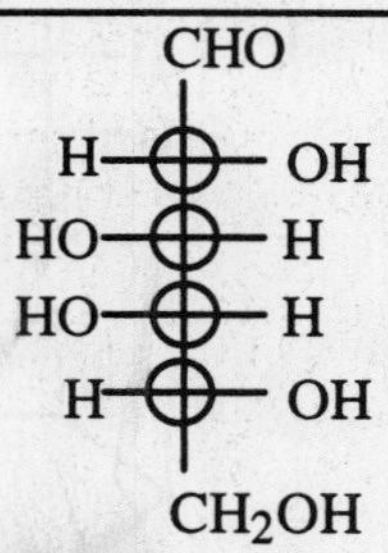

a. aldohexose
b. $2^n = 2^4 = 16$
c. D-galactose

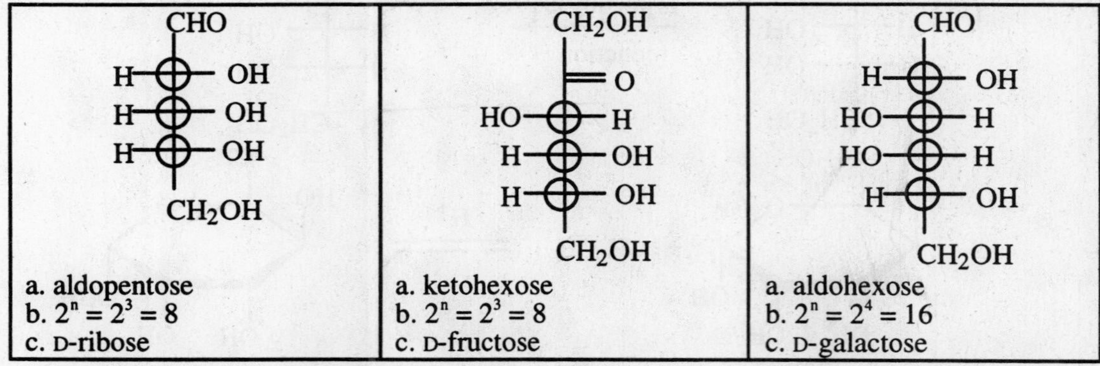

18.9 a.

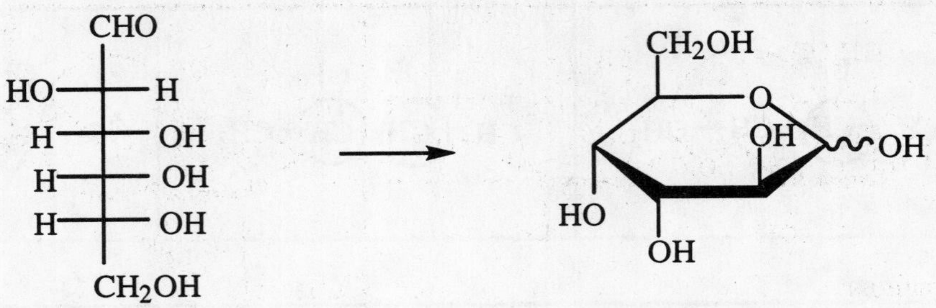

CHO
HO—H
H—OH
H—OH
H—OH
CH_2OH

→

CH_2OH
O
OH ∿OH
HO
OH

b.

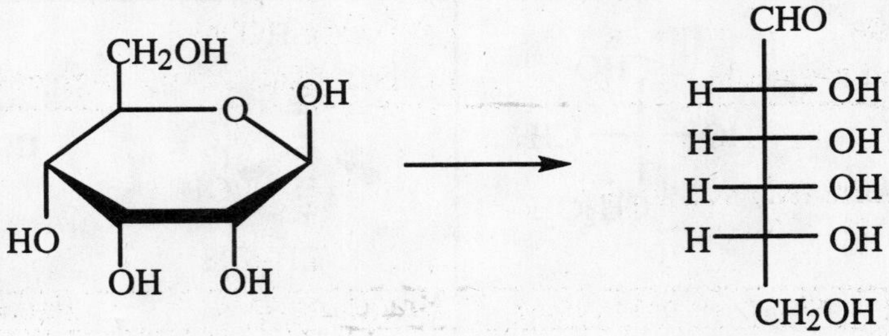

CH_2OH
O OH
HO
OH OH

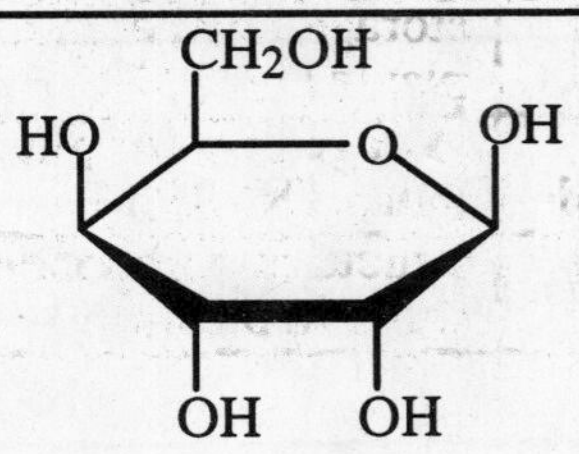

→

CHO
H—OH
H—OH
H—OH
H—OH
CH_2OH

18.10

CH_2OH HO O OH OH OH a. beta b. hemiacetal	CH_2OH HO O OH OH OH a. alpha b. hemiacetal	CH_2OH O CH_2OH HO HO OH a. alpha b. hemiketal

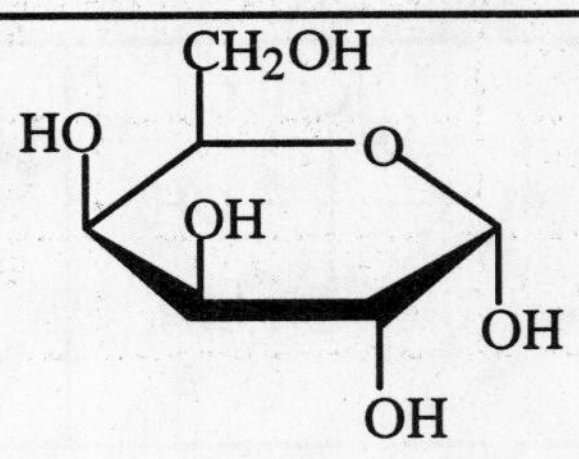

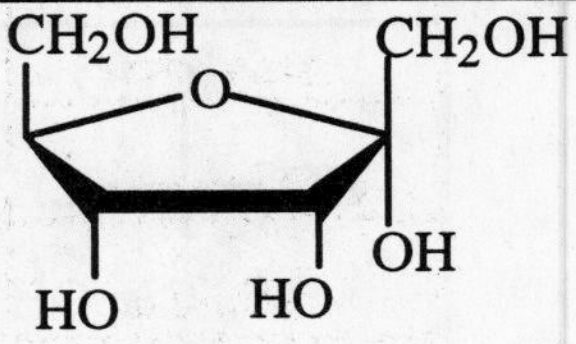

18.11 a.

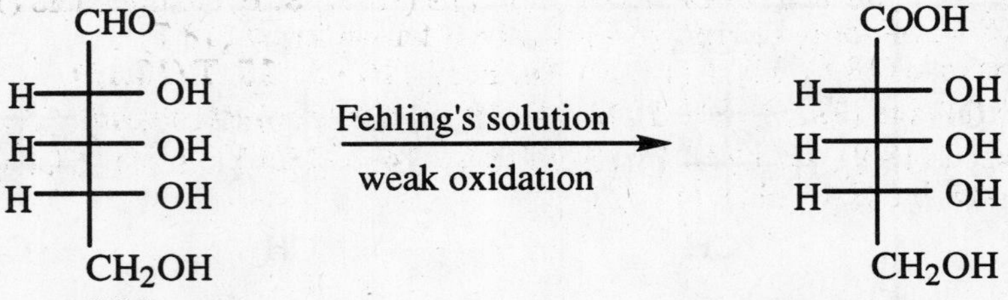

CHO
H—OH
H—OH
H—OH
CH_2OH

Fehling's solution
weak oxidation
→

COOH
H—OH
H—OH
H—OH
CH_2OH

b.

CHO
H—OH
H—OH
H—OH
CH_2OH

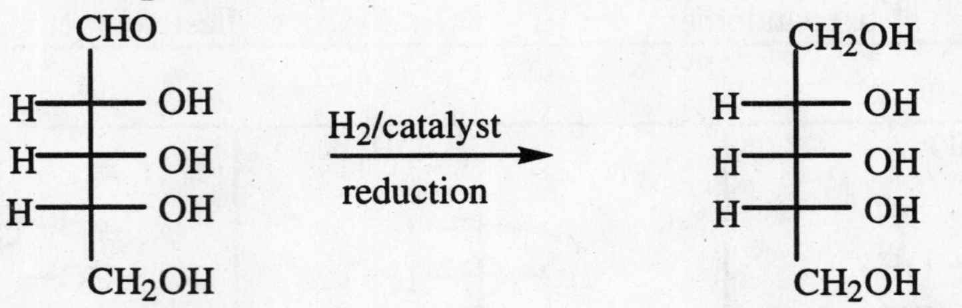

H_2/catalyst
reduction
→

CH_2OH
H—OH
H—OH
H—OH
CH_2OH

c.

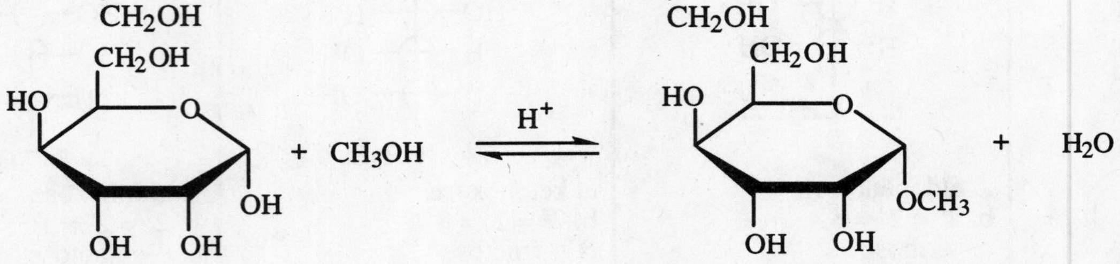

CH_2OH
HO O
OH
OH OH

+ CH_3OH $\overset{H^+}{\rightleftharpoons}$

CH_2OH
HO O
OCH_3
OH OH

+ H_2O

18.12

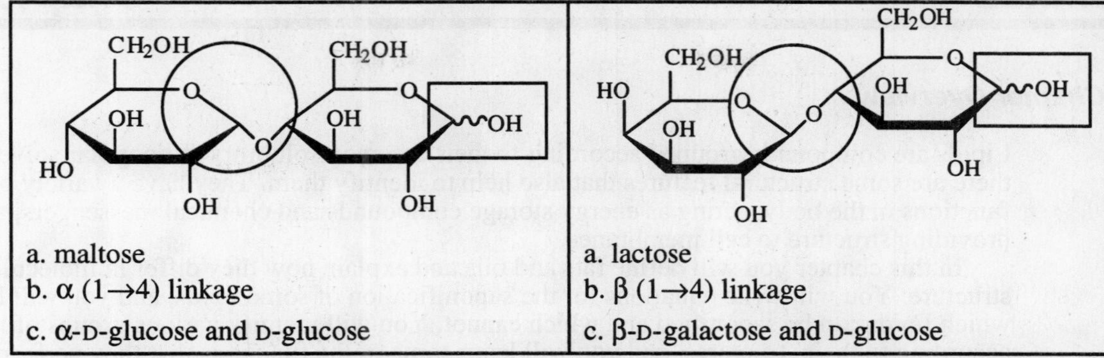

a. maltose	a. lactose
b. α (1→4) linkage	b. β (1→4) linkage
c. α-D-glucose and D-glucose	c. β-D-galactose and D-glucose

18.13 Both disaccharides are reducing sugars.

18.14

Polysaccharide	Linkage	Size (amu)	Branching	Function
glycogen	α(1→4) α(1→6)	3,000,000	very highly branched	storage form of glucose in animals
amylose	α(1→4)	50,000	straight-chain	storage form of glucose in plants (15-20%)
amylopectin	α(1→4)	300,000	highly branched	storage form of glucose in plants (80-85%)
cellulose	β(1→4)	900,000	straight-chain	structural component of cell walls in plants

Answers to Self-Test

The numbers in parentheses refer to sections in your textbook:
1. T (18.1) **2**. T (18.2) **3**. F; two to ten (18.3) **4**. F; an achiral object (18.4)
5. F; no (18.4) **6**. F; vertical lines (18.6) **7**. F; are (18.5) **8**. F; enantiomers (18.9)
9. T (18.7) **10**. F; (+) or (–) (18.6, 18.7) **11**. F; often different (18.7)
12. F; a ketohexose (18.8) **13**. T (18.10) **14**. F; H_2 (18.12) **15**. T (18.13)
16. F; β(1→4) linkage (18.14) **17**. T (18.14) **18**. F; heteropolysaccharide (18.15) **19**. b (18.2)
20. c (18.4) **21**. d (18.9) **22**. a (18.9) **23**. d (18.9) **24**. b (18.13, 18.14) **25**. c (18.13)
26. a (18.14) **27**. b (18.4) **28**. a (18.13)

Lipids

Chapter 19

Chapter Overview

Lipids are compounds grouped according to their common solubility in nonpolar solvents, but there are some structural features that also help to identify them. They have a variety of functions in the body, acting as energy storage compounds and chemical messengers, and providing structure to cell membranes.

In this chapter you will define fats and oils and explain how they differ in molecular structure. You will write equations for the saponification of some lipids, and you will learn which lipids can be saponified and which cannot. You will identify various groups of lipids according to their structures, and you will learn some of the functions that they perform in biological systems.

Practice Exercises

19.1 **Saponifiable lipids** (19.1) can be hydrolyzed under alkaline conditions to yield salts of **fatty acids** (Sec. 19.2), naturally occurring carboxylic acids that contain long, unbranched hydrocarbon chains 12 to 26 carbon atoms in length. Fatty acids are classified as **saturated** (no carbon-carbon double bonds), **monounsaturated** (one carbon-carbon double bond) and **polyunsaturated** (Sec. 19.2) (two or more carbon-carbon double bonds). In the omega classification system, the number of the first double bond on the carbon chain is counted from the methyl end of the chain.

Using information from the tables in section 19.2 of your textbook, classify the fatty acids below according to their length, the number of double bonds, and the position of the first carbon-carbon double bond using the omega classification system.

Name of fatty acid	Number of carbon atoms	Number of carbon-carbon double bonds	Omega classification system
arachidonic acid			
linolenic acid			
palmitic acid			
linoleic acid			
oleic acid			
stearic acid			

19.2 **Fats and oils** (Sec. 19.4) are **triacylglycerols** (Sec. 19.4) produced by the reaction of three fatty acid molecules with a glycerol molecule.

a. Draw the triacylglycerol prepared from glycerol and three molecules of linolenic acid.
b. Draw the product of the complete hydrogenation of the triacylglycerol in part a.

a.	b.

19.3 Fats have a high percentage of saturated fatty acids; oils have a high percentage of unsaturated fatty acids. Tell whether each of the compounds in Practice Exercise 19.2 would be a fat or an oil.

a. ______________________________
b. ______________________________

How many molecules of H_2 would be required to hydrogenate the compound in part a. to form the compound in part b.? ______________________________

19.4 The saponification (alkaline hydrolysis reaction) of fats and oils produces glycerol and fatty acid salts. Write the equation for the saponification of the triacylglycerol produced in Practice Exercise 19.2, part b. Name the products formed in the reaction.

19.5 Triacylglycerols are the most common of the saponifiable lipids. Other important saponifiable lipids are the **phosphoacylglycerols** (Sec. 19.6), the **waxes** (Sec. 19.7), and the **sphingolipids** (Sec. 19.8). These lipids differ in structure, containing different acid and alcohol parts.

In the table below give the general types of compounds from which each of these lipid groups is formed.

Name of lipid	Compounds from which lipid forms
triacylglycerols	glycerol and three fatty acids
phosphoacylglycerols	
waxes	
sphingolipids	

19.6 The two types of **nonsaponifiable lipids** (Sec. 19.1) that are considered in this chapter are the **steroids** (Sec. 19.10) and the **eicosanoids** (Sec. 19.11). Match the name of each of the nonsaponifiable lipids in the table below with the number of its function in biological systems.

Answer	Name of lipid	Function of lipid
	thromboxanes	1. control reproduction, secondary sex characteristics
	mineralocorticoids	2. raise body temperature, enhance inflammation
	steroid hormones	3. starting material for synthesis of steroid hormones
	glucocorticoids	4. control glucose metabolism, reduce inflammation
	cholesterol	5. emulsifying agents produced by gall bladder
	bile salts	6. found in white blood cells, hypersensitivity response
	leukotrienes	7. promote formation of blood clots
	prostaglandins	8. control the balance of Na^+ and K^+ ions in cells

19.7 After each name in the table above, write either A. if the compound is a steroid or B. if the compound is an eicosanoid.

Self-Test

True-false: Indicate whether the following statements are true or false. If the statement is false, give the word or phrase that may be substituted for the underlined portion to make the statement true.

1. Fats contain a high proportion of <u>unsaturated fatty acid</u> chains.
2. Fats and oils become rancid when the double bonds on triacylglycerol side chains are <u>oxidized</u>.
3. Waxes are esters of long-chain fatty acids and <u>aromatic alcohols</u>.
4. Rancid fats contain <u>low-molecular-mass acids</u>.
5. <u>Sphingolipids</u> are complex lipids found in the brain and nerves.
6. In a plasma membrane, the <u>nonpolar tails</u> of the phospholipids are on the outside surfaces of the lipid bilayer.
7. The presence of cholesterol molecules in the lipid bilayer of a plasma membrane makes the bilayer <u>more fluid</u>.
8. Essential fatty acids <u>cannot</u> be synthesized within the human body from other substances.

9. Most naturally occurring triacylglycerols are formed with three identical fatty acid molecules.
10. Partial hydrogenation of vegetable oils produces a product of higher melting point than the original oil.
11. Phosphoacylglycerols are important components of cell membranes.
12. Some synthetic derivatives of adrenocortical hormones are used to control inflammatory diseases.
13. Cholesterol cannot be synthesized within the human body.
14. The eicosanoids are hormone-like substances that are transported in the blood stream to the site of action.
15. Eicosanoids are derivatives of polyunsaturated fatty acid that have 20 carbon atoms.
16. The lipid molecules in the lipid bilayer of a cell membrane are held to one another by covalent bonds.
17. Active transport involves movement of a substance across a membrane against a concentration gradient with the expenditure of cellular energy.

Multiple choice:

18. A triacylglycerol is prepared by combining glycerol and:
    a) long-chain alcohols
    b) fatty acids
    c) unsaturated hydrocarbons
    d) saturated hydrocarbons
    e) none of these
19. Which of the following lipid groups is nonsaponifiable?
    a) lecithins
    b) waxes
    c) glucocorticoids
    d) sphingomyelins
    e) none of these
20. An example of a phosphoacylglycerol is:
    a) prostaglandin
    b) progesterone
    c) a lecithin
    d) glycerol tristearate
    e) none of these
21. An example of a steroidal hormone is:
    a) progesterone
    b) cholesterol
    c) insulin
    d) serotonin
    e) none of these
22. Which of the following groups are not steroids?
    a) anabolic agents
    b) androgens
    c) mineralocorticoids
    d) cerebrosides
    e) none of these
23. The physiological effects of eicosanoids include the mediation of:
    a) development of secondary sex characteristics
    b) synthesis of vitamin D
    c) digestion of fats and oil
    d) regulation of smooth muscle contraction
    e) none of these

Answers to Practice Exercises

19.1

Name	Number of carbon atoms	Number of double bonds	Omega classification system
arachidonic acid	20	4	omega-6
linolenic acid	18	3	omega-3
palmitic acid	16	0	--
linoleic acid	18	2	omega-6
oleic acid	18	1	omega-9
stearic acid	18	0	--

19.2

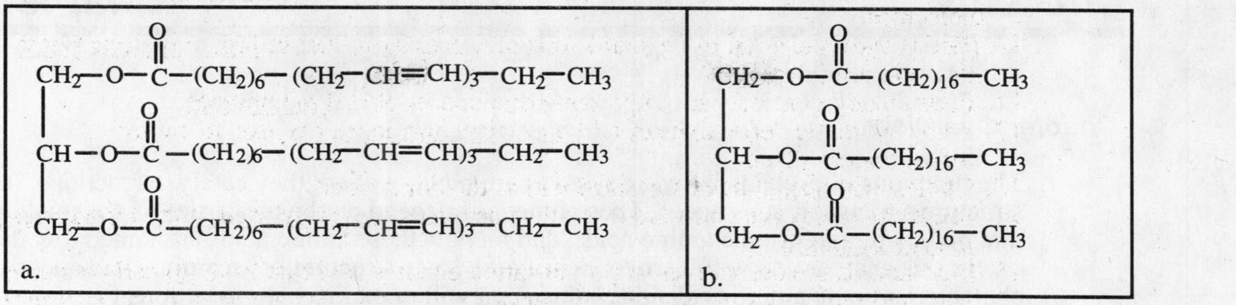

a.	b.
$CH_2-O-\overset{O}{\overset{\Vert}{C}}-(CH_2)_6-(CH_2-CH{=}CH)_3-CH_2-CH_3$	$CH_2-O-\overset{O}{\overset{\Vert}{C}}-(CH_2)_{16}-CH_3$
$CH-O-\overset{O}{\overset{\Vert}{C}}-(CH_2)_6-(CH_2-CH{=}CH)_3-CH_2-CH_3$	$CH-O-\overset{O}{\overset{\Vert}{C}}-(CH_2)_{16}-CH_3$
$CH_2-O-\overset{O}{\overset{\Vert}{C}}-(CH_2)_6-(CH_2-CH{=}CH)_3-CH_2-CH_3$	$CH_2-O-\overset{O}{\overset{\Vert}{C}}-(CH_2)_{16}-CH_3$

19.3 a. oil; b. fat; nine molecules of H_2 would be required.

19.4

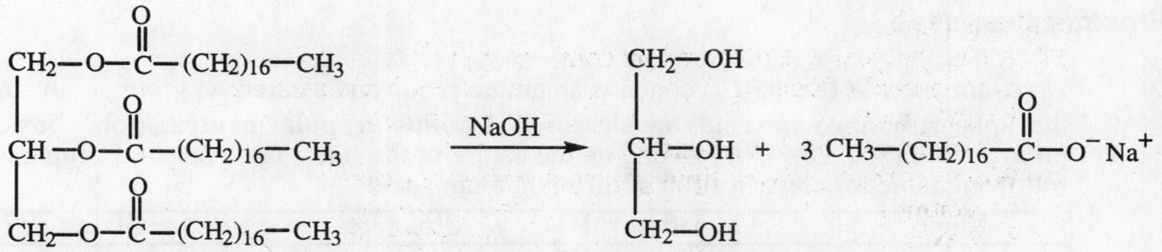

$$\begin{array}{l} CH_2-O-\overset{O}{\overset{\Vert}{C}}-(CH_2)_{16}-CH_3 \\ CH-O-\overset{O}{\overset{\Vert}{C}}-(CH_2)_{16}-CH_3 \\ CH_2-O-\overset{O}{\overset{\Vert}{C}}-(CH_2)_{16}-CH_3 \end{array} \xrightarrow{NaOH} \begin{array}{l} CH_2-OH \\ CH-OH \\ CH_2-OH \end{array} + 3\ CH_3-(CH_2)_{16}-\overset{O}{\overset{\Vert}{C}}-O^-\ Na^+$$

Glycerol and sodium stearate are the products of the saponification.

19.5

Name of lipid	Compounds from which lipid forms
triacylglycerols	glycerol and three fatty acids
phosphoacylglycerols	glycerol, two fatty acids, phosphoric acid, an alcohol
waxes	long-chain monohydroxy alcohol, fatty acid
sphingolipids	sphingosine, fatty acid, either phosphoric acid and choline or a carbohydrate

19.6

Answer	Name of lipid	Function of lipid
7.	thromboxanes	1. control reproduction, secondary sex characteristics
8.	mineralocorticoids	2. raise body temperature, enhance inflammation
1.	steroid hormones	3. starting material for synthesis of steroid hormones
4.	glucocorticoids	4. control glucose metabolism, reduce inflammation
3.	cholesterol	5. emulsifying agents produced by gall bladder
5.	bile salts	6. found in white blood cells, hypersensitivity response
6.	leukotrienes	7. promote formation of blood clots
2.	prostaglandins	8. control the balance of Na^+ and K^+ ions in cells

19.7 thromboxanes – B; mineralocorticoids – A; steroid hormones – A; glucocorticoids – A; cholesterol – A; bile salts – A; leukotrienes – B; prostaglandins – B

Answers to Self-Test

The numbers in parentheses refer to sections in your textbook:
1. F; saturated fatty acid (19.4) **2**. T (19.4) **3**. F; long-chain alcohols (19.7)
4. T (19.5) **5**. T (19.8) **6**. F; polar heads (19.12) **7**. F; more rigid (19.12) **8**. T (19.2)
9. F; a mixture of (19.4) **10**. T (19.5) **11**. T (19.6) **12**. T (19.10) **13**. F; can (19.10)
14. F; are synthesized at their site of action (19.11) **15**. T (19.11)
16. F; dipole-dipole attractions (19.12) **17**. T (19.13) **18**. b (19.4) **19**. c (19.9) **20**. c (19.6)
21. a (19.10) **22**. d (19.10) **23**. d (19.11)

Proteins — Chapter 20

Chapter Overview

The functions of proteins in living systems are highly varied; they catalyze reactions, form structures, transport substances. These functions depend on the properties of the small units that make up proteins, the amino acids, and on how these amino acids are joined together.

In this chapter you will learn to identify the basic structure of an amino acid and characterize some amino acid side chains. You will draw Fischer projections for amino acids and explain why an amino acid exists as a zwitterion. You will define the peptide bond and write abbreviated names for some peptides. You will compare the primary, secondary, tertiary, and quaternary structures of proteins.

Practice Exercises

20.1 An **α-amino acid** (Sec. 20.2) contains an amino group and a carboxyl group, both attached to the alpha carbon. Amino acids are classified as **nonpolar, polar neutral, polar basic, or polar acidic** (Sec. 20.2), depending on the nature of the side chain present. Complete the following table for classification of different amino acids.

Name	Abbreviation	Structure	Classification
leucine			
aspartic acid			
lysine			
serine			
histidine			
asparagine			

20.2 The α-carbon atom of an α-amino acid is a chiral center (except in glycine); naturally occurring amino acids are L isomers. The Fischer projection for an amino acid is drawn by putting the –COOH group at the top of the projection and the $-NH_2$ group to the left for the L isomer.

a. Draw the Fischer projection for L-aspartic acid. b. Draw the enantiomer of L-aspartic acid using a Fischer projection.

$HO-\overset{O}{\overset{\|\|}{C}}-CH_2-\underset{NH_2}{\underset{\|}{CH}}-\overset{O}{\overset{\|\|}{C}}-OH$ aspartic acid	a. Fischer projection of L-aspartic acid	b. enantiomer of L-aspartic acid

20.3 Since an amino acid has both an acidic group (the carboxyl group) and a basic group (the amino group), it exists as a **zwitterion** (Sec. 20.4), a molecule that has a positive charge on one atom and a negative charge on another atom. Draw the structural formula for neutral leucine and the zwitterionic structure of leucine.

Structural formula of leucine	Zwitterionic structure of leucine

20.4 In solution, three different amino acid forms can exist in equilibrium. The amount of each form present depends on the pH of the solution.

Draw the structure of leucine at each of the following pH's.

pH = 1	pH = 7	pH = 12

20.5 Referring to the structures in Practice Exercise 20.4, predict the direction (if any) of migration toward the positively or negatively charged electrode for each structure. The pH at which no migration occurs is called the **isoelectric point** (Sec. 20.4). Write "isoelectric" if no migration occurs.

pH = 1 ____________________

pH = 7 ____________________

pH = 12 ____________________

20.6 The **peptide bond** (Sec. 20.5) is an amide linkage joining amino acids to form **peptides** (Sec. 20.5). The peptide bond forms between the carboxyl group of one amino acid and the amino group of another.

a. Draw the structural formulas of the two different dipeptides that could form from the amino acids aspartic acid (Asp) and lysine (Lys). b. Circle the peptide bond in each structure. c. Label the N-terminal end and C-terminal end.

20.7 Peptides that contain the same amino acids but in different order are different molecules with different properties.

a. Using abbreviations for the amino acids, draw all possible tripeptides that could be formed from one molecule each of the following amino acids: Pro, Trp, and Lys.

b. Draw the abbreviated peptide formulas for all possible tripeptides that could be formed from two molecules of Pro and one molecule of Lys.

20.8 **Proteins** (Sec. 20.6) and polypeptides can undergo hydrolysis of their peptide bonds. Sometimes only a few of the bonds are hydrolyzed, yielding a mixture of smaller peptides. How many different di- and tripeptides could be present in a solution of partially hydrolyzed Arg-Arg-Cys-Lys?

20.9 The sequence of the amino acids in a protein's peptide chain is called its **primary structure** (Sec. 20.7). The shape of a protein molecule is the result of forces acting between various parts of the peptide chain. These forces determine the **secondary, tertiary, and quaternary structure of the protein** (Sec. 20.8, 20.9, and 20.10).

Complete the following table by writing the type of protein structure (primary, secondary, tertiary, or quaternary) most readily associated with the following terms.

Structural term	Type of protein structure
amino acid sequence	
beta-pleated sheet	
disulfide bonds	
alpha-helix	
globular protein (Sec. 20.11)	
hemoglobin (tetramer)	
collagen (triple helix)	
fibrous protein (Sec. 20.11)	
hydrophobic attractions	

20.10 Determine the primary structure of a heptapeptide containing six different amino acids, if the following smaller peptides are among the partial-hydrolysis products: Tyr-Trp, Glu-Asp-Tyr, Trp-Asp-Val, Val-Pro, Tyr-Trp-Asp. The C-terminal end is proline.

Structure of heptapeptide ________________________________

20.11 **Glycoproteins** (Sec. 20.15) are **conjugated proteins** (Sec. 20.12) containing carbohydrates as well as amino acids. They are important in several ways in biological systems. Match each term below with the statement that describes its role in a process involving glycoproteins.

Answer	Term	Statement
	collagen	1. glycoprotein produced as a protective response
	antigens	2. foreign substances that invade the body
	antibodies	3. aggregates caused by carbohydrate cross linking
	immunoglobulin	4. premilk substance containing immunoglobulins
	colostrum	5. triple helix of chains of amino acids
	collagen fibrils	6. molecules that counteract specific antigens

Self-Test

True-false: Indicate whether the following statements are true or false. If the statement is false, give the word or phrase that may be substituted for the underlined portion to make the statement true.

1. Naturally occurring amino acids are generally in the <u>L form</u>.
2. A peptide bond is the bond formed between the amino group of an amino acid and the carboxyl group of <u>the same</u> amino acid.
3. Since amino acids have both an acidic group and a basic group, they are able to undergo <u>internal acid-base reactions</u>.
4. When the pH of an amino acid solution is lowered, the amino acid zwitterion forms more of the <u>positively</u> charged species.
5. A peptide bond <u>differs slightly from</u> an amide bond.
6. A tripeptide has a COO^- group <u>at each end</u> of the molecule.
7. <u>Vasopressin</u> is a peptide in the human body that regulates uterine contractions and lactation.
8. The function of a protein is controlled by the protein's <u>primary</u> structure.

9. The alpha helix structure is a part of the primary structure of some proteins.
10. Denaturation of a protein involves changes in the protein's primary structure.
11. The quaternary structure of a protein involves associations among separate polypeptide chains.
12. An antibody is a foreign substance that invades the body.
13. The peptides in a beta-pleated sheet are held in place by hydrogen bonds.

Multiple choice:

14. An example of a polar basic amino acid is:
    a) lysine
    b) serine
    c) tryptophan
    d) leucine
    e) none of these
15. A tripeptide is formed from two alanine molecules and one glycine molecule. The maximum number of different tripeptides that could be formed from this combination is:
    a) two
    b) three
    c) four
    d) five
    e) none of these
16. The peptide that regulates the excretion of water by the kidneys is:
    a) endorphin
    b) insulin
    c) vasopressin
    d) leucine enkephalin
    e) none of these
17. The partial hydrolysis of the pentapeptide Val-Ala-Ala-Gly-Ser could yield which of the following peptides? (The amino acid on the left is the N-terminal amino acid.)
    a) Val-Ala-Gly
    b) Ser-Gly-Ala
    c) Ala-Gly-Ser
    d) Ala-Ala-Ser
    e) none of these
18. Which of the following amino acids would have a net charge of zero at a pH of 7?
    a) arginine
    b) phenylalanine
    c) aspartic acid
    d) lysine
    e) none of these
19. Which of the following attractive interactions does **not** affect the formation of a protein's tertiary structure?
    a) disulfide bonds
    b) salt bridges
    c) hydrogen bonds
    d) peptide bonds
    e) none of these
20. Proteins may be denatured by:
    a) heat
    b) acid
    c) ethanol
    d) all of the above (a, b, and c)
    e) a) and b) only
21. Which of the following proteins is **not** a conjugated protein?
    a) hemoglobin
    b) insulin
    c) collagen
    d) immunoglobulin
    e) lipoprotein
22. Which of these is true of globular proteins?
    a) stringlike molecules
    b) water-insoluble
    c) shape is roughly spherical
    d) structural function in body
    e) none of these

Answers to Practice Exercises

20.1

Name	Abbreviation	Structure	Classification
leucine	Leu	$CH_3-CH(CH_3)-CH_2-CH(NH_2)-C(=O)-OH$	nonpolar

aspartic acid	Asp	$HO-\overset{O}{\overset{\|}{C}}-CH_2-\underset{NH_2}{\underset{\|}{CH}}-\overset{O}{\overset{\|}{C}}-OH$	polar acidic
lysine	Lys	$H_2N-(CH_2)_4-\underset{NH_2}{\underset{\|}{CH}}-\overset{O}{\overset{\|}{C}}-OH$	polar basic
serine	Ser	$HO-CH_2-\underset{NH_2}{\underset{\|}{CH}}-\overset{O}{\overset{\|}{C}}-OH$	polar neutral
histidine	His	(imidazolyl, HN / N ring)$-CH_2-\underset{NH_2}{\underset{\|}{CH}}-\overset{O}{\overset{\|}{C}}-OH$	polar basic
asparagine	Asn	$H_2N-\overset{O}{\overset{\|}{C}}-CH_2-\underset{NH_2}{\underset{\|}{CH}}-\overset{O}{\overset{\|}{C}}-OH$	polar neutral

20.2

$HO-\overset{O}{\overset{\|}{C}}-CH_2-\underset{NH_2}{\underset{\|}{CH}}-\overset{O}{\overset{\|}{C}}-OH$	COOH H_2N —+— H CH_2 COOH	COOH H —+— NH_2 CH_2 COOH
aspartic acid	a. Fischer projection of L- aspartic acid	b. enantiomer of L-aspartic acid

20.3

Structural formula of leucine	Zwitterion structure of leucine
$CH_3-\underset{CH_3}{\underset{\|}{CH}}-CH_2-\underset{NH_2}{\underset{\|}{CH}}-\overset{O}{\overset{\|}{C}}-OH$	$CH_3-\underset{CH_3}{\underset{\|}{CH}}-CH_2-\underset{\overset{+}{NH_3}}{\underset{\|}{CH}}-\overset{O}{\overset{\|}{C}}-O^-$

20.4

$CH_3-\underset{CH_3}{\underset{\|}{CH}}-CH_2-\underset{\overset{+}{NH_3}}{\underset{\|}{CH}}-\overset{O}{\overset{\|}{C}}-OH$	$CH_3-\underset{CH_3}{\underset{\|}{CH}}-CH_2-\underset{\overset{+}{NH_3}}{\underset{\|}{CH}}-\overset{O}{\overset{\|}{C}}-O^-$	$CH_3-\underset{CH_3}{\underset{\|}{CH}}-CH_2-\underset{NH_2}{\underset{\|}{CH}}-\overset{O}{\overset{\|}{C}}-O^-$
pH = 1	pH = 7	pH = 12

20.5 pH = 1: toward negative electrode
pH = 7: isoelectric
pH = 12: toward positive electrode

20.6

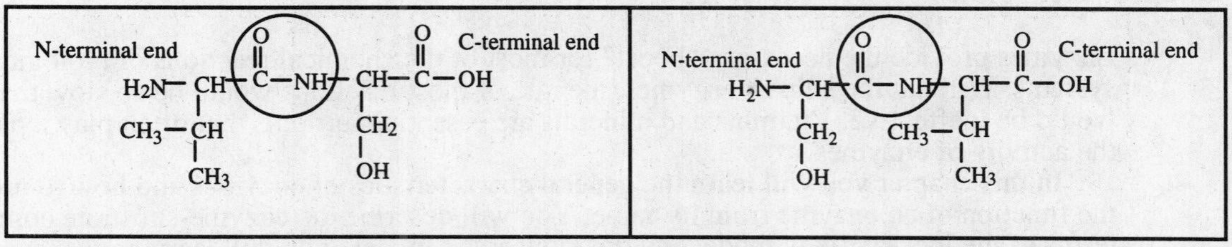

N-terminal end; C-terminal end $H_2N-CH-\overset{O}{\overset{\parallel}{C}}-NH-CH-\overset{O}{\overset{\parallel}{C}}-OH$ first CH: $CH_3-CH-CH_3$; second CH: CH_2-OH	N-terminal end; C-terminal end $H_2N-CH-\overset{O}{\overset{\parallel}{C}}-NH-CH-\overset{O}{\overset{\parallel}{C}}-OH$ first CH: CH_2-OH; second CH: $CH_3-CH-CH_3$

20.7 a. Pro-Trp-Lys, Pro-Lys-Trp, Trp-Pro-Lys, Trp-Lys-Pro, Lys-Trp-Pro, Lys-Pro-Trp.
b. Pro-Pro-Lys, Pro-Lys-Pro, Lys-Pro-Pro

20.8 Three dipeptides (Arg-Arg, Arg-Cys, Cys-Lys) and two tripeptides (Arg-Arg-Cys and Arg-Cys-Lys)

20.9

Structural Term	Classification of protein structure
amino acid sequence	primary
beta-pleated sheet	secondary
disulfide bonds	tertiary
alpha-helix	secondary
globular proteins	tertiary
hemoglobin (tetramer)	quaternary
collagen (triple helix)	secondary
fibrous protein	secondary
hydrophobic attractions	tertiary

20.10 Structure of heptapeptide: Glu-Asp-Tyr-Trp-Asp-Val-Pro

20.11

Answer	Term	Statement
5.	collagen	1. glycoprotein produced as a protective response
2.	antigens	2. foreign substances that invade the body
6.	antibodies	3. aggregates caused by carbohydrate cross linking
1.	immunoglobulin	4. premilk substance containing immunoglobulins
4.	colostrum	5. triple helix of chains of amino acids
3.	collagen fibrils	6. molecules that counteract specific antigens

Answers to Self-Test

The numbers in parentheses refer to sections in your textbook:
1. T (20.3) **2**. F; another (20.5) **3**. T (20.4) **4**. T (20.4) **5**. F; is the same as (20.5)
6. F; at one end and an amino group at the other end (20.5) **7**. F; oxytocin (20.5)
8. F; primary, secondary, tertiary, and quaternary (20.6) **9**. F; secondary (20.8)
10. F; secondary, tertiary, and quaternary (20.14) **11**. T (20.10) **12**. F; antigen (20.15)
13. T (20.8) **14**. a (20.2) **15**. b (20.5) **16**. c (20.5) **17**. c (20.13) **18**. b (20.5)
19. d (20.9) **20**. d (20.14) **21**. b (20.12) **22**. c (20.11)

Enzymes, Vitamins, and Minerals — Chapter 21

Chapter Overview

Enzymes provide the necessary "boost" for most of the chemical reactions of biological systems. Without the help of enzymes, the rate of most reactions would be so slow that they would be ineffective. Vitamins and minerals are essential nutrients that often play a part in the activity of enzymes.

In this chapter you will learn the general characteristics of enzymes and how to predict the function of an enzyme from its name. You will describe the enzyme-substrate complex in terms of the lock-and-key model and the induced-fit model. You will learn how enzyme action can be inhibited and how it is controlled in biologic systems, and you will learn the roles of vitamins and minerals in enzyme activity.

Practice Exercises

21.1 **Enzymes** (Sec. 21.1) are commonly named by taking the name of the **substrate** (Sec. 21.2), the compound undergoing change, and adding the ending *-ase*. Sometimes the enzyme name gives the type of reaction being catalyzed.

In the table below, predict the function for a given enzyme name or suggest a name for the enzyme that catalyzes the given type of reaction.

Name of enzyme	Type of reaction
α-amylase	
dehydrogenase	
	removal of a carboxyl group from a substrate
	transfer of an amino group from one molecule to another
peptidase	
oxidase	
	hydrolysis of ester linkages in lipids

21.2 Enzymes can be divided into two classes: **simple enzymes** and **conjugated enzymes** (Sec. 21.3). Other common terms used in describing enzymes are defined in section 21.3 of your textbook. Match the terms below with the correct description.

Answer	Term	Description
	coenzyme	1. nonprotein portion of a conjugated enzyme
	cofactor	2. obtained from dietary minerals
	inorganic ion cofactor	3. protein portion of a conjugated enzyme
	apoenzyme	4. small organic molecule that serves as a cofactor

21.3 Before an enzyme-catalyzed reaction takes place, an **enzyme-substrate complex** (Sec. 21.4) is formed: the substrate binds to the **active site** (Sec. 21.4) of the enzyme.

a. What are two models that account for the specific way an enzyme selects a substrate?

b. What is the main difference between these two models?

21.4 **Enzyme activity** (Sec. 21.6) is a measure of the rate at which an enzyme converts substrate to products.

a. Name four factors that affect enzyme activity.

b. Describe the effects on rate of reaction of an increase in each of the four factors named above.

21.5 An **enzyme inhibitor** (Sec. 21.7) is a substance that slows or stops the normal catalytic function of an enzyme by binding to it. Enzyme activity can also be changed by regulators produced within a cell. After reading the sections on enzyme inhibition and regulation (sections 21.7–21.10) in your textbook, match the following terms and their descriptions.

Answer	Term	Description
	penicillin	1. binds to a site other than the active site
	competitive inhibitor	2. contains both active and regulator sites
	zymogen	3. competes with substrate for active site
	sulfa drugs	4. forms a covalent bond at the active site
	allosteric enzyme	5. inhibits cell wall formation in bacteria
	irreversible inhibitor	6. inactive precursor of an enzyme
	noncompetitive inhibitor	7. inhibitors of folic acid production

21.6 **Vitamins** (Sec. 21.12) are organic chemicals that must be obtained in the diet and that are essential in trace amounts for proper functioning of the human body. Match each of the vitamins named below with its function in the human body.

Answer	Vitamin name	Function in human body
	Vitamin A	1. prevents oxidation of polyunsaturated fatty acids
	Vitamin C	2. helps form compounds that regulate blood clotting
	Vitamin D	3. group of vitamins that are enzyme cofactors
	Vitamin B	4. essential for formation of collagen
	Vitamin E	5. promotes normal bone mineralization
	Vitamin K	6. keeps skin and mucous membranes healthy

21.7 Vitamins are divided into two groups: water-soluble and fat-soluble. After each vitamin name in Practice Exercise 21.6, put a W if the vitamin is water-soluble or an F if it is fat-soluble.

21.8 In addition to the elements that are found in the common organic compounds (C, H, O, N, P, and S), there are other elements that must be supplied in the diet. Inorganic compounds that contain these elements are called **minerals** (Sec. 21.15). Match each of the mineral elements below with its function in the human body. Pairs of minerals listed are those that function in similar, though not identical, areas.

Answer	Mineral	Function of mineral in the human body
	iodine	1. nerve impulse transmission, electrolytes
	calcium, magnesium	2. part of thyroxine, regulator of basal metabolic rate
	zinc	3. bone building, muscle contraction, electrolytes
	sodium, potassium	4. cofactor for many enzymes
	iron	5. transport of oxygen (in hemoglobin)

Self-Test

True-false: Indicate whether the following statements are true or false. If the statement is false, give the word or phrase that may be substituted for the underlined portion to make the statement true.

1. In an enzyme catalyzed reaction, the compound that undergoes a chemical change is called a substrate.
2. Enzyme names are usually based on the structure of the enzyme.
3. The protein portion of a conjugated enzyme is called the coenzyme.
4. The active site of an enzyme is the small part of an enzyme where catalysis takes place.
5. According to the lock-and-key model of enzyme action, the active site of the enzyme is flexible in shape.
6. A carboxypeptidase is an enzyme that is specific for one group of compounds.
7. A substance that binds to an enzyme's active site and is not released is called a noncompetitive enzyme inhibitor.
8. Enzymes undergo all the reactions of proteins except denaturation.
9. A cofactor is a protein part of an enzyme necessary for the enzyme's function.
10. Large doses of water-soluble vitamins may be toxic, since they can be retained in the body in excess of need.
11. Vitamin K is necessary for the formation of prothrombin used in blood clotting.
12. Some vitamins act as cofactors in conjugated enzymes.
13. Tissue plasminogen activator is used in the diagnosis of heart attacks.

Multiple choice:

14. Enzymes assist chemical reactions by:
    a) increasing the rate of the reactions
    b) increasing the temperature of the reactions
    c) being consumed during the reactions
    d) all of these
    e) none of these
15. The enzyme that catalyzes the reaction of an alcohol to form an aldehyde would be:
    a) an oxidase
    b) a decarboxylase
    c) a dehydratase
    d) a reductase
    e) none of these
16. A competitive inhibitor of an enzymatic reaction:
    a) distorts the shape of the enzyme molecule
    b) attaches to the substrate
    c) blocks an enzyme's active site
    d) weakens the enzyme-substrate complex
    e) none of these

17. When the enzyme-substrate complex forms, the actual bond-breaking and/or bond formation take place:
    a) at the active site
    b) at the regulator bonding site
    c) on the zymogen
    d) on the positive regulator
    e) none of these
18. α-Amylase does not catalyze the hydrolysis of β-glycosidic bonds because the enzyme α-amylase is:
    a) difficult to activate
    b) specific
    c) a proenzyme
    d) inhibited
    e) none of these
19. The antibiotic penicillin acts by:
    a) activating zymogens for bacterial enzymes
    b) cleaving peptide bonds in bacterial proteins
    c) catalyzing the hydrolysis of bacterial cell walls
    d) inhibiting bacterial enzymes that help form cell walls
    e) none of these
20. The vitamin that protects polyunsaturated fatty acids from oxidation is:
    a) vitamin B
    b) vitamin A
    c) vitamin D
    d) vitamin E
    e) none of these

Answers to Practice Exercises

21.1

Name of enzyme	Type of reaction
α-amylase	hydrolysis of α-linkage in starch molecules
dehydrogenase	removal of hydrogen from a substrate
decarboxylase	removal of a carboxyl group from a substrate
transaminase	transfer of an amino group from one molecule to another
peptidase	hydrolysis of peptide linkages
oxidase	oxidation of a substrate
lipase	hydrolysis of ester linkages in lipids

21.2

Answer	Term	Description
4.	coenzyme	1. nonprotein portion of a conjugated enzyme
1.	cofactor	2. obtained from dietary minerals
2.	inorganic ion cofactor	3. protein portion of a conjugated enzyme
3.	apoenzyme	4. small organic molecule that serves as a cofactor

21.3
a. The two models are the lock-and-key model and the induced-fit model.
b. The main difference between the two models is that according to the lock-and-key model the active site of the enzyme is fixed and rigid, but in the induced-fit model the active site can change its shape slightly to accommodate the shape of the substrate.

21.4 a. Four factors that affect the rate of enzyme activity are temperature, pH, substrate concentration, and enzyme concentration.
b. Increase in temperature: Temperature increases the rate of a reaction, but if the temperature is high enough to denature the protein enzyme, the rate will decrease.
Increase in pH: Each enzyme has an optimum pH for action; if the pH increases above this point the reaction will slow down.
Increase in substrate concentration: Increased rate of reaction until maximum enzyme capacity is reached; after this no further increase.
Increase in enzyme concentration: Reaction rate increases.

21.5

Answer	Term	Description
5.	penicillin	1. binds to a site other than the active site
3.	competitive inhibitor	2. contains both active and regulator sites
6.	zymogen	3. competes with substrate for active site
7.	sulfa drugs	4. forms a covalent bond at the active site
2.	allosteric enzyme	5. inhibits cell wall formation in bacteria
4.	irreversible inhibitor	6. inactive precursor of an enzyme
1.	noncompetitive inhibitor	7. inhibitor of folic acid production

21.6 and **21.7**

Answer	Vitamin name	Function in human body
6.	Vitamin A -- F	1. prevents oxidation of polyunsaturated fatty acids
4.	Vitamin C -- W	2. helps form compounds that regulate blood clotting
5.	Vitamin D -- F	3. group of vitamins that are enzyme cofactors
3.	Vitamin B -- W	4. essential for formation of collagen
1.	Vitamin E -- F	5. promotes normal bone mineralization
2.	Vitamin K -- F	6. keeps skin and mucous membranes healthy

21.8

Answer	Mineral	Function of mineral in the human body
2.	iodine	1. nerve impulse transmission, electrolytes
3.	calcium, magnesium	2. part of thyroxine, regulator of basal metabolic rate
4.	zinc	3. bone building, muscle contraction, electrolytes
1.	sodium, potassium	4. cofactor for many enzymes
5.	iron	5. transport of oxygen (in hemoglobin)

Answers to Self-Test

The numbers in parentheses refer to sections in your textbook:
1. T (21.2) **2**. F; function (21.2) **3**. F; apoenzyme (21.3) **4**. T (21.4)
5. F; fixed and rigid (21.4) **6**. T (21.5) **7**. F; an irreversible (21.7)
8. F; including (21.1) **9**. F; nonprotein part (21.3) **10**. F; fat-soluble vitamins (21.12)
11. T (21.14) **12**. T (21.12) **13**. F; treatment (21.11) **14**. a (21.1) **15**. a (21.2)
16. c (21.7) **17**. a (21.4) **18**. b (21.5) **19**. d (21.10) **20**. d (21.14)

Nucleic Acids

Chapter 22

Chapter Overview

Nucleic acids are the molecules of heredity. Every inherited trait of every living organism is coded in these huge molecules. The complexity of their structure has been unraveled in fairly recent times, and this new knowledge of the transmission of genetic information has led to the exciting field of recombinant DNA technology.

In this chapter you will name and identify the structures of nucleotides and nucleic acids. You will write shorthand forms for nucleotide sequences in segments of DNA and RNA. You will identify the amino acid sequence coded by a given segment of DNA, and describe the processes of replication, transcription, and translation leading to protein synthesis. You will learn the basic ideas of recombinant DNA technology and gene therapy.

Practice Exercises

22.1 **Nucleotides** (Sec. 22.2) are the structural units from which the polymeric **nucleic acids** (Sec. 22.2) are formed. A nucleotide is composed of a pentose sugar bonded to a phosphate group and a nitrogen-containing heterocyclic base.

The information needed to complete the table below can be found in your textbook in Table 22.1.

Name of nucleotide	Abbreviation	Base	Sugar
deoxyadenosine 5′-monophosphate			
	dTMP		
		guanine	ribose
	CMP		

22.2 The formation of nucleotides from three constituent molecules involves condensation, with the formation of a water molecule. The nucleotide can undergo hydrolysis (addition of water) to yield the three molecules from which it was formed. Write the names for the products of the hydrolysis of each nucleotide below.

a. dCMP

b. UMP

22.3 Nucleotide units can be linked to each other through sugar–phosphate bonds. Draw the structural formula for the dinucleotide that forms between dAMP and dTMP, so that dAMP is the 5′ end and to the left in your drawing, and dTMP is the 3′ end and to the right in your drawing.

22.4 DNA (deoxyribonucleic acid) molecules contain deoxyribose molecules as their sugar units and four different bases: adenine (A), guanine (G), cytosine (C), and thymine (T). The DNA molecule consists of two strands linked by hydrogen bonds between two pairs of **complementary bases** (Sec. 22.4): A–T and G–C. The relative amounts of these base pairs are constant for a given life form.

If, in a DNA molecule, the percentage of the base adenine is 20% of the total bases present, what would be the percentages of the bases thymine, cytosine, and guanine?

22.5 In the DNA double helix, the two complementary strands run in opposite directions, one in the 5′ to 3′ direction, the other 3′ to 5′. Complete the following segment of a DNA double helix. Write symbols for the missing bases. Indicate the correct number of hydrogen bonds between the bases in each pair.

```
5′ T – C –    – C –    – G –    – A  3′
   ||  |||
3′ A – G – T –    – T –    – A –     5′
```

22.6 During **DNA replication** (Sec. 22.5), the DNA molecule makes an exact duplicate of itself. The two strands unwind, and free nucleotides line up along each strand, with complementary base pairs attracted to one another by hydrogen bonding. Polymerization of the new strand takes place. The daughter strand (3′ to 5′) is in the opposite direction to the parent strand (5′ to 3′).

Write the sequence of bases for the replication of the DNA strand below:

5′ T–A–A–G–C–G–T–G–G 3′

22.7 During **transcription** (Sec. 22.6), one strand of a DNA molecule acts as the template for the formation of a molecule of RNA. The nucleotides that line up next to the DNA strand have ribose as a sugar; the same bases are present except that uracil is substituted for thymine.

Using the new DNA strand formed in Practice Exercise 22.6, write the sequence of bases in the new RNA strand formed by transcription. Hint: Rewrite the answer in Practice Exercise 22.6, with 5′ on the left side, or starting point.

22.8 What would be the sequence of bases in a DNA strand that produces the following **messenger RNA** (Sec. 22.7) strand? Give your answer starting with the 5′ end.

5′ A–U–U–U–G–C–C–G–A 3′

22.9 A **codon** (Sec. 22.9) is a sequence of three nucleotides in an mRNA molecule that codes for a specific amino acid.

a. Complete the tables below with correct amino acid names and codons. The information can be obtained from Table 22.2 in your textbook.

Codon	Amino acid
UCA	
	asparagine
GAC	

Codon	Amino acid
	methionine
GAU	
	tryptophan

b. Are there any synonyms among the codons in the table in part a.?

c. Why is ATC not listed in Table 22.2 as one of the codon sequences?

22.10 a. Write a base sequence for mRNA that codes for the tripeptide Gly-Pro-Leu

b. Will there be only one answer? Explain.

22.11 Determine the amino acid sequence of a tetrapeptide if the DNA base sequence is that shown below. (Hint: Find the mRNA sequence first; then reverse it to put the 5′ end on the left.)

5′ G–C–C–A–C–T–T–C–G–G–G–A 3′ DNA

mRNA

tetrapeptide

22.12 **Mutations** (Sec. 22.12) are changes in the base sequence. This change in genetic information can cause a change in the amino acid sequence in protein synthesis. Consider the following segment of mRNA:

5′ G–C–C–U–A–C–A–A–U–G–C–G 3′

a. What is the amino acid sequence formed by **translation** (Sec. 22.6)?

b. What amino acid sequence would result if adenine was substituted for the first uracil?

Self-Test

True-false: Indicate whether the following statements are true or false. If the statement is false, give the word or phrase that may be substituted for the underlined portion to make the statement true.

1. The sugar unit found in DNA molecules is <u>ribose</u>.
2. In DNA the amount of adenine is equal to the amount of <u>guanine</u>.
3. In DNA strands, a phosphate ester bridge connects hydroxyl groups on the 3′ and 5′ positions of the <u>sugar</u> units.
4. DNA molecules <u>are the same</u> for individuals of the same species.
5. <u>Transfer RNA</u> molecules carry the genetic code from DNA to the ribosomes.

6. A codon is a series of three adjacent bases that carry the code for a specific amino acid.
7. A gene is an individual DNA molecule bound to a group of proteins.
8. The two strands of a DNA molecule are connected to each other by hydrogen bonds between the base units.
9. The process by which a DNA molecule forms an exact duplicate of itself is called transcription.
10. Primary transcript RNA is edited under the direction of enzymes and joined together to form messenger RNA.
11. Two different codons that specify the same amino acid are called synonyms.
12. Different species of organisms usually have the same genetic code for an amino acid.
13. A single mRNA molecule can serve as a codon sequence for the synthesis of one protein molecule at a time.
14. Mutagens are agents that cause a change in the structure of a DNA molecule.
15. Viruses are tiny disease-causing agents composed of a protein coat and a glycogen core.
16. Viruses can reproduce in water with dissolved organic nutrients.

Multiple choice:

17. The codon 5′ UGC 3′ would have as its anticodon:
   a) 5′ GCA 3′
   b) 5′ UAG 3′
   c) 5′ GAC 3′
   d) 5′ AUC 3′
   e) none of these
18. Sections of DNA that carry noncoding base sequences are called:
   a) introns
   b) exons
   c) codons
   d) anticodons
   e) none of these
19. Fifteen nucleotide units in a DNA molecule can contain the code for no more than:
   a) 3 amino acids
   b) 5 amino acids
   c) 10 amino acids
   d) 15 amino acids
   e) none of these
20. Which of the following types of molecules does not carry information for protein synthesis?
   a) DNA
   b) ribosomal RNA
   c) messenger RNA
   d) transfer RNA
   e) none of these
21. A sequence of three nucleotides in an mRNA molecule is a(n):
   a) exon
   b) intron
   c) codon
   d) anticodon
   e) none of these
22. The intermediary molecules that deliver amino acids to the ribosomes are:
   a) ptRNA
   b) tRNA
   c) rRNA
   d) mRNA
   e) none of these
23. The codon that initiates protein synthesis when it occurs as the first codon in an amino acid sequence is:
   a) GTA
   b) UGA
   c) GAC
   d) AUG
   e) none of these
24. The process of inserting recombinant DNA into a host cell is:
   a) translation
   b) transformation
   c) transcription
   d) transmission
   e) none of these
25. Cells that have descended from a single cell and have identical DNA are called:
   a) mutagens
   b) mutations
   c) clones
   d) plasmids
   e) none of these

Answers to Practice Exercises

22.1

Name of nucleotide	Abbreviation	Base	Sugar
deoxyadenosine 5′-monophosphate	dAMP	adenine	deoxyribose
deoxythymidine 5′-monophosphate	dTMP	thymine	deoxyribose
guanosine 5′-monophosphate	GMP	guanine	ribose
cytidine 5′-monophosphate	CMP	cytosine	ribose

22.2 a. dCMP -- cytosine, phosphate, deoxyribose

b. UMP -- uracil, phosphate, ribose

22.3

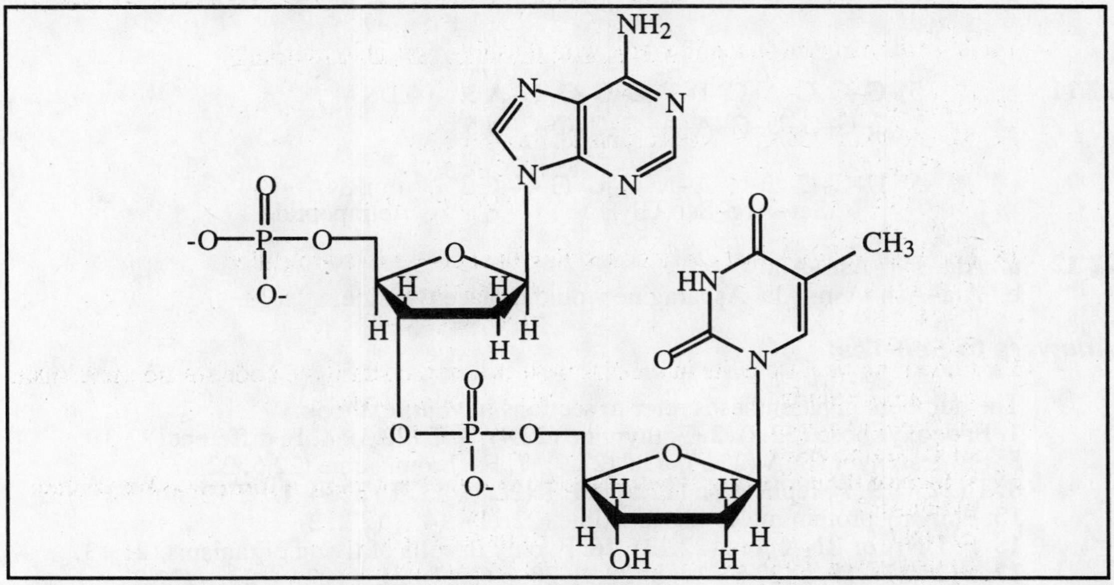

NH_2 N N N N O -O P O O- O H H H H H O O P O O- O CH$_3$ HN O N O H H H H OH H

22.4 We know that %A = %T and %G = %C, since these bases are paired in DNA.
If %A = 20%, then %T = 20%. %A + %T = 40%,
so %G + %C = 60% and %G = %C = 30%

22.5

5′ T – C – A – C – A – G – T – A 3′
|| ||| || ||| || ||| || ||
3′ A – G – T – G – T – C – A – T 5′

22.6 3′ A–T–T–C–G–C–A–C–C 5′ or 5′ C–C–A–C–G–C–T–T–A 3′

22.7 5′ C–C–A–C–G–C–T–T–A 3′ DNA
3′ G–G–U–G–C–G–A–A–U 5′ RNA

22.8 mRNA: 5′ A–U–U–U–G–C–C–G–A 3′
DNA: 3′ T–A–A–A–C–G–G–C–T 5′ or 5′ T–C–G-G–C–A–A–A–T 3′

22.9 a.

Codon	Amino acid
UCA	serine
AAU and AAC	asparagine
GAC	aspartic acid

Codon	Amino acid
AUG	methionine
GAU	aspartic acid
UGG	tryptophan

b. Yes; GAU and GAC both code for aspartic acid, and AAU and AAC both code for asparagine.

c. ATC is not listed as a codon because RNA does not contain thymine.

22.10 a. 5′ G–G–U–C–C–C–C–U–U 3′ is one possible answer.
b. No, because there is more than one codon for most amino acids.

22.11

5′ G–C–C–A–C–T–T–C–G–G–G–A 3′ DNA
3′ C–G–G–U–G–A–A–G–C–C–C–U 5′

5′ U–C–C–C–G–A–A–G–U–G–G–C 3′ mRNA
Ser–Arg–Ser–Gly tetrapeptide

22.12 a. Ala–Tyr–Asn–Ala
b. Ala–Asn–Asn–Ala Asparagine would replace tyrosine.

Answers to Self-Test

The numbers in parentheses refer to sections in your textbook:
1. F; deoxyribose (22.2) **2**. F; thymine (22.4) **3**. T (22.3) **4**. F; different (22.3)
5. F; Messenger RNA (22.7) **6**. T (22.9) **7**. F; chromosome (22.5, 22.8)
8. T (22.4) **9**. F; replication (22.5) **10**. T (22.7) **11**. T (22.9) **12**. T (22.9)
13. F; many protein molecules at a time (22.11) **14**. T (22.12)
15. F; DNA or RNA core (22.13) **16**. F; only in cells of living organisms (22.13)
17. a (22.11) **18**. a (22.8) **19**. b (22.9) **20**. b (22.7) **21**. c (22.9) **22**. b (22.7)
23. d (22.9) **24**. b (22.14) **25**. c (22.14)

Biochemical Energy Production — Chapter 23

Chapter Overview

The most important job of the body's cells is the production of energy to be utilized in carrying out all the complex processes known as life. The production and use of energy by living organisms involves an important intermediate called ATP.

In this chapter you will study the formation of acetyl CoA from the products of the digestion of food, and the further oxidation of acetyl CoA during the individual steps of the citric acid cycle. You will study the function and processes of the electron transport chain, and the important role of ATP in energy transfer and release.

Practice Exercises

23.1 During the processes of **metabolism** (Sec. 23.1) by which food is converted into energy, several compounds function as key intermediates: ATP, which contains two high energy phosphate bonds, and the coenzymes FAD, NAD^+, and coenzyme A.

a. Complete the equations below showing some of the reactions of these **high energy compounds** (Sec. 23.4) by filling in the excluded products or reactants.

$$ATP + H_2O \rightarrow \;?\; + HPO_4^{2-} + H^+ + \text{energy}$$

$$? + H_2O \rightarrow AMP + HPO_4^{2-} + H^+ + \text{energy}$$

$$FAD + 2e^- + 2H^+ \rightarrow$$

$$NAD^+ + 2e^- + 2H^+ \rightarrow$$

b. In the equations for part a., which molecules contained high energy bonds that were hydrolyzed?

c. Which molecules gained electrons (were reduced)?

d. What is the reduced form of FAD?

23.2 The **catabolism** (Sec. 23.1) of food begins with digestion and continues as the food is further broken down to release its stored energy. Complete the following table for the different stages of biochemical energy production by the catabolism of food.

Stage of catabolism	Where process occurs	Products
1. digestion		
2. acetyl group formation		
3. citric acid cycle		
4. electron transport chain and oxidative phosphorylation		

23.3 The **citric acid cycle** (Sec. 23.6) is the series of reactions in which the acetyl group of acetyl CoA is oxidized to CO_2, and $FADH_2$ and NADH are produced. Complete the following table summarizing the steps of the citric acid cycle. Use the discussion and equations in section 23.6 of your textbook.

Step	Type of reaction	Final product(s)	Energy-rich compound formed
1.	condensation		none
2.		isocitrate	
3.			NADH
4.			
5.			
6.			
7.			
8.			

23.4 Four of the steps summarized in Practice Exercise 23.3 involve oxidation. For each of these steps, give the name and/or symbol for the substance that was oxidized and for the one that was reduced.

Step	Substance oxidized	Substance reduced
3.		
4.	α-ketoglutarate	
6.		FAD (to $FADH_2$)
8.		

23.5 The **electron transport chain** (Sec. 23.7) is a series of reactions in which electrons and hydrogen ions from NADH and $FADH_2$ are passed to intermediate carriers and ultimately react with molecular oxygen to produce water.

Use the summary of the electron transport chain given in section 23.7 in your textbook to complete the following table.

Step	Substance oxidized	Substance reduced
1.		
2.		
3.		
4.		
5.		
6.		
7.		
8.		
final step		

23.6 **Oxidative phophorylation** (Sec. 23.8) is the process by which ATP is synthesized from ADP using energy released in the electron transport chain. Each NADH from the citric acid cycle produces 3 ATP molecules, each $FADH_2$ produces 2 ATP molecules, and each GTP produces 1 ATP molecule. Using Practice Exercise 23.3, summarize the number of ATP molecules produced in one turn of the citric acid cycle (steps 1 through 8).

Step	Energy-rich compound formed	Number of ATP's produced
3.		
4.		
5.		
6.		
8.		
	Total:	

Self-Test

True-false: Indicate whether the following statements are true or false. If the statement is false, give the word or phrase that may be substituted for the underlined portion to make the statement true.

1. Catabolic reactions usually release energy.
2. Bacterial cells do not contain a nucleus and so are classified as eukaryotic.
3. Ribosomes are organelles that have a central role in the production of energy.
4. The active portion of the FAD molecule is the flavin subunit.
5. NAD^+ is the oxidized form of nicotine adenine dinucleotide.
6. The citric acid cycle occurs in the cytoplasm of cells.
7. In the first step of the citric acid cycle, oxaloacetate reacts with glucose.
8. The electron transport chain receives electrons and hydrogen ions from NAD^+ and FAD.
9. The electrons that pass through the electron transport chain gain energy in each transfer along the chain.
10. Cytochromes contain iron atoms that are reversibly oxidized and reduced.
11. ATP is produced from ADP using energy released in the electron transport chain.

Multiple choice:

12. Which of the following is **not** an organelle?
    a) mitochondrion
    b) hemoglobin
    c) ribosome
    d) lysosome
    e) none of these
13. Which of these molecules is **not** part of the electron transport chain?
    a) coenzyme Q
    b) acetyl CoA
    c) cytochrome b
    d) ATP
    e) none of these
14. Energy used in cells is obtained directly from:
    a) oxidation of NADH
    b) activation of acetyl CoA
    c) oxidation of $FADH_2$
    d) hydrolysis of ATP
    e) none of these
15. The final acceptor of electrons in the electron transport chain is:
    a) NAD^+
    b) water
    c) oxygen
    d) FAD
    e) none of these

Use the discussion of the citric acid cycle in section 23.6 of your textbook as you answer the following questions:

16. Which step in the citric acid cycle is involved with the removal of a CO_2 molecule?
    a) 1
    b) 3
    c) 5
    d) 7
    e) 8

17. Which step in the citric acid cycle involves the removal of hydrogen atoms from the carbon chain to form an alkene?
 a) 1
 b) 4
 c) 6
 d) 7
 e) none of these
18. Which step in the citric acid cycle involves isomerization by dehydration of a molecule and then hydration?
 a) 1
 b) 2
 c) 6
 d) 7
 e) none of these
19. Which step in the citric acid cycle involves the removal of hydrogen atoms by FAD?
 a) 3
 b) 4
 c) 7
 d) 8
 e) none of these
20. Which step in the citric acid cycle involves the oxidation of a secondary alcohol to a ketone?
 a) 2
 b) 4
 c) 6
 d) 7
 e) 8

Answers to Practice Exercises

23.1 a. $ATP + H_2O \rightarrow ADP + HPO_4^{2-} + H^+ + \text{energy}$

$ADP + H_2O \rightarrow AMP + HPO_4^{2-} + H^+ + \text{energy}$

$FAD + 2e^- + 2H^+ \rightarrow FADH_2$

$NAD^+ + 2e^- + 2H^+ \rightarrow NADH + H^+$

b. ATP and ADP
c. FAD and NAD^+
d. $FADH_2$

23.2

Stage of catabolism	Where process occurs	Products
1. digestion	mouth, stomach, small intestine	glucose, fatty acids and glycerol, amino acids
2. acetyl group formation	cytoplasm and mitochondria	acetyl CoA and NADH
3. citric acid cycle	mitochondria	CO_2, water, energy (of electrons carried by NADH and $FADH_2$)
4. electron transport chain and oxidative phosphorylation	mitochondria	energy for ATP production

23.3

Step	Type of reaction	Final product(s)	Energy-rich compound
1.	condensation	citrate, CoA-SH	none
2.	isomerization	isocitrate	none
3.	oxidation, decarboxylation	α-ketoglutarate, CO_2	NADH
4.	oxidation, decarboxylation	succinyl CoA, CO_2	NADH
5.	phosphorylation	succinate, CoA-SH	GTP
6.	oxidation (dehydrogenation)	fumarate	$FADH_2$
7.	hydration	L-malate	none
8.	oxidation (dehydrogenation)	oxaloacetate	NADH

23.4

Step	Substance oxidized	Substance reduced
3.	isocitrate	NAD^+ (to NADH)
4.	α-ketoglutarate	NAD^+ (to NADH)
6.	succinate	FAD (to $FADH_2$)
8.	L-malate	NAD^+ (to NADH)

23.5

Step	Substance oxidized	Substance reduced
1.	NADH	FMN
2.	$FMNH_2$	Fe(III)
3.	Fe(II) (or $FADH_2$)	CoQ
4.	$CoQH_2$	Fe(III) cytochrome b
5.	Fe(II) cytochrome b	Fe(III) cytochrome c_1
6.	Fe(II) cytochrome c_1	Fe(III) cytochrome c
7.	Fe(II) cytochrome c	Fe(III) cytochrome a
8.	Fe(II) cytochrome a	Fe(III) cytochrome a_3
final step	Fe(II) cytochrome a_3	O_2

23.6

Step	Energy-rich compound formed	Number of ATP's produced
3.	NADH	3
4.	NADH	3
5.	GTP	1
6.	$FADH_2$	2
8.	NADH	3
	Total:	12

Answers to Self-Test

The numbers in parentheses refer to sections in your textbook:
1. T (23.1) **2**. F; prokaryotic (23.2) **3**. F; mitochondria (23.2) **4**. T (23.3) **5**. T (23.3)
6. F; mitochondria (23.5) **7**. F; acetyl CoA (23.6) **8**. F; NADH and $FADH_2$ (23.6)
9. F; lose energy (23.7) **10**. T (23.7) **11**. T (23.8) **12**. b (23.2) **13**. b (23.7)
14. d (23.9) **15**. c (23.7) **16**. b (23.6) **17**. c (23.6) **18**. b (23.6) **19**. e (23.6)
20. e (23.6)

Carbohydrate Metabolism — Chapter 24

Chapter Overview

The complete oxidation of glucose supplies the energy needed by cells to carry out their many vital functions. This oxidation takes place in a number of steps that conserve the energy contained in the chemical bonds of glucose and transfer it efficiently.

In this chapter you will study the reactions of glycolysis to produce pyruvate, the pathways of pyruvate under aerobic and anaerobic conditions, and the pentose phosphate pathway. You will study the ways that glycogen is synthesized in the body and broken down, and the hormones that control these processes.

Practice Exercises

24.1 **Glycolysis** (Sec. 24.2) is the metabolic pathway by which glucose is converted into two molecules of pyruvate. Using the steps of glycolysis discussed in section 24.2 of your textbook, give the number(s) of the reaction steps for each of the following:

a. Where are ATP's produced? ______________________

b. Where are ATP's used? ______________________

c. Where is NAD^+ reduced? ______________________

d. Where is the carbon chain split? ______________________

e. Where is a ketone isomerized to an aldehyde? ______________________

f. Where are phosphate groups added to sugar molecules? ______________________

g. Where are phosphate groups removed from sugar molecules? ______________________

h. Where is water lost? ______________________

24.2 The pyruvate produced by glycolysis reacts further in several different ways according to the conditions and the type of organism. Complete the following table on the fates of pyruvate produced by glycolysis.

Conditions	Name of process	Name of product	Number of NADH used or produced
1. aerobic			
2. anaerobic (humans)			
3. anaerobic (yeasts)			

24.3 The very different metabolic processes of **glycogenesis** (Sec. 24.4), **glycogenolysis** (Sec. 24.4), and **gluconeogenesis** (Sec. 24.6) have like-sounding names. Fill in the following table showing the differences between these processes.

Name of process	What the process accomplishes	Where the process takes place	High energy phosphate molecules used per glucose molecule
glycogenesis			
glycogenolysis			
gluconeogenesis			

24.4 Another pathway by which glucose may be degraded is the **pentose phosphate pathway** (Sec. 24.8).

a. What are the two main functions of the pentose phosphate pathway?

1.

2.

b. In the table below, compare the coenzyme NADPH to NADH.

Coenzyme	Oxidized form	Function in human body
NADH		
NADPH		

24.5 Three hormones affect carbohydrate metabolism. Complete the table below comparing these hormones.

Hormone	Source	Effect
insulin		
glucagon		
epinephrine		

24.6 Review the following terms, introduced in this chapter, by matching each with its correct description.

Answer	Term	Description
	fermentation	1. lactate → pyruvate → glucose
	glycogenesis	2. stimulates: glycogen(liver) → glucose
	gluconeogenesis	3. glucose → glycogen
	glycolysis	4. stimulates: glycogen(muscle) → glucose-6-phosphate
	insulin	5. glucose → ethanol
	epinephrine	6. glycogen → glucose or glucose-6-phosphate
	glycogenolysis	7. stimulates: glucose(blood) → glucose(cells)
	glucagon	8. glucose → pyruvate

Self-Test

True-false: Indicate whether the following statements are true or false. If the statement is false, give the word or phrase that may be substituted for the underlined portion to make the statement true.

1. The main site for carbohydrate digestion is <u>the small intestine</u>.
2. Glycolysis takes place in the <u>cytoplasm</u> of cells.
3. The process by which glucose is degraded to ethanol is called <u>fermentation</u>.
4. The enzymes maltase, sucrase, and lactase convert <u>starch to disaccharides</u>.
5. Substrate-level phosphorylation is the direct transfer of <u>phosphate ions in solution</u> to ADP molecules to produce ATP.
6. Fructose and galactose are converted, <u>in the liver</u>, to intermediates that enter the glycolysis pathway.
7. In human metabolism, under anaerobic conditions, pyruvate is reduced to <u>acetyl CoA</u>.
8. The complete oxidation of glucose in skeletal muscle and nerve cells yields <u>36</u> ATP molecules per glucose molecule.

9. When glycogen stored in muscle and liver tissues is depleted by strenuous exercise, glucose molecules can be synthesized by the process of gluconeogenesis.
10. Glucagon speeds up the rate of glycogenolysis in the muscle cells.
11. The breakdown of glycogen in order to maintain normal glucose levels in the bloodstream is called glycolysis.
12. The pentose phosphate pathway metabolizes glucose to produce ribose and other sugars needed for biosynthesis.

Multiple choice:
13. The metabolic pathway converting glucose to two molecules of pyruvate is:
    a) glycogenolysis
    b) glycogenesis
    c) glycolysis
    d) gluconeogenesis
    e) none of these
14. The hormone that lowers blood glucose levels is:
    a) glucagon
    b) insulin
    c) epinephrine
    d) norepinephrine
    e) none of these
15. Carbohydrate digestion produces:
    a) glucose
    b) galactose
    c) fructose
    d) a), b) and c)
    e) none of these
16. The lactate produced during strenuous exercise is converted to pyruvate in the:
    a) kidneys
    b) liver
    c) muscles
    d) small intestine
    e) none of these
17. One of the control mechanisms of glycolysis is feedback inhibition of hexokinase by:
    a) glucose
    b) glucose 6-phosphate
    c) glucose 1-phosphate
    d) fructose 6-phosphate
    e) none of these
18. In order for the electrons from NADH produced during glycolysis to enter the electron transport chain, they must:
    a) react with acetyl CoA and then enter the citric acid cycle
    b) reduce FAD, which then passes through the mitochondrial membrane
    c) be shuttled by an intermediate through the mitochondrial membrane to FAD
    d) pass energy directly to ATP in the cytoplasm
    e) none of these

Refer to the steps of glycolysis discussed in section 24.2 of your textbook as you answer the next three questions:
19. In glycolysis, which steps involve the removal of a phosphate group from ATP?
    a) 1 and 3
    b) 2 and 5
    c) 4 and 3
    d) 1 and 5
    e) 5 and 7
20. In glycolysis, the formation of ATP from ADP occurs during which step?
    a) 1
    b) 4
    c) 5
    d) 6
    e) 7
21. In glycolysis, which step involves the isomerization of an aldehyde to a ketone?
    a) 1
    b) 2
    c) 5
    d) 7
    e) none of the above

Answers to Practice Exercises

24.1 a. Steps 7 and 10; b. Steps 1 and 3; c. Step 6; d. Step 4
e. Step 5; f. Steps 1, 3, and 6; g. Steps 7 and 10; h. Step 9

24.2

Conditions	Name of process	Product	Number of NADH
1. aerobic	oxidation	acetyl CoA	1 produced
2. anaerobic (humans)	reduction	lactate	1 used
3. anaerobic (yeasts)	fermentation (reduction)	ethanol	1 used

24.3

Name of process	What process accomplishes	Where process takes place	Phosphate molecules used/glucose
glycogenesis	synthesis of glycogen from glucose	muscle and liver tissue	2 ATP
glycogenolysis	breakdown of glycogen to glucose	muscle and brain (glucose-6-phosphate) and liver (free glucose)	none
gluconeogenesis	glucose synthesis from non-carbohydrates	liver	4 ATP, 2 GTP

24.4 a. 1. synthesis of NADPH
2. production of ribose-5-phosphate for synthesis of nucleic acids and coenzymes

b.

Coenzyme	Oxidized form	Function in human body
NADH	NAD^+	promotes common metabolic pathways of energy transfer
NADPH	$NADP^+$	promotes biosynthetic reactions of lipids and nucleic acids

24.5

Hormone	Source	Effect
insulin	beta cells of pancreas	increases uptake and utilization of glucose by cells; lowers blood glucose
glucagon	alpha cells of pancreas	increases blood glucose by speeding up glycogenolysis in liver
epinephrine	adrenal glands	stimulates glycogenolysis in muscle cells; gives quick energy to muscle cells

24.6

Answer	Term	Description
5.	fermentation	1. lactate → pyruvate → glucose
3.	glycogenesis	2. stimulates: glycogen(liver) → glucose
1.	gluconeogenesis	3. glucose → glycogen
8.	glycolysis	4. stimulates: glycogen(muscle) → glucose-6-phosphate
7.	insulin	5. glucose → ethanol
4.	epinephrine	6. glycogen → glucose or glucose-6-phosphate
6.	glycogenolysis	7. stimulates: glucose(blood) → glucose(cells)
2.	glucagon	8. glucose → pyruvate

Answers to Self-Test

The numbers in parentheses refer to sections in your textbook:
1. T (24.1) **2**. T (24.2) **3**. T (24.3) **4**. F; disaccharides to monosaccharides (24.1)
5. F; high-energy phosphate groups from substrate molecules (24.2) **6**. T (24.2)
7. F; lactate (24.3) **8**. T (24.4) **9**. T (24.6) **10**. F; epinephrine (24.9)
11. F; glycogenolysis (24.5) **12**. T (24.8) **13**. c (24.2) **14**. b (24.9) **15**. d (24.1)
16. b (24.3) **17**. b (24.2) **18**. c (24.4) **19**. a (24.2) **20**. e (24.2) **21**. b (24.2)

Lipid Metabolism — Chapter 25

Chapter Overview

Lipids are the most efficient energy storage compounds of the body. They are also important materials in membranes. This chapter discusses the biosynthesis and storage of lipids and their degradation to produce energy.

In this chapter you will study the processes of digestion of triacyglycerols and their absorption into the bloodstream, their storage, and their degradation by means of the fatty acid spiral. You will define ketone bodies and learn the conditions under which they are produced. You will compare fatty acid synthesis to fatty acid oxidation, and you will study the biosynthesis of cholesterol.

Practice Exercises

25.1 Most dietary lipids are triacylglycerols (TAGs). Complete the following diagram showing the stages of digestion and absorption of TAGs. In each box, write the processes that take place at that location.

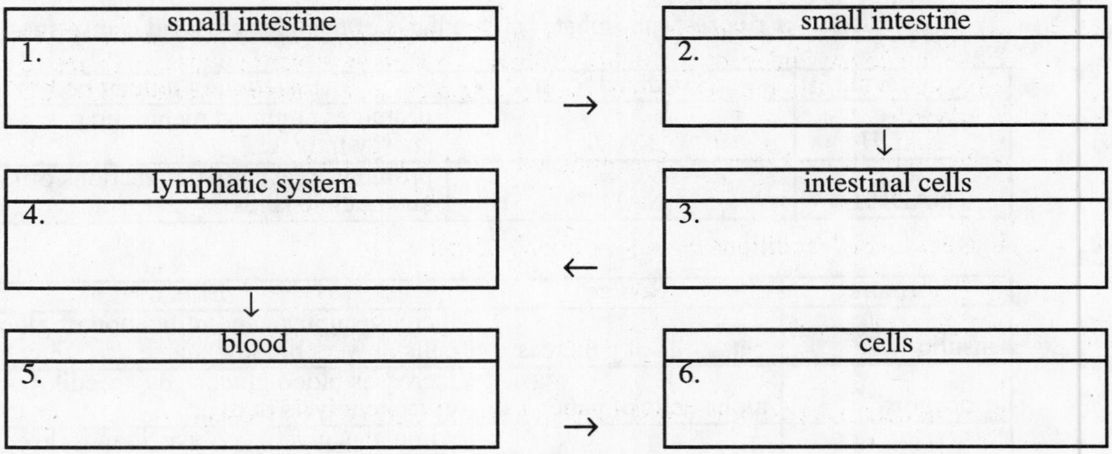

small intestine		small intestine
1.	→	2.
		↓
lymphatic system		**intestinal cells**
4.	←	3.
↓		
blood		**cells**
5.	→	6.

25.2 TAG molecules can be stored in adipose tissue until needed for energy production.
a. What is the process of **triacylglycerol mobilization** (Sec. 25.2)?

b. What happens to the products of triacylglycerol mobilization?

25.3 The **fatty acid spiral** (Sec. 25.4) is a four-step process in which two carbon atoms are cleaved from a fatty acid for each turn of the spiral.
a. Summarize the type of reaction in each of the following steps of the fatty acid spiral by completing the table below. (Use the discussion of the fatty acid spiral in section 25.4 of your textbook.)

Step	Type of reaction	Coenzyme oxidizing agent
1. alkane → alkene		
2. alkene → secondary alcohol		
3. secondary alcohol → ketone		
4. ketone → shortened fatty acid		

b. Give the numbers of the reaction steps for these parts of the process.

1. Where is FAD reduced? ______________________
2. Where is NAD^+ reduced? ______________________
3. Where is CoA-SH added? ______________________
4. Where is water added to the carbon chain? ____________
5. Where is acetyl CoA removed from the carbon chain? ____________

25.4 a. Calculate the total number of acetyl CoA molecules, NADH molecules, and $FADH_2$ molecules produced by the aerobic catabolism of capric acid, a saturated 10-carbon fatty acid.

b. Determine the total number of molecules of ATP produced for energy use by the total oxidation of capric acid to carbon dioxide and water.

25.5 Acetyl CoA from the fatty acid spiral is usually processed further through the citric acid cycle; however, under some conditions there is not enough oxaloacetate produced to react with acetyl CoA in the first step of the citric acid cycle, and **ketone bodies** (Sec. 25.6) are produced.

a. Name the three ketone bodies produced in the human body.

b. Under what conditions do ketone bodies form?

c. What is ketosis?

d. Which ketone bodies could produce a lowered blood pH (acidosis)?

25.6 **Lipogenesis** (Sec. 25.7) is the synthesis of fatty acids from acetyl CoA. It is not simply the reverse of the fatty acid spiral for degradation of fatty acids. Complete the table below comparing these two processes.

Comparisons	Degradation of fatty acids	Lipogenesis
reaction site in cell		
carbons lost/gained per turn		
intermediate carriers		
types of enzymes		
coenzymes for energy transfer		

25.7 Four reactions occurring in a cyclic pattern constitute the chain elongation process of lipogenesis. Complete the table below summarizing the types of reactions that take place in each turn of the cycle.

Step	Type of reaction	Coenzyme reducing agent
1.		
2.		
3.		
4.		

25.8 In the biosynthesis of capric acid, a 10-carbon saturated fatty acid, from acetyl CoA molecules:
a. How many rounds of the fatty acid biosynthesis pathway are needed? ____________
b. How many molecules of malonyl ACP must be formed? ____________
c. How many high-energy ATP bonds are consumed? ____________
d. How many NADPH molecules are needed? ____________

25.9 Biosynthesis of cholesterol takes place in the liver. There are at least 27 steps in the process, but these can be considered to occur in biosynthetic stages. Fill in the missing parts of the diagram below summarizing the stages of cholesterol biosynthesis.

1. 3 acetyl CoA + 3ATP	→	1 isoprene unit
2.	→	1 squalene molecule
3.	→	
4. lanosterol	→	

25.10 Acetyl CoA is a key intermediate for many metabolic processes. Fill in the diagram below showing the products or reactants in these processes.

[ ] (degradation) → acetyl CoA → (biosynthesis) [ ]
↓ (energy production)
oxidation in citric acid cycle

Self-Test

True-false: Indicate whether the following statements are true or false. If the statement is false, give the word or phrase that may be substituted for the underlined portion to make the statement true.

1. Bile is released into the small intestine where it acts as <u>a hydrolytic enzyme</u> for lipids.
2. A meal high in triacylglycerols will cause the concentration of <u>fatty acid micelles</u> in the blood and lymph to peak in four to six hours.
3. Adipocytes are triacylglycerol storage cells found mainly in <u>liver tissue</u>.
4. Chylomicrons consist of lipoproteins combined with <u>fatty acids</u> in the intestinal cells.
5. Glycerol is metabolized to <u>dihydroxyacetone phosphate</u> before entering glycolysis.
6. A stearic acid molecule produces <u>more than 4 times</u> as much energy as a glucose molecule.
7. An important intermediate in the synthesis of cholesterol is <u>isoprene</u>.
8. Acetyl CoA molecules can be further oxidized in the <u>fatty acid spiral</u>.
9. Cholesterol synthesis takes place in the <u>liver</u>.
10. When glucose is not available to the body, acetyl CoA molecules are used to manufacture <u>pyruvate</u>.
11. The liver uses <u>fatty acids</u> as the preferred fuel.
12. The last turn of the fatty acid spiral produces <u>two</u> acetyl CoA molecules.

Multiple choice:

13. Triacylglycerol mobilization is the process in which triacylglycerols are:
    a) oxidized
    b) hydrolyzed
    c) synthesized
    d) hydrogenated
    e) none of these
14. Acetyl CoA may not be used to:
    a) synthesize ketone bodies
    b) synthesize fatty acids
    c) synthesize glucose
    d) synthesize cholesterol
    e) none of these
15. Diabetic ketosis results in:
    a) ketonemia
    b) ketonuria
    c) metabolic acidosis
    d) all of these
    e) none of these

16. One round of the fatty acid spiral produces:
    a) one NADH and one FAD
    b) two NADH and one FAD
    c) one NADH and two $FADH_2$
    d) one NADH and one $FADH_2$
    e) none of these
17. Which of the following does not aid in the digestion of lipids?
    a) pancreatic lipases
    b) cholecytokinin
    c) salivary enzymes
    d) churning action of stomach
    e) all of the above are used
18. A fatty acid micelle produced by hydrolysis of lipids contains:
    a) triacylglycerols
    b) diacylglycerols
    c) monoacylglycerols
    b) lipoproteins
    e) none of these
19. A 20 carbon saturated fatty acid will produce 10 acetyl CoA molecules from the fatty acid spiral after:
    a) 1 turn
    b) 9 turns
    c) 10 turns
    d) 20 turns
    e) none of these
20. The total number of ATPs produced in the body by complete oxidation of a 20-carbon saturated fatty acid is:
    a) 170
    b) 168
    c) 165
    d) 163
    e) none of these

Answers to Practice Exercises

25.1

small intestine		small intestine
1. emulsification by bile, hydrolysis of TAGs to form glycerol, monoacylglycerols and fatty acids	→	2. formation of fatty acid micelles, which can be absorbed into the intestinal cells
		↓
lymphatic system		**intestinal cells**
4. chylomicrons enter lymphatic vessels and are transported to the thoracic duct, where fluid enters a vein	←	3. TAGs are reassembled and combined with water-soluble protein in chylomicrons
↓		
blood		**cells**
5. TAGs are hydrolyzed to fatty acids and glycerol, which are absorbed by the cells	→	6. fatty acids and glycerol are broken down for energy or stored as TAGs

25.2 a. Triacylglycerol mobilization is the hydrolysis of stored TAGs and the release of the fatty acids and glycerol produced into the bloodstream.

b. The fatty acids and glycerol are oxidized to produce energy, the fatty acids through the fatty acid spiral and the glycerol in a two-step process that produces dihydroxyacetone phosphate.

25.3 a.

Step	Type of reaction	Oxidizing agent
1. alkane → alkene	oxidation (dehydrogenation)	FAD
2. alkene → secondary alcohol	hydration	none
3. secondary alcohol → ketone	oxidation (dehydrogenation)	NAD^+
4. ketone → shortened fatty acid	chain cleavage	none

b. 1. step one; 2. step three; 3. step four; 4. step two; 5. step four

25.4 a. In four times through the fatty acid spiral a 10-carbon fatty acid produces: 5 acetyl CoA molecules, 4 NADH, and 4 $FADH_2$

b. Citric acid cycle: 5 acetyl CoA x 12 ATP/ 1 acetyl CoA = 60 ATP
Electron transport chain: 4 NADH x 3 ATP/ 1 NADH = 12 ATP
4 $FADH_2$ x 2 ATP/ 1 $FADH_2$ = 8 ATP
Activation of fatty acid spiral: –2 ATP
Total: 78 ATP

25.5 a. Acetoacetate, β-hydroxybutyrate, acetone
b. Ketone bodies form from acetyl CoA when there is insufficient oxaloacetate being formed from pyruvate. This would occur when dietary intakes are high in fat and low in carbohydrates, when the body cannot adequately process glucose (as in diabetes), and during prolonged fasting or starvation.
c. Ketosis is the accumulation of ketone bodies in the blood and urine.
d. Acetoacetate and β-hydroxybutyrate are acids and could produce a lowered blood pH.

25.6

Comparisons	Fatty acid degradation	Lipogenesis
reaction site in cell	mitochondrial matrix	cytoplasm
carbons lost/gained per turn	two carbons lost	two carbons gained
intermediate carriers	coenzyme A	ACP (acyl carrier protein)
types of enzymes	independent enzymes	fatty acid synthetase complex
coenzymes -- energy transfer	FAD and NAD^+	NADPH

25.7

Step	Type of reaction	Coenzyme reducing agent
1.	condensation	none
2.	hydrogenation	NADPH
3.	dehydration	none
4.	hydrogenation	NADPH

25.8 a. four rounds; b. four malonyl ACP molecules; c. four ATP bonds; d. eight NADPH molecules

25.9

1.	3 acetyl CoA + 3ATP	→	1 isoprene unit
2.	6 isoprene units	→	squalene
3.	squalene	→	lanosterol (4-ring system)
4.	lanosterol	→	cholesterol

25.10

fatty acids, glucose, glycerol (degradation) → acetyl CoA → (biosynthesis) fatty acids, ketone bodies, cholesterol

↓ (energy production)
oxidation in citric acid cycle

Answers to Self-Test

The numbers in parentheses refer to sections in your textbook:
1. F; an emulsifier (25.1) **2**. F; chylomicrons (25.1) **3**. F; adipose tissue (25.2)
4. F; triacylglycerols (25.1) **5**. T (25.3) **6**. T (25.5) **7**. T (25.8)
8. F; citric acid cycle (25.4) **9**. T (25.8) **10**. F; ketone bodies (25.6) **11**. T (25.5)
12. T (25.4) **13**. b (25.2) **14**. c (25.9) **15**. d (25.6) **16**. d (25.4) **17**. c (25.1)
18. c (25.1) **19**. b (25.4) **20**. d (25.5)

Chapter Overview

Like other organic compounds, proteins and amino acids are metabolized by the body. Although their primary uses are in biosynthesis, they can also be used as sources of energy for the body.

In this chapter you will study the digestion of proteins and the absorption of amino acids, the breakdown of amino acids in the body, and the entry of the products of catabolism into the metabolic pathways of carbohydrates and fatty acids. You will study the biosynthesis of the nonessential amino acids, and learn how the urea cycle is used to rid the body of toxic ammonium ion. You will learn how the body uses the degradation products of hemoglobin.

Practice Exercises

26.1 The digestion of proteins produces amino acids, which enter the blood stream for distribution throughout the body. This **amino acid pool** (Sec. 26.2) is utilized in four different ways.

a. Complete the diagram below for the digestion of proteins and absorption of amino acids.

b. What are the four ways that the amino acids from the amino acid pool are utilized in the human body?

1.
2.
3.
4.

26.2 The degradation of amino acids takes place in two stages, **transamination** (Sec. 26.3) and **oxidative deamination** (Sec. 26.3). Complete the equations below by drawing structural formulas for the missing products and reactants. (Equations a. and b. are transamination reactions; equations c. and d. are oxidative deamination reactions.)

a. $CH_3-CH(NH_3^+)-COO^- + {}^-OOC-CH_2-C(=O)-COO^- \xrightarrow{\text{trans-aminase}} CH_3-C(=O)-COO^- + \ ?$

b. $OH-CH(CH_3)-CH(NH_3^+)-COO^- + CH_3-C(=O)-COO^- \xrightarrow{\text{trans-aminase}} OH-CH(CH_3)-C(=O)-COO^- + \ ?$

c. $? \ + NAD^+ + H_2O \longrightarrow NH_4^+ + CH_3-C(=O)-COO^- + NADH + H^+$

d. ${}^-OOC-CH_2-CH(NH_3^+)-COO^- + \ ? \longrightarrow NH_4^+ + \ ? \ + NADH + H^+$

26.3 The ammonium ion produced by oxidative deamination is toxic; it is removed from the human body as urea in the **urea cycle** (Sec. 26.4). Using the steps of the urea cycle shown in your textbook, answer the following questions:

a. Where is urea removed from the cycle? __________
b. Where is carbamoyl phosphate added? __________
c. Where is water added? __________
d. Where is fumarate produced? __________
e. Where is ATP used? __________
f. How much ATP is consumed in the production of one urea molecule? __________

26.4 The carbon skeletons produced by transamination or oxidative deamination of amino acids undergo a sequence of degradations. The seven degradation products of these series of reactions are further metabolized, and the original amino acids are classified by the pathways taken by their degradation products.

Classification of original amino acids	Products of degradation sequences	Further metabolic reactions
glucogenic (Sec. 26.5)		
ketogenic (Sec. 26.5)		
glucogenic and ketogenic		

26.5 The essential amino acids must be consumed by humans, but the 11 nonessential amino acids can be synthesized in the human body from intermediates of glycolysis and the citric acid cycle. Complete the equations below for the three amino acids that can be biosynthesized by transamination of the appropriate α-keto acid. Name the amino acid produced.

a. $^{-}OOC-CH_2-\overset{O}{\overset{\|}{C}}-COO^{-} \xrightarrow{\text{transaminase}}$

b. $CH_3-\overset{O}{\overset{\|}{C}}-COO^{-} \xrightarrow{\text{transaminase}}$

c. $^{-}OOC-CH_2-CH_2-\overset{O}{\overset{\|}{C}}-COO^{-} \xrightarrow{\text{transaminase}}$

26.6 Hemoglobin is the conjugated protein responsible for the oxygen-carrying ability of red blood cells. Hemoglobin catabolism takes place in the spleen and liver. Complete the diagram below showing the degradation products of hemoglobin.

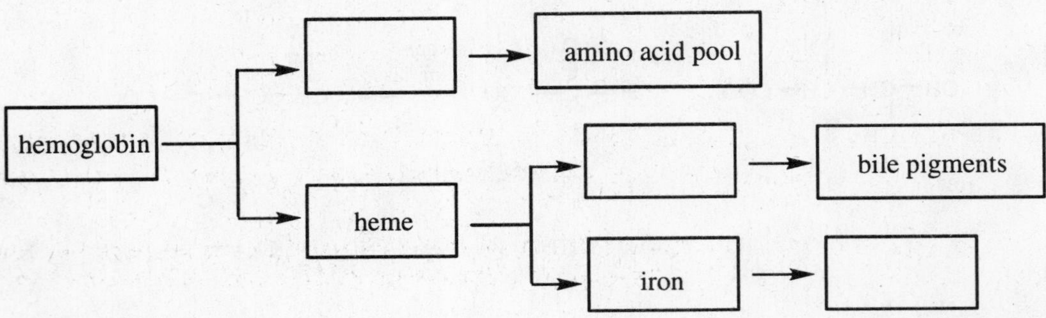

26.7 Complete the table below summarizing some of the metabolism of the three groups of nutrients you have been studying.

Class of nutrient	Basic structural unit	Storage compounds	Degradation pathway
carbohydrates			
lipids			
proteins			

Self-Test

True-false: Indicate whether the following statements are true or false. If the statement is false, give the word or phrase that may be substituted for the underlined portion to make the statement true.

1. Protein digestion begins in the <u>stomach</u>.
2. Proteins are <u>denatured</u> in the stomach by hydrochloric acid.
3. The pH of the pancreatic juice that aids in protein digestion is between <u>1.5 and 2.0</u>.
4. <u>Transamination</u> of an amino acid releases the amino group as an ammonium ion.
5. Transaminases require the presence of the coenzyme <u>pyridoxal phosphate</u>.
6. Oxidative deamination is the conversion of an amino acid into a keto acid with the release of <u>urea</u>.
7. <u>Glucogenic</u> amino acids are converted to acetyl CoA.
8. <u>All</u> of the nonessential amino acids can be synthesized in the body.
9. <u>Two</u> of the essential amino acids can be synthesized in the body.
10. A <u>positive</u> nitrogen balance means that more nitrogen is excreted than is taken in.
11. Phenylketonuria (PKU) is a lack of the enzyme needed to <u>synthesize</u> phenylalanine.
12. During negative nitrogen balance, the body uses amino acids obtained from <u>the amino acid pool</u>.
13. Only two amino acids are purely ketogenic: <u>leucine and lysine</u>.

Multiple choice:

14. The urea cycle is important because it:
    a) removes ammonium ion from the body
    b) regenerates α-ketoglutarate by oxidative deamination
    c) results in transamination of amino acids
    d) all of these
    e) none of these
15. Before entering the urea cycle, ammonia is converted to carbamoyl phosphate. This requires how much ATP?
    a) one molecule
    b) two molecules
    c) three molecules
    d) four molecules
    e) none of these
16. Which of the following amino acids can be synthesized by the human body?
    a) valine
    b) lysine
    c) alanine
    d) tryptophan
    e) phenylalanine
17. The degradation products of the globin portion of hemoglobin contribute to the formation of:
    a) bile pigments
    b) ferritin molecules
    c) tetrapyrrole
    d) the amino acid pool
    e) none of these
18. The net effect of the transamination of amino acids is to produce a single amino acid:
    a) glutamate
    b) aspartate
    c) pyridoxine
    d) oxaloacetate
    e) none of these

19. The amino acids whose carbon skeletons are degraded to citric acid cycle intermediates are classified as:
 a) essential amino acids
 b) nonessential amino acids
 c) ketogenic amino acids
 d) glucogenic amino acids
 e) both glucogenic and ketogenic
20. Jaundice can occur because of:
 a) a negative nitrogen balance
 b) a low rate of protein turnover
 c) a low rate of heme degradation
 d) a low rate of bilirubin excretion by the liver
 e) none of these

Answers to Practice Exercises

26.1 a.

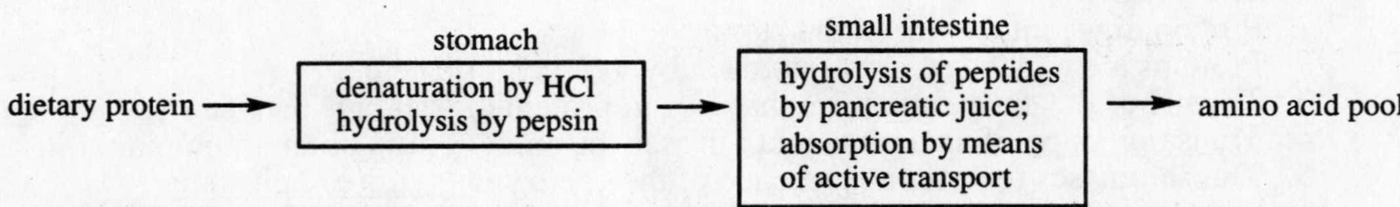

b. Four ways that the amino acids from the amino acid pool are utilized:
 1. energy production
 2. biosynthesis of functional proteins
 3. biosynthesis of nonessential amino acids
 4. biosynthesis of nonprotein nitrogen-containing compounds

26.2 a.

$$CH_3-CH(\overset{+}{N}H_3)-COO^- + {}^-OOC-CH_2-\overset{O}{\overset{\|}{C}}-COO^- \xrightarrow{\text{trans-aminase}} CH_3-\overset{O}{\overset{\|}{C}}-COO^- + {}^-OOC-CH_2-CH(\overset{+}{N}H_3)-COO^-$$

b.

$$OH-CH(CH_3)-CH(\overset{+}{N}H_3)-COO^- + CH_3-\overset{O}{\overset{\|}{C}}-COO^- \xrightarrow{\text{trans-aminase}} OH-CH(CH_3)-\overset{O}{\overset{\|}{C}}-COO^- + CH_3-CH(\overset{+}{N}H_3)-COO^-$$

c.

$$CH_3-CH(\overset{+}{N}H_3)-COO^- + NAD^+ + H_2O \longrightarrow \overset{+}{N}H_4 + CH_3-\overset{O}{\overset{\|}{C}}-COO^- + NADH + H^+$$

d.

$${}^-OOC-CH_2-CH(\overset{+}{N}H_3)-COO^- + NAD^+ + H_2O \longrightarrow \overset{+}{N}H_4 + {}^-OOC-CH_2-\overset{O}{\overset{\|}{C}}-COO^- + NADH + H^+$$

26.3 a. step 4
 b. step 1
 c. step 4
 d. step 3
 e. in the formation of carbamoyl phosphate (2) and step 2 (2 high energy bonds)
 f. 4 ATP molecules

26.4

Classification of original amino acids	Products of degradation sequences	Further metabolic reactions
glucogenic	citric acid cycle intermediates	gluconeogenesis, ATP production
ketogenic	acetyl CoA, acetoacetyl CoA	production of ketone bodies or fatty acids, ATP production
glucogenic and ketogenic	pyruvate	any of the above

26.5 a. $^{-}OOC{-}CH_2{-}\overset{O}{\overset{\|}{C}}{-}COO^{-} \xrightarrow{\text{transaminase}} {}^{-}OOC{-}CH_2{-}\overset{\overset{+}{N}H_3}{\overset{|}{C}H}{-}COO^{-}$ aspartate

b. $CH_3{-}\overset{O}{\overset{\|}{C}H}{-}COO^{-} \xrightarrow{\text{transaminase}} CH_3{-}\overset{\overset{+}{N}H_3}{\overset{|}{C}H}{-}COO^{-}$ alanine

c. $^{-}OOC{-}CH_2{-}CH_2{-}\overset{O}{\overset{\|}{C}}{-}COO^{-} \xrightarrow{\text{transaminase}} {}^{-}OOC{-}CH_2{-}CH_2{-}\overset{\overset{+}{N}H_3}{\overset{|}{C}H}{-}COO^{-}$ glutamate

26.6

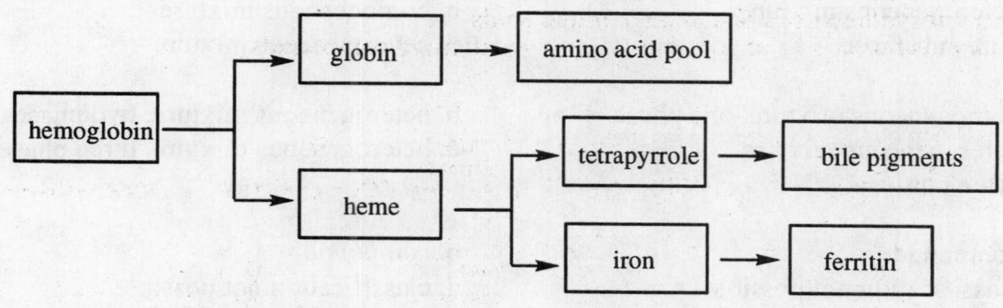

26.7

Nutrient class	Basic structural unit	Storage compounds	Degradation pathway
carbohydrates	monosaccharides	glycogen	glycolysis, citric acid cycle
lipids	fatty acids, glycerol	triacylglycerols	fatty acid spiral, citric acid cycle
proteins	amino acids	functional proteins (amino acids are not "stored")	transamination, oxidative deamination, citric acid cycle

Answers to Self-Test

The numbers in parentheses refer to sections in your textbook:
1. T (26.1) **2**. T (26.1) **3**. F; 7 and 8 (26.1) **4**. F; oxidative deamination (26.3) **5**. T (26.3) **6**. F; ammonium ion (26.3) **7**. F; ketogenic (26.5) **8**. T (26.6) **9**. F; none (26.6) **10**. F; negative (26.2) **11**. F; oxidize (26.6) **12**. F; degradation of functional proteins (26.2) **13**. T (26.5) **14**. a (26.4) **15**. b (26.4) **16**. c (26.6) **17**. d (26.7) **18**. a (26.3) **19**. d (26.3) **20**. d (26.7)

Basic Concepts About Matter — Chapter 1

Solutions to Exercises

1.1 a. matter b. matter c. energy d. energy e. matter f. matter

1.3 a. indefinite shape versus definite shape b. indefinite volume versus definite volume

1.5 a. no b. no c. yes d. yes

1.7 a. physical b. chemical c. chemical d. physical

1.9 a. chemical b. physical c. chemical d. physical

1.11 a. chemical b. physical c. physical d. chemical

1.13 a. physical b. physical c. chemical d. physical

1.15 a. false b. true c. false d. true

1.17 a. heterogeneous mixture b. homogeneous mixture c. pure substance d. heterogeneous mixture

1.19 a. homogeneous mixture, one phase b. heterogeneous mixture, two phases c. heterogeneous mixture, three phases d. heterogeneous mixture, three phases

1.21 a. compound b. compound c. classification not possible d. classification not possible

1.23 a. A, classification not possible; B, classification not possible; C, compound
b. D, compound; E, classification not possible; F, classification not possible; G, classification not possible

1.25 a. true b. false c. false d. false

1.27 a. false b. true c. false d. true

1.29 a. silver b. gold c. calcium d. sodium e. phosphorus f. sulfur

1.31 a. Sn b. Cu c. Al d. B e. Ba f. Ar

1.33 a. no b. yes c. yes d. no

1.35 a. true b. false; triatomic molecules must contain at least one kind of atom c. true d. false; both homoatomic and heteroatomic molecules may contain three or more atoms

1.37 a.  b.  c.  d.

1.39 a. compound b. compound c. element d. compound e. compound f. compound
g. element h. element

1.41 a. $C_8H_{10}N_4O_2$ b. $C_{12}H_{22}O_{11}$ c. HCN d. H_2SO_4

1.43 a. 3 elements, 2 H atoms, 1 C atom, 3 O atoms
b. 4 elements, 1 N atom, 4 H atoms, 1 Cl atom, 4 O atoms
c. 3 elements, 1 Ca atom, 1 S atom, 4 O atoms
d. 2 elements, 4 C atoms, 10 H atoms

1.45 a. solid is pulverized, granules are heated
b. discoloration, bursts into flame and burns

1.46 a. element b. compound c. mixture d. compound

1.47 a. homogeneous mixture b. heterogeneous mixture c. compound
d. compound

1.48 a. element b. mixture c. mixture d. compound

1.49 a. B-Ar-Ba-Ra b. Eu-Ge-Ne c. He-At-H-Er d. Al-La-N

1.50 a. same, both 4 b. more, 6 and 5 c. same, both 5 d. fewer, 13 and 15

1.51 a. $1 + 2 + x = 6$; $x = 3$ b. $2 + 3 + 3x = 17$; $x = 4$ c. $1 + x + x = 5$; $x = 2$
d. $x + 2x + x = 8$; $x = 2$

1.52 a. 2 (N_2, NH_3) b. 4 (N, H, C, Cl) c. 110; 5(2 + 6 + 4 + 5 + 5) d. 56; 4(4 + 3 + 4 + 3)

1.53 a. 2 b. 1, 4, 5, 6 c. 2, 3 d. 1, 3, 4, 5, 6

1.54 a. 2, 3, 4 b. 6 c. 3, 5 d. 1, 4

1.55 a. 6 b. 4, 5 c. 2, 5 d. all choices

1.56 a. 2 b. 4, 5 c. 1, 4 d. (3, 6)

Measurement in Chemistry — Chapter 2

Solutions to Exercises

2.1 a. kilo b. milli c. micro d. deci

2.3 a. centimeter b. kiloliter c. microliter d. nanogram

2.5 a. nanogram, milligram, centigram
b. kilometer, megameter, gigameter
c. picoliter, microliter, deciliter
d. microgram, milligram, kilogram

2.7 a. 0.1°C b. 0.01 mL c. 1 mL d. 0.1 mm

2.9 a. 4 b. 2 c. 4 d. 3 e. 5 f. 4

2.11 a. same b. different c. same d. same

2.13 a. 0.351 b. 653,900 c. 22.556 d. 0.2777

2.15 a. 2 b. 2 c. 2 d. 2

2.17 a. 0.0080 b. 0.0143 c. 14 d. 0.182 e. 1.1 f. 5.72

2.19 a. 162 b. 9.3 c. 1261 d. 20.0

2.21 a. 1.207×10^{2} b. 3.4×10^{-3} c. 2.3100×10^{2} d. 2.3×10^{4} e. 2.00×10^{-1}
f. 1.011×10^{-1}

2.23 a. 10^{8} b. 10^{2} c. 10^{-8} d. 10^{8} e. 10^{-8} f. 10^{-2}

2.25 a. 5.50×10^{12} b. 4.14×10^{-2} c. 1.5×10^{4} d. 2.0×10^{-7} e. 1.5×10^{11} f. 1.2×10^{6}

2.27 a. $\frac{1\text{ kg}}{10^{3}\text{ g}}$ and $\frac{10^{3}\text{ g}}{1\text{ kg}}$ b. $\frac{1\text{ nm}}{10^{-9}\text{ m}}$ and $\frac{10^{-9}\text{ m}}{1\text{ nm}}$

c. $\frac{1\text{ mL}}{10^{-3}\text{ L}}$ and $\frac{10^{-3}\text{ L}}{1\text{ mL}}$ d. $\frac{1.00\text{ lb}}{454\text{ g}}$ and $\frac{454\text{ g}}{1.00\text{ lb}}$

e. $\frac{1.00\text{ km}}{0.621\text{ mi}}$ and $\frac{0.621\text{ mi}}{1.00\text{ km}}$ f. $\frac{1.00\text{ L}}{0.265\text{ gal}}$ and $\frac{0.265\text{ gal}}{1.00\text{ L}}$

2.29 a. $1.6 \times 10^3 \text{ dm} \times \left(\frac{10^{-1} \text{ m}}{1 \text{ dm}}\right) = 1.6 \times 10^2 \text{ m}$

b. $24 \text{ nm} \times \left(\frac{10^{-9} \text{ m}}{1 \text{ nm}}\right) = 2.4 \times 10^{-8} \text{ m}$

c. $0.003 \text{ km} \times \left(\frac{10^3 \text{ m}}{1 \text{ km}}\right) = 3 \text{ m}$

d. $3.0 \times 10^8 \text{ mm} \times \left(\frac{10^{-3} \text{ m}}{1 \text{ mm}}\right) = 3.0 \times 10^5 \text{ m}$

2.31 $2500 \text{ mL} \times \left(\frac{10^{-3} \text{ L}}{1 \text{ mL}}\right) = 2.5 \text{ L}$

2.33 $1550 \text{ g} \times \left(\frac{1.00 \text{ lb}}{454 \text{ g}}\right) = 3.41 \text{ lb}$

2.35 $25 \text{ mL} \times \left(\frac{10^{-3} \text{ L}}{1 \text{ mL}}\right) \times \left(\frac{0.265 \text{ gal}}{1.00 \text{ L}}\right) = 0.0066 \text{ gal}$

2.37 $83.2 \text{ kg} \times \left(\frac{2.20 \text{ lb}}{1.00 \text{ kg}}\right) = 183 \text{ lb}$

$1.92 \text{ m} \times \left(\frac{39.4 \text{ in.}}{1.00 \text{ m}}\right) \times \left(\frac{1 \text{ ft}}{12 \text{ in.}}\right) = 6.30 \text{ ft}$ (6 ft 4 in.)

2.39 $\frac{524.5 \text{ g}}{38.72 \text{ cm}^3} = 13.55 \frac{\text{g}}{\text{cm}^3}$

2.41 $20.0 \text{ g} \times \left(\frac{1 \text{ mL}}{0.791 \text{ g}}\right) = 25.3 \text{ mL}$

2.43 $236 \text{ mL} \times \left(\frac{1.03 \text{ g}}{1 \text{ mL}}\right) = 243 \text{ g}$

2.45 $\frac{5}{9} (525° - 32°) = 274°\text{C}$

2.47 $\frac{9}{5} (-38.9°) + 32.0° = -38.0°\text{F}$

2.49 $\frac{9}{5}(-10°) + 32° = 14°F$; $-10°C$ is higher

2.51 $0.63 \frac{\text{cal}}{\text{g°C}} \times \left(\frac{4.184 \text{ J}}{1 \text{ cal}}\right) = 2.6 \frac{\text{J}}{\text{g°C}}$

2.53 $\frac{18.6 \text{ cal}}{12.0 \text{ g} \times 10.0°\text{C}} = 0.155 \frac{\text{cal}}{\text{g°C}}$

2.55 a. $0.057 \frac{\text{cal}}{\text{g°C}} \times (42.0 \text{ g}) \times (20.0°\text{C}) = 48 \text{ cal}$

b. $1.00 \frac{\text{cal}}{\text{g°C}} \times (42.0 \text{ g}) \times (20.0°\text{C}) = 8.40 \times 10^2 \text{ cal}$

c. $0.21 \frac{\text{cal}}{\text{g°C}} \times (42.0 \text{ g}) \times (20.0°\text{C}) = 180 \text{ cal}$

2.57 a. 4.72051 b. 4.7205 c. 4.721 d. 4.7

2.58 a. 3.00×10^{-3} b. 9.4×10^5 c. 2.35×10^1 d. 4.50000×10^8

2.59 a. smaller, 10^3 b. larger, 10^9 c. smaller, 10^8 d. smaller, 10^8

2.60 a. 4 significant figures b. 4 significant figures c. 3 significant figures d. exact

2.61 a. $\frac{1.0 \text{ g}}{2.0 \text{ cm}^3} = 5.0 \times 10^{-1} \frac{\text{g}}{\text{cm}^3}$ b. $\frac{1.000 \text{ g}}{2.00 \text{ cm}^3} = 5.00 \times 10^{-1} \frac{\text{g}}{\text{cm}^3}$

c. $\frac{1.0000 \text{ g}}{2.0000 \text{ cm}^3} = 5.0000 \times 10^{-1} \frac{\text{g}}{\text{cm}^3}$ d. $\frac{1.000 \text{ g}}{2.0000 \text{ cm}^3} = 5.000 \times 10^{-1} \frac{\text{g}}{\text{cm}^3}$

2.62 a. $75.0 \text{ g} \times \frac{1 \text{ mL}}{0.56 \text{ g}} = 1.3 \times 10^2 \text{ mL}$

b. $75.0 \text{ g} \times \frac{1 \text{ cm}^3}{0.93 \text{ g}} \times \frac{1 \text{ mL}}{1 \text{ cm}^3} = 81 \text{ mL}$

c. $75.0 \text{ g} \times \frac{1 \text{ L}}{0.759 \text{ g}} \times \frac{1 \text{ mL}}{10^{-3} \text{ L}} = 9.88 \times 10^4 \text{ mL}$

d. $75.0 \text{ g} \times \frac{1 \text{ mL}}{13.6 \text{ g}} = 5.51 \text{ mL}$

2.63 a. $\frac{4.5 \text{ mg}}{1 \text{ mL}} \times \frac{1 \text{ mL}}{10^{-3} \text{ L}} = 4.5 \times 10^3 \text{ mg/L}$

b. $\frac{4.5 \text{ mg}}{1 \text{ mL}} \times \frac{10^{-3} \text{ g}}{1 \text{ mg}} \times \frac{1 \text{ pg}}{10^{-12} \text{ g}} = 4.5 \times 10^9 \text{ pg/mL}$

c. $\frac{4.5 \text{ mg}}{1 \text{ mL}} \times \frac{10^{-3} \text{ g}}{1 \text{ mg}} \times \frac{1 \text{ mL}}{10^{-3} \text{ L}} = 4.5 \text{ g/L}$

d. $\frac{4.5 \text{ mg}}{1 \text{ mL}} \times \frac{10^{-3} \text{ g}}{1 \text{ mg}} \times \frac{1 \text{ kg}}{10^3 \text{ g}} \times \frac{1 \text{ mL}}{1 \text{ cm}^3} \times \frac{1 \text{ cm}^3}{(10^{-2} \text{ m})^3} = 4.5 \text{ kg/m}^3$

2.64 $7 \text{ pounds} \times \frac{1 \text{ franc}}{0.102 \text{ pounds}} \times \frac{1 \text{ dollar}}{5.94 \text{ francs}} = 11.6 \text{ dollars}$

2.65 a. 3, 5 b. 1, 4, 6 c. (1, 5), (3, 4) d. (1, 4), (3, 5), (4, 6)

2.66 a. 1, 4, 6 b. 2, 3, 5 c. 1, 4 d. all choices

2.67 a. 2, 5 b. 1, 2, 3 c. 2, 5 d. all choices

2.68 a. 1 b. 2 c. 2, 3, 4, 5 d. 4, 5

Atomic Structure And The Periodic Table *Chapter 3*

Solutions to Exercises

3.1 a. electron b. neutron c. proton d. proton

3.3 a. false b. false c. false d. true

3.5 a. true b. false c. false d. false

3.7 a. atomic number = 2, mass number = 4 b. atomic number = 4, mass number = 9
c. atomic number = 5, mass number = 9 d. atomic number = 28, mass number = 58

3.9 a. 8 protons, 8 neutrons, and 8 electrons b. 8 protons, 10 neutrons, and 8 electrons
c. 20 protons, 24 neutrons, and 20 electrons d. 100 protons, 157 neutrons, and 100 electrons

3.11 a. S, Cl, Ar, K b. Ar, K, Cl, S c. S, Cl, Ar, K d. S, Cl, K, Ar

3.13 a. 24, 29, 24, 53, 77 b. 101, 155, 101, 256, 357 c. 30, 37, 30, 67, 97 d. 20, 20, 20, 40, 60

3.15 $^{90}_{40}Zr$, $^{91}_{40}Zr$, $^{92}_{40}Zr$, $^{94}_{40}Zr$, $^{96}_{40}Zr$

3.17 a. false b. false c. true d. true

3.19 a. not the same b. same c. not the same d. same

3.21 8.91 x 12.0 amu = 106.92 amu = 107 amu (3 significant figures)

3.23 a. 0.0742 x 6.01 amu = 0.446 amu
0.9258 x 7.02 amu = <u>6.50 amu</u>
6.946 amu = 6.95 amu (answer is limited to the hundredths place)

b. 0.7899 x 23.99 amu = 18.95 amu
0.1000 x 24.99 amu = 2.499 amu
0.1101 x 25.98 amu = <u>2.860 amu</u>
24.309 amu = 24.31 amu (answer is limited to the hundredths place)

3.25 The exact number 12 applies only to ^{12}C. The number 12.011 is an average obtained by considering *all* naturally-occurring isotopes of C.

3.27 a. Ca b. Mo c. Li d. Sn

3.29 a. K, Rb b. P, As c. F, I d. Na, Cs

3.31 a. group b. periodic law c. periodic law d. group

3.33 a. alkali metal b. alkali metal c. noble gas d. alkaline earth metal e. halogen
f. noble gas g. alkaline earth metal h. halogen

3.35 a. no b. no c. yes d. yes

3.37 a. S b. P c. I d. Cl

3.39 a. orbital b. orbital c. shell d. shell

3.41 a. true b. true c. false d. true

3.43 a. 2 b. 2 c. 6 d. 18

3.45 a. $1s^22s^22p^2$ b. $1s^22s^22p^63s^1$ c. $1s^22s^22p^63s^23p^4$ d. $1s^22s^22p^63s^23p^6$

3.47 a. $1s^22s^22p^63s^23p^5$ b. $1s^22s^22p^63s^23p^64s^23d^{10}4p^65s^24d^7$ c. $1s^22s^22p^63s^23p^64s^2$
d. $1s^22s^22p^63s^23p^64s^23d^1$

3.49 a. (↑↓) (↑↓) (↑)(↑)( )

b. (↑↓) (↑↓) (↑↓)(↑↓)(↑↓) (↑↓)

c. (↑↓) (↑↓) (↑↓)(↑↓)(↑↓) (↑↓) (↑)(↑)(↑)

d. (↑↓) (↑↓) (↑↓)(↑↓)(↑↓) (↑↓) (↑↓)(↑↓)(↑↓) (↑↓) (↑)(↑)(↑)(↑)(↑)

3.51 a. 3 b. 0 c. 1 d. 5

3.53 a. no b. yes c. no d. yes

3.55 a. p^1 b. d^3 c. s^2 d. p^6

3.57 a. representative b. noble gas c. transition d. inner transition

3.59 a. noble gas b. representative c. transition d. representative

3.61 a. ${}^{44}_{20}Ca$ b. ${}^{211}_{86}Rn$ c. ${}^{110}_{47}Ag$ d. ${}^{9}_{4}Be$

3.62 a. same number of neutrons, 7 b. same number of neutrons, 10
c. same total number of subatomic particles, 54 d. same number of electrons, 17

3.63 a. ${}^{57}_{24}Cr$ b. ${}^{50}_{24}Cr$ c. ${}^{55}_{24}Cr$ d. ${}^{65}_{24}Cr$

3.64 9[(12 x 6) + (22 x 1) + (11 x 8)] = 1638 electrons

3.65 a. Be, Al b. Be, Al, Ag, Au (the metals) c. N, Be, Ar, Al, Ag, Au d. Ag, Au

3.66 the same

3.67 a. $1s^22s^22p^1$ (B) b. $1s^22s^22p^63s^23p^1$ (Al) c. $1s^22s^1$ (Li) d. $1s^22s^22p^63s^23p^3$ (P)

3.68 a. ${}_{8}O$ b. ${}_{10}Ne$ c. ${}_{30}Zn$ d. ${}_{12}Mg$

3.69 a. all choices b. 2, 3, 5, 6 c. 1, 4 d. (3, 4), (5, 6)

3.70 a. 1, 2 b. 3, 5 c. 1,2 d. (3, 4), (5, 6)

3.71 a. 2 b. 1, 4 c. 3, 4, 5, 6 d. 1, 2, 3, 4, 5

3.72 a. 1, 4, 6 b. 5, 6 c. 5, 6 d. 2

Chemical Bonding: The Ionic Bond Model Chapter 4

Solutions to Exercises

4.1 a. 2 b. 2 c. 3 d. 4

4.3 a. 1 b. 8 c. 2 d. 7

4.5 a. $1s^22s^22p^2$ b. $1s^22s^22p^5$ c. $1s^22s^22p^63s^2$ d. $1s^22s^22p^63s^23p^3$

4.7 a. Mg b. K c. P d. Kr

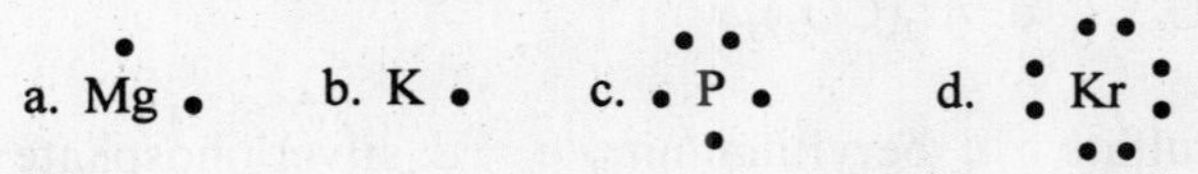

4.9 a. Li b. F c. Be d. N

4.11 a. O^{2-} b. Mg^{2+} c. F^- d. Al^{3+}

4.13 a. Ca^{2+} b. O^{2-} c. Na^+ d. Al^{3+}

4.15 a. 15p, 18e b. 7p, 10e c. 12p, 10e d. 3p, 2e

4.17 a. 2+ b. 3− c. 1+ d. 1−

4.19 a. 2 lost b. 1 gained c. 2 lost d. 2 gained

4.21 a. $1s^22s^22p^63s^23p^1$ b. $1s^22s^22p^6$

4.23 a. Be O b. Mg S c. K K K N d. Ca F F

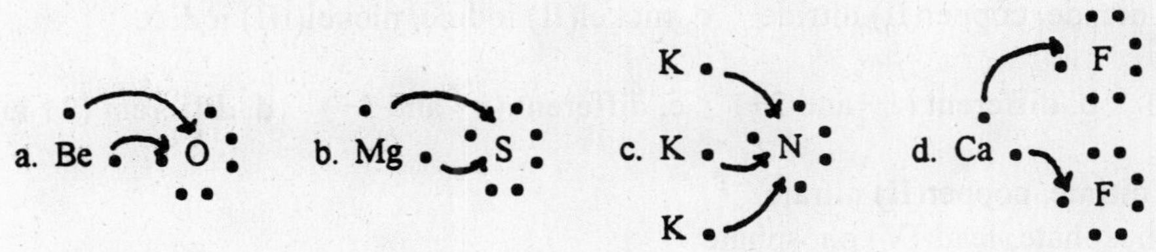

4.25 a. $BaCl_2$ b. $BaBr_2$ c. Ba_3N_2 d. BaO

4.27 a. MgF_2 b. BeF_2 c. LiF d. AlF_3

4.29 a. Na_2S b. CaI_2 c. Li_3N d. $AlBr_3$

4.31 a. potassium iodide b. beryllium oxide c. aluminum fluoride d. sodium phosphide

4.33 a. +1 b. +2 c. +4 d. +2

4.35 a. iron(II) oxide b. gold(III) oxide c. copper(II) sulfide d. cobalt(II) bromide

4.37 a. gold(I) chloride b. potassium chloride c. silver chloride d. copper(II) chloride

4.39 a. KBr b. Ag_2O c. BeF_2 d. Ba_3P_2

4.41 a. CoS b. Co_2S_3 c. SnI_4 d. Pb_3N_2

4.43 a. SO_4^{2-} b. ClO_3^- c. OH^- d. CN^-

4.45 a. PO_4^{3-} and HPO_4^{2-} b. NO_3^- and NO_2^- c. H_3O^+ and OH^- d. CrO_4^{2-} and $Cr_2O_7^{2-}$

4.47 a. $NaClO_4$ b. $Fe(OH)_3$ c. $Ba(NO_3)_2$ d. $Al_2(CO_3)_3$

4.49 a. magnesium carbonate b. zinc sulfate c. beryllium nitrate d. silver phosphate

4.51 a. iron(II) hydroxide b. copper(II) carbonate c. gold(I) cyanide
d. manganese(II) phosphate

4.53 a. $KHCO_3$ b. $Au_2(SO_4)_3$ c. $AgNO_3$ d. $Cu_3(PO_4)_2$

4.55 a. Na^+ b. F^- c. S^{2-} d. Ca^{2+}

4.56 a. XZ_2 b. X_2Z c. XZ d. XZ

4.57 a. S b. Mg c. P d. Al

4.58 a. K^+, Cl^- b. Ca^{2+}, S^{2-} c. Be^{2+}, two F^- d. two Al^{3+}, three S^{2-}

4.59 a. tin(IV) chloride, tin(II) chloride b. iron(II) sulfide, iron(III) sulfide
c. copper(I) nitride, copper(II) nitride d. nickel(II) iodide, nickel(III) iodide

4.60 a. same (3+) b. different (1+ and 2+) c. different (1+ and 3+) d. different (2+ and 1+)

4.61 a. copper(I) nitrate, copper(II) nitrate
b. lead(II) phosphate, lead(IV) phosphate
c. manganese(III) cyanide, manganese(II) cyanide
d. cobalt(II) chlorate, cobalt(III) chlorate

4.62 a. Na_2S b. Na_2SO_4 c. Na_2SO_3 d. $Na_2S_2O_3$

4.63 a. 1, 2 b. 4 c. (1, 3), (2, 3), (5, 6) d. (1, 5), (2, 5), (3, 6)

4.64 a. 1, 2 b. 1, 3 c. 1, 2, 4 d. 2, 5, 6

4.65 a. 4, 6 b. 1, 2, 6 c. 3 d. 3

4.66 a. 1, 2, 4, 5 b. 3, 4 c. 1, 3, 5, 6 d. 1, 5

Chemical Bonding: The Covalent Bond Model Chapter 5

Solutions to Exercises

5.1

```
       ••  ••            ••            ••  ••             ••  ••
a.  : Br : Br :   b. H : I :     c. : I : Br :     d.  : Br : F :
       ••  ••            ••            ••  ••             ••  ••
```

5.3 a. 2 b. 2 c. 0 d. 6

5.5

```
                                                               ••  ••
a. :N ≡ N:   b. H – C = C – H   c. H – C – H    d. H – O – O – H
                    |   |               ||             ••  ••
                    H   H             : O
                                        ••
```

5.7 a. NF_3 b. Cl_2O c. H_2S d. CH_4

5.9 a. N b. C c. N d. C

5.11 oxygen forms three bonds instead of the normal two

5.13

```
                                                    ••
                                                  : Br :
           ••             ••   ••   ••              ••             ••   ••   ••
a. H : P : H    b.  : Cl : P : Cl :     c.  : Br : Si : Br :   d. : F : O : F :
           ••             ••   ••   ••        ••    ••   ••        ••   ••   ••
           H                : Cl :                : Br :
                              ••                    ••
```

5.15

```
                                      ••
                                    : I :
                                      ••
     ••   ••   ••              ••           ••           ••    ••   ••                ••
a. : F : S : F :      b.  : I : C : I :        c. : Br : N : Br :     d. H : Se : H
     ••   ••   ••              ••           ••           ••    ••   ••                ••
                                      ••                    : Br :
                                    : I :                     ••
                                      ••
```

5.17

```
                                                  ••   ••        ••   ••
a. H : C :: C ::: C : H                b.  : F : N :: N : F :
       ••          ••                         ••                   ••
       H           H

       H                                      H
       ••                                     ••
c. H : C : C ::: N :                   d. H : C : C ::: C : H
       ••                                     ••
       H                                      H
```

5.19 a.
```
[  ••        ]–
[ :O : H     ]
[  ••        ]
```

b.
```
[      H      ]2–
[      ••     ]
[ H : Be : H  ]
[      ••     ]
[      H      ]
```

c.
```
[          ••          ]–
[        : Cl :        ]
[          ••          ]
[   ••            ••   ]
[ : Cl : Al : Cl :     ]
[   ••            ••   ]
[          ••          ]
[        : Cl :        ]
[          ••          ]
```

d.
```
[  ••          ••   ]–
[ :O : N : O :      ]
[  ••          ••   ]
[        ••         ]
[        ••         ]
[       :O          ]
[        ••         ]
```

5.21 a. Na^+ [:C ::: N:]$^-$

b. 3[K^+]
```
[          ••          ]3–
[        : O :         ]
[          ••          ]
[   ••            ••   ]
[ : O :   P   : O :    ]
[   ••            ••   ]
[          ••          ]
[        : O :         ]
[          ••          ]
```

5.23 a. angular b. angular c. angular d. linear

5.25 a. trigonal pyramidal b. trigonal planar c. tetrahedral d. tetrahedral

5.27 a. trigonal pyramidal b. tetrahedral c. angular d. angular

5.29 a. trigonal planar about each carbon atom
b. tetrahedral about carbon atom and angular about oxygen atom

5.31 a. Na, Mg, Al, P b. I, Br, Cl, F c. Al, P, S, O d. Ca, Mg, C, O

5.33 a. Br, Cl, N, O, F b. K, Na, Ca, Li c. Cl, N, O, F d. 0.5

5.35 a. $\overset{\delta+}{B} - \overset{\delta-}{N}$ b. $\overset{\delta+}{Cl} - \overset{\delta-}{F}$ c. $\overset{\delta-}{N} - \overset{\delta+}{C}$ d. $\overset{\delta-}{F} - \overset{\delta+}{O}$

5.37 a. H – Br, H – Cl, H – O
b. O – F, P – O, Al – O
c. Br – Br, H – Cl, B – N
d. P – N, S – O, Br – F

5.39 a. polar covalent b. ionic c. nonpolar covalent d. polar covalent

5.41 a. nonpolar b. polar c. polar d. polar

5.43 a. polar b. polar c. nonpolar d. polar

5.45 a. sulfur tetrafluoride b. tetraphosphorus hexoxide
c. chlorine dioxide d. hydrogen sulfide

5.47 a. ICl b. N_2O c. NCl_3 d. HBr

5.49 a. H_2O_2 b. CH_4 c. NH_3 d. PH_3

5.51 a. 26 b. 24 c. 14 d. 8

5.52 a. same, both single b. different, double and single c. same, both triple
d. same, both single

5.53 a. not enough electron dots b. not enough electron dots
c. improper placement of a correct number of electron dots d. too many electron dots

5.54 a. tetrahedral; tetrahedral b. tetrahedral; tetrahedral c. trigonal planar; angular
d. trigonal planar; trigonal planar

5.55 a. BrI b. SO_2 c. NF_3 d. H_3CF

5.56 DA, DC, CA, CB

5.57
```
        H                  H                      ..
        |                  |     ..             : F :
a. H —  C — H        H — C — F :                  |     ..
        |                  |     ..           H — C — F :
        H                  H                      |     ..
                                                  H

        ..                   ..
      : F :                : F :
        |     ..       ..    |     ..
  H —  C — F :        : F — C — F :
        |     ..       ..    |     ..
      : F :                : F :
        ..                   ..
```

b. all are tetrahedral
c. nonpolar, polar, polar, polar, nonpolar

5.58 a. sodium chloride b. bromine monochloride c. potassium sulfide
d. dichlorine monoxide

5.59 a. 4 b. 5 c. (1, 4), (2, 4), (3, 4) d. (1, 2), (1, 3), (2, 3)

5.60 a. 1, 3, 6 b. 1, 4, 5, 6 c. 1, 3, 6 d. (1, 6), (2, 4), (2, 5), (4, 5)

5.61 a. 2, 5 b. 1, 2, 6 c. 5 d. 1, 2, 4, 5

5.62 a. 2, 4 b. 5 c. 4 d. 1, 2, 4, 5

Chemical Calculations: Formula Masses, Moles, And Chemical Equations — Chapter 6

Solutions to Exercises

6.1 a. $[12(12.01) + 22(1.01) + 11(16.00)]$ amu = 342.34 amu
b. $[7(12.01) + 16(1.01)]$ amu = 100.23 amu
c. $[7(12.01) + 5(1.01) + 14.01 + 3(16.00) + 32.07]$ amu = 183.20 amu
d. $[2(14.01) + 8(1.01) + 32.07 + 4(16.00)]$ amu = 132.17 amu

6.3 a. $2.00 \text{ moles } H_2O \times \left(\frac{6.02 \times 10^{23} \text{ molecules } H_2O}{1 \text{ mole } H_2O}\right) = 1.20 \times 10^{24} \text{ molecules } H_2O$

b. $3.25 \text{ moles } CO_2 \times \left(\frac{6.02 \times 10^{23} \text{ molecules } CO_2}{1 \text{ mole } CO_2}\right) = 1.96 \times 10^{24} \text{ molecules } CO_2$

c. $0.356 \text{ mole CO} \times \left(\frac{6.02 \times 10^{23} \text{ molecules CO}}{1 \text{ mole CO}}\right) = 2.14 \times 10^{23} \text{ molecules CO}$

d. Avogadro's number = 6.02×10^{23} molecules CO

6.5 a. $0.542 \text{ mole Cu} \times \left(\frac{6.02 \times 10^{23} \text{ atoms Cu}}{1 \text{ mole Cu}}\right) = 3.26 \times 10^{23} \text{ atoms Cu}$

b. $0.542 \text{ mole Fe} \times \left(\frac{6.02 \times 10^{23} \text{ atoms Fe}}{1 \text{ mole Fe}}\right) = 3.26 \times 10^{23} \text{ atoms Fe}$

c. $0.542 \text{ mole } HNO_3 \times \left(\frac{6.02 \times 10^{23} \text{ molecules } HNO_3}{1 \text{ mole } HNO_3}\right) = 3.26 \times 10^{23} \text{ molecules } HNO_3$

d. $0.542 \text{ mole } C_9H_8O_4 \times \left(\frac{6.02 \times 10^{23} \text{ molecules } C_9H_8O_4}{1 \text{ mole } C_9H_8O_4}\right) = 3.26 \times 10^{23} \text{ molecules } C_9H_8O_4$

6.7 a. $[12.01 + 16.00]$ g = 28.0 g
b. $[12.01 + 2(16.00)]$ g = 44.0 g
c. $[22.99 + 35.45]$ g = 58.4 g
d. $[12(12.01) + 22(1.01) + 11(16.00)]$ g = 342 g

6.9 a. $0.034 \text{ mole Au} \times \left(\frac{196.97 \text{ g Au}}{1 \text{ mole Au}}\right) = 6.7 \text{ g Au}$

b. $0.034 \text{ mole Ag} \times \left(\frac{107.87 \text{ g Ag}}{1 \text{ mole Ag}}\right) = 3.7 \text{ g Ag}$

c. $3.00 \text{ moles O} \times \left(\frac{16.00 \text{ g O}}{1 \text{ mole O}}\right) = 48.0 \text{ g O}$

d. $3.00 \text{ moles } O_2 \times \left(\frac{32.00 \text{ g } O_2}{1 \text{ mole } O_2}\right) = 96.0 \text{ g } O_2$

6.11 a. $5.00 \text{ g CO} \times \left(\frac{1 \text{ mole CO}}{28.01 \text{ g CO}}\right) = 0.179 \text{ mole CO}$

b. $5.00 \text{ g } CO_2 \times \left(\frac{1 \text{ mole } CO_2}{44.01 \text{ g } CO_2}\right) = 0.114 \text{ mole } CO_2$

c. $5.00 \text{ g } B_4H_{10} \times \left(\frac{1 \text{ mole } B_4H_{10}}{53.34 \text{ g } B_4H_{10}}\right) = 0.0937 \text{ mole } B_4H_{10}$

d. $5.00 \text{ g U} \times \left(\frac{1 \text{ mole U}}{238 \text{ g U}}\right) = 0.0210 \text{ mole U}$

6.13 a. $\frac{2 \text{ moles H}}{1 \text{ mole } H_2SO_4}$, $\frac{1 \text{ mole } H_2SO_4}{2 \text{ moles H}}$, $\frac{1 \text{ mole S}}{1 \text{ mole } H_2SO_4}$, $\frac{1 \text{ mole } H_2SO_4}{1 \text{ mole S}}$,

$\frac{4 \text{ moles O}}{1 \text{ mole } H_2SO_4}$, $\frac{1 \text{ mole } H_2SO_4}{4 \text{ moles O}}$

b. $\frac{1 \text{ mole P}}{1 \text{ mole } POCl_3}$, $\frac{1 \text{ mole } POCl_3}{1 \text{ mole P}}$, $\frac{1 \text{ mole O}}{1 \text{ mole } POCl_3}$, $\frac{1 \text{ mole } POCl_3}{1 \text{ mole O}}$,

$\frac{3 \text{ moles Cl}}{1 \text{ mole } POCl_3}$, $\frac{1 \text{ mole } POCl_3}{3 \text{ moles Cl}}$

6.15 a. $2.00 \text{ moles } SO_2 \times \left(\frac{1 \text{ mole S}}{1 \text{ mole } SO_2}\right) = 2.00 \text{ moles S}$

$2.00 \text{ moles } SO_2 \times \left(\frac{2 \text{ moles O}}{1 \text{ mole } SO_2}\right) = 4.00 \text{ moles O}$

b. $2.00 \text{ moles } SO_3 \times \left(\frac{1 \text{ mole S}}{1 \text{ mole } SO_3}\right) = 2.00 \text{ moles S}$

$2.00 \text{ moles } SO_3 \times \left(\frac{3 \text{ moles O}}{1 \text{ mole } SO_3}\right) = 6.00 \text{ moles O}$

c. $3.00 \text{ moles } NH_3 \times \left(\frac{1 \text{ mole N}}{1 \text{ mole } NH_3}\right) = 3.00 \text{ moles N}$

$3.00 \text{ moles } NH_3 \times \left(\frac{3 \text{ moles H}}{1 \text{ mole } NH_3}\right) = 9.00 \text{ moles H}$

d. $3.00 \text{ moles } N_2H_4 \times \left(\frac{2 \text{ moles N}}{1 \text{ mole } N_2H_4}\right) = 6.00 \text{ moles N}$

$3.00 \text{ moles } N_2H_4 \times \left(\frac{4 \text{ moles H}}{1 \text{ mole } N_2H_4}\right) = 12.0 \text{ moles H}$

6.17 a. $10.0 \text{ g B} \times \left(\frac{1 \text{ mole B}}{10.81 \text{ g B}}\right) \times \left(\frac{6.02 \times 10^{23} \text{ atoms B}}{1 \text{ mole B}}\right) = 5.57 \times 10^{23} \text{ atoms B}$

b. $32.0 \text{ g Ca} \times \left(\frac{1 \text{ mole Ca}}{40.08 \text{ g Ca}}\right) \times \left(\frac{6.02 \times 10^{23} \text{ atoms Ca}}{1 \text{ mole Ca}}\right) = 4.81 \times 10^{23} \text{ atoms Ca}$

c. $2.0 \text{ g Ne} \times \left(\frac{1 \text{ mole Ne}}{20.18 \text{ g Ne}}\right) \times \left(\frac{6.02 \times 10^{23} \text{ atom Ne}}{1 \text{ mole Ne}}\right) = 6.0 \times 10^{22} \text{ atoms Ne}$

d. $7.0 \text{ g N} \times \left(\frac{1 \text{ mole N}}{14.01 \text{ g N}}\right) \times \left(\frac{6.02 \times 10^{23} \text{ atoms N}}{1 \text{ mole N}}\right) = 3.0 \times 10^{23} \text{ atoms N}$

6.19 a. 6.02×10^{23} atoms Cu x $\left(\frac{1 \text{ mole Cu}}{6.02 \times 10^{23} \text{ atoms Cu}}\right) \times \left(\frac{63.55 \text{ g Cu}}{1 \text{ mole C}}\right) = 63.6 \text{ g Cu}$

b. 3.01×10^{23} atoms Cu x $\left(\frac{1 \text{ mole Cu}}{6.02 \times 10^{23} \text{ atoms Cu}}\right) \times \left(\frac{63.55 \text{ g Cu}}{1 \text{ mole C}}\right) = 31.8 \text{ g Cu}$

c. 557 atoms Cu x $\left(\frac{1 \text{ mole Cu}}{6.02 \times 10^{23} \text{ atoms Cu}}\right) \times \left(\frac{63.55 \text{ g Cu}}{1 \text{ mole Cu}}\right) = 5.88 \times 10^{-20} \text{ g Cu}$

d. 1 atom Cu x $\left(\frac{1 \text{ mole Cu}}{6.02 \times 10^{23} \text{ atoms Cu}}\right) \times \left(\frac{63.55 \text{ g Cu}}{1 \text{ mole Cu}}\right) = 1.06 \times 10^{-22} \text{ g Cu}$

6.21 a. 10.0 g He x $\left(\frac{1 \text{ mole He}}{4.00 \text{ g He}}\right) = 2.50 \text{ moles He}$

b. 10.0 g N_2O x $\left(\frac{1 \text{ mole } N_2O}{44.02 \text{ g } N_2O}\right) = 0.227 \text{ mole } N_2O$

c. 4.0×10^{10} atoms P x $\left(\frac{1 \text{ mole P}}{6.02 \times 10^{23} \text{ atoms P}}\right) = 6.6 \times 10^{-14} \text{ mole P}$

d. 4.0×10^{10} atoms Be x $\left(\frac{1 \text{ mole Be}}{6.02 \times 10^{23} \text{ atoms Be}}\right) = 6.6 \times 10^{-14} \text{ mole Be}$

6.23 a. $10.0\text{ g } H_2SO_4 \times \left(\frac{1\text{ mole } H_2SO_4}{98.09\text{ g } H_2SO_4}\right) \times \left(\frac{1\text{ mole S}}{1\text{ mole } H_2SO_4}\right) \times \left(\frac{6.02 \times 10^{23}\text{ atoms S}}{1\text{ mole S}}\right)$
$= 6.14 \times 10^{22}$ atoms S

b. $20.0\text{ g } SO_3 \times \left(\frac{1\text{ mole } SO_3}{80.07\text{ g } SO_3}\right) \times \left(\frac{1\text{ mole S}}{1\text{ mole } SO_3}\right) \times \left(\frac{6.02 \times 10^{23}\text{ atoms S}}{1\text{ mole S}}\right)$
$= 1.50 \times 10^{23}$ atoms S

c. $30.0\text{ g } Al_2S_3 \times \left(\frac{1\text{ mole } Al_2S_3}{150.17\text{ g } Al_2S_3}\right) \times \left(\frac{3\text{ moles S}}{1\text{ mole } Al_2S_3}\right) \times \left(\frac{6.02 \times 10^{23}\text{ atoms S}}{1\text{ mole S}}\right)$
$= 3.61 \times 10^{23}$ atoms S

d. $2.00\text{ moles } S_2O \times \left(\frac{2\text{ moles S}}{1\text{ mole } S_2O}\right) \times \left(\frac{6.02 \times 10^{23}\text{ atoms S}}{1\text{ mole S}}\right) = 2.41 \times 10^{24}$ atoms S

6.25 a. 3.01×10^{23} molecules S_2O x

$\left(\frac{1\text{ mole } S_2O}{6.02 \times 10^{23}\text{ molecules } S_2O}\right) \times \left(\frac{2\text{ moles S}}{1\text{ mole } S_2O}\right) \times \left(\frac{32.07\text{ g S}}{1\text{ mole S}}\right) = 32.1\text{ g S}$

b. $3\text{ molecules } S_4N_4 \times \left(\frac{1\text{ mole } S_4N_4}{6.02 \times 10^{23}\text{ molecules } S_4N_4}\right) \times \left(\frac{4\text{ moles S}}{1\text{ mole } S_4N_4}\right) \times \left(\frac{32.07\text{ g S}}{1\text{ mole S}}\right)$

$= 6.39 \times 10^{-22}$ g S

c. $2.00\text{ moles } SO_2 \times \left(\frac{1\text{ mole S}}{1\text{ mole } SO_2}\right) \times \left(\frac{32.07\text{ g S}}{1\text{ mole S}}\right) = 64.1\text{ g S}$

d. $4.50\text{ moles } S_8 \times \left(\frac{8\text{ moles S}}{1\text{ mole } S_8}\right) \times \left(\frac{32.07\text{ g S}}{1\text{ mole S}}\right) = 1150\text{ g S}$

6.27 a. $2H_2 + O_2 \rightarrow 2H_2O$
b. $2NO + O_2 \rightarrow 2NO_2$
c. $2Fe_2O_3 \rightarrow 4Fe + 3\ O_2$
d. $NH_4NO_2 \rightarrow N_2 + 2H_2O$

6.29 a. $2Na + 2H_2O \rightarrow 2NaOH + H_2$
b. $2Na + ZnSO_4 \rightarrow Na_2SO_4 + Zn$
c. $2NaBr + Cl_2 \rightarrow 2NaCl + Br_2$
d. $2ZnS + 3\ O_2 \rightarrow 2ZnO + 2SO_2$

6.31 a. $CH_4 + 2\ O_2 \rightarrow CO_2 + 2H_2O$
b. $2C_6H_6 + 15\ O_2 \rightarrow 12CO_2 + 6H_2O$
c. $C_4H_8O_2 + 5\ O_2 \rightarrow 4CO_2 + 4H_2O$
d. $C_5H_{10}O + 7\ O_2 \rightarrow 5CO_2 + 5H_2O$

6.33 a. $3PbO + 2NH_3 \rightarrow 3Pb + N_2 + 3H_2O$ b. $2Fe(OH)_3 + 3H_2SO_4 \rightarrow Fe_2(SO_4)_3 + 6H_2O$

6.35 $\dfrac{2 \text{ moles } Ag_2CO_3}{4 \text{ moles } Ag}$, $\dfrac{2 \text{ moles } Ag_2CO_3}{2 \text{ moles } CO_2}$, $\dfrac{2 \text{ moles } Ag_2CO_3}{1 \text{ mole } O_2}$, $\dfrac{4 \text{ moles } Ag}{2 \text{ moles } CO_2}$, $\dfrac{4 \text{ moles } Ag}{1 \text{ mole } O_2}$,

$\dfrac{2 \text{ moles } CO_2}{1 \text{ mole } O_2}$, the other six are reciprocals of these six factors

6.37 a. $2.00 \text{ moles } C_7H_{16} \times \left(\dfrac{7 \text{ moles } CO_2}{1 \text{ mole } C_7H_{16}}\right) = 14.0 \text{ moles } CO_2$

b. $2.00 \text{ moles HCl} \times \left(\dfrac{1 \text{ mole } CO_2}{2 \text{ moles HCl}}\right) = 1.00 \text{ mole } CO_2$

c. $2.00 \text{ moles } Na_2SO_4 \times \left(\dfrac{2 \text{ moles } CO_2}{1 \text{ mole } Na_2SO_4}\right) = 4.00 \text{ moles } CO_2$

d. $2.00 \text{ moles } Fe_3O_4 \times \left(\dfrac{1 \text{ mole } CO_2}{1 \text{ mole } Fe_3O_4}\right) = 2.00 \text{ moles } CO_2$

6.39 a. $20.0 \text{ g } N_2 \times \left(\dfrac{1 \text{ mole } N_2}{28.02 \text{ g } N_2}\right) \times \left(\dfrac{4 \text{ moles } NH_3}{2 \text{ moles } N_2}\right) \times \left(\dfrac{17.04 \text{ g } NH_3}{1 \text{ mole } NH_3}\right) = 24.3 \text{ g } NH_3$

b. $20.0 \text{ g } N_2 \times \left(\dfrac{1 \text{ mole } N_2}{28.02 \text{ g } N_2}\right) \times \left(\dfrac{1 \text{ mole } (NH_4)_2Cr_2O_7}{1 \text{ mole } N_2}\right) \times \left(\dfrac{252.10 \text{ g } (NH_4)_2Cr_2O_7}{1 \text{ mole } (NH_4)_2Cr_2O_7}\right)$

$= 1.80 \times 10^2 \text{ g } (NH_4)_2Cr_2O_7$

c. $20.0 \text{ g } N_2 \times \left(\dfrac{1 \text{ mole } N_2}{28.02 \text{ g } N_2}\right) \times \left(\dfrac{1 \text{ mole } N_2H_4}{1 \text{ mole } N_2}\right) \times \left(\dfrac{32.06 \text{ g } N_2H_4}{1 \text{ mole } N_2H_4}\right) = 22.9 \text{ g } N_2H_4$

d. $20.0 \text{ g } N_2 \times \left(\dfrac{1 \text{ mole } N_2}{28.02 \text{ g } N_2}\right) \times \left(\dfrac{2 \text{ moles } NH_3}{1 \text{ mole } N_2}\right) \times \left(\dfrac{17.04 \text{ g } NH_3}{1 \text{ mole } NH_3}\right) = 24.3 \text{ g } NH_3$

6.41 $3.50 \text{ g } CO_2 \times \left(\dfrac{1 \text{ mole } CO_2}{44.01 \text{ g } CO_2}\right) \times \left(\dfrac{2 \text{ moles } O_2}{1 \text{ mole } CO_2}\right) \times \left(\dfrac{32.00 \text{ g } O_2}{1 \text{ mole } O_2}\right) = 5.09 \text{ g } O_2$

6.43 $25.0 \text{ g CO} \times \left(\frac{1 \text{ mole CO}}{28.01 \text{ g CO}}\right) \times \left(\frac{1 \text{ mole } O_2}{2 \text{ moles CO}}\right) \times \left(\frac{32.00 \text{ g } O_2}{1 \text{ mole } O_2}\right) = 14.3 \text{ g } O_2$

6.45 $10.0 \text{ g } SO_2 \times \left(\frac{1 \text{ mole } SO_2}{64.07 \text{ g } SO_2}\right) \times \left(\frac{2 \text{ moles } H_2O}{1 \text{ mole } SO_2}\right) \times \left(\frac{18.02 \text{ g } H_2O}{1 \text{ mole } H_2O}\right) = 5.63 \text{ g } H_2O$

6.47 $[3(12.01) + y(1.01) + 32.07] \text{ amu} = 76.18 \text{ amu}$
$1.01\,y = 8.08$
$y = 8.00$

6.48 a. 1.00 mole S_8; it contains eight times as many atoms as 1.00 mole S
b. 28.0 g Al; the molar mass of Al is 27.0 g, so 28.0 g of Al is more than 1.00 mole
c. 30.0 g Mg; the molar mass of Mg is 24.3 g, so 30.0 g of Mg is more than 1.00 mole and the molar mass of Si is 28.1 g so 28.1 g of Si is 1.00 mole
d. 6.02×10^{23} atoms He; this is 1.00 mole of He and 2.00 g of Na is less than 1.00 mole of Na since the molar mass of Na is 23.0 g

6.49 a. $1.000 \text{ g Si} \times \frac{1 \text{ mole Si}}{28.09 \text{ g Si}} \times \frac{1 \text{ mole } SiH_4}{1 \text{ mole Si}} = 0.03560 \text{ mole } SiH_4$

b. $1.000 \text{ g Si} \times \frac{1 \text{ mole Si}}{28.09 \text{ g Si}} \times \frac{1 \text{ mole } SiO_2}{1 \text{ mole Si}} \times \frac{60.09 \text{ g } SiO_2}{1 \text{ mole } SiO_2} = 2.139 \text{ g } SiO_2$

c. $1.000 \text{ g Si} \times \frac{1 \text{ mole Si}}{28.09 \text{ g Si}} \times \frac{1 \text{ mole } (CH_3)_3SiCl}{1 \text{ mole Si}} \times \frac{6.022 \times 10^{23} \text{ molecules } (CH_3)_3SiCl}{1 \text{ mole } (CH_3)_3SiCl}$
$= 2.144 \times 10^{22} \text{ molecules } (CH_3)_3SiCl$

d. $1.000 \text{ g Si} \times \frac{1 \text{ mole Si}}{28.09 \text{ g Si}} \times \frac{6.022 \times 10^{23} \text{ atoms Si}}{1 \text{ mole Si}} = 2.144 \times 10^{22} \text{ atoms Si}$

6.50 $2.10 \text{ moles Ar} \times \frac{1 \text{ mole Si}}{1 \text{ mole Ar}} \times \frac{28.09 \text{ g Si}}{1 \text{ mole Si}} = 59.0 \text{ g Si}$

6.51 Since the oxygen is balanced with 22 atoms on each side of the equation, the compound butyne contains only C and H.

$$2C_xH_y + 11\,O_2 \rightarrow 8CO_2 + 6H_2O$$

carbon balance: $2x = 8$ $x = 4$
hydrogen balance: $2y = 6(2)$ $y = 6$
butyne has the formula C_4H_6

6.52 $75.0 \text{ g } (NH_4)_2Cr_2O_7 \times \left(\frac{1 \text{ mole } (NH_4)_2Cr_2O_7}{252 \text{ g } (NH_4)_2Cr_2O_7}\right) \times \left(\frac{1 \text{ mole } N_2}{1 \text{ mole } (NH_4)_2Cr_2O_7}\right) \times \left(\frac{28.0 \text{ g } N_2}{1 \text{ mole } N_2}\right) = 8.33 \text{ g } N_2$

$75.0 \text{ g } (NH_4)_2Cr_2O_7 \times \left(\frac{1 \text{ mole } (NH_4)_2Cr_2O_7}{252 \text{ g } (NH_4)_2Cr_2O_7}\right) \times \left(\frac{1 \text{ mole } H_2O}{1 \text{ mole } (NH_4)_2Cr_2O_7}\right) \times \left(\frac{18.0 \text{ g } H_2O}{1 \text{ mole } H_2O}\right) = 21.4 \text{ g } H_2O$

$75.0 \text{ g } (NH_4)_2Cr_2O_7 \times \left(\frac{1 \text{ mole } (NH_4)_2Cr_2O_7}{252 \text{ g } (NH_4)_2Cr_2O_7}\right) \times \left(\frac{4 \text{ moles } Cr_2O_3}{1 \text{ mole } (NH_4)_2Cr_{2_7}}\right) \times \left(\frac{152 \text{ g } Cr_2O_3}{1 \text{ mole } Cr_2O_3}\right) = 45.2 \text{ g } Cr_2O_3$

6.53 $125 \text{ g } Ag_2S \times \left(\frac{1 \text{ mole } Ag_2S}{248 \text{ g } Ag_2S}\right) \times \left(\frac{2 \text{ moles Ag}}{1 \text{ mole } Ag_2S}\right) \times \left(\frac{108 \text{ g Ag}}{1 \text{ mole Ag}}\right) = 109 \text{ g Ag}$

$125 \text{ g } Ag_2S \times \left(\frac{1 \text{ mole } Ag_2S}{248 \text{ g } Ag_2S}\right) \times \left(\frac{1 \text{ mole S}}{1 \text{ mole } Ag_2S}\right) \times \left(\frac{32.1 \text{ g S}}{1 \text{ mole S}}\right) = 16.2 \text{ g S}$

6.54 $45.0 \text{ g } N_2 \times \left(\frac{1 \text{ mole } N_2}{28.0 \text{ g } N_2}\right) \times \left(\frac{3 \text{ moles Be}}{1 \text{ mole } N_2}\right) \times \left(\frac{9.01 \text{ g Be}}{1 \text{ mole Be}}\right) = 43.4 \text{ g Be}$

6.55 a. (2, 3), (5, 6) b. (2, 3), (5, 6) c. (5, 6) d. (5, 6)

6.56 a. 1, 3, 5, 6 b. 2, 6 c. 1 d. 6

6.57 a. 1, 3, 4 b. 2, 6 c. 2, 3, 4, 5, 6 d. 1, 4, 6

6.58 a. 2, 3 b. 2, 3, 6 c. 2, 3, 4, 6 d. 2, 3

Solutions to Exercises

7.1 a. velocity increases with increasing temperature b. potential energy
c. disruptive force magnitude increases with increasing temperature d. gaseous state

7.3 a. cohesive forces are strong enough that the space between particles changes very little with a temperature increase
b. particles of a gas are widely separated because disruptive forces are greater than cohesive forces

7.5 a. $735 \text{ mm Hg} \times \left(\frac{1 \text{ atm}}{760 \text{ mm Hg}}\right) = 0.967 \text{ atm}$

b. $0.530 \text{ atm} \times \left(\frac{760 \text{ mm Hg}}{1 \text{ atm}}\right) = 403 \text{ mm Hg}$

c. $0.530 \text{ atm} \times \left(\frac{760 \text{ torr}}{1 \text{ atm}}\right) = 403 \text{ torr}$

d. $12.0 \text{ psi} \times \left(\frac{1 \text{ atm}}{14.7 \text{ psi}}\right) = 0.816 \text{ atm}$

7.7 $P_2 = 3.0 \text{ atm} \times \left(\frac{6.0 \text{ L}}{2.5 \text{ L}}\right) = 7.2 \text{ atm}$

7.9 $V_2 = 3.00 \text{ L} \times \left(\frac{655 \text{ mm Hg}}{725 \text{ mm Hg}}\right) = 2.71 \text{ L}$

7.11 $V_2 = 2.73 \text{ L} \times \left(\frac{400 \text{ K}}{300 \text{ K}}\right) = 3.64 \text{ L}$

7.13 $T_2 = 298 \text{ K} \times \left(\frac{525 \text{ mL}}{375 \text{ mL}}\right) = 417 \text{ K}$

$417 \text{ K} - 273 = 144° \text{ C}$

7.15 a. $T_1 = T_2 \times \frac{P_1}{P_2} \times \frac{V_1}{V_2}$

b. $P_2 = P_1 \times \frac{V_1}{V_2} \times \frac{T_2}{T_1}$

c. $V_1 = V_2 \times \frac{P_2}{P_1} \times \frac{T_1}{T_2}$

7.17 a. $V_2 = 15.2 \text{ L} \times \left(\frac{1.35 \text{ atm}}{3.50 \text{ atm}}\right) \times \left(\frac{308 \text{ K}}{306 \text{ K}}\right) = 5.90 \text{ L}$

b. $P_2 = 1.35 \text{ atm} \times \left(\frac{15.2 \text{ L}}{10.0 \text{ L}}\right) \times \left(\frac{315 \text{ K}}{306 \text{ K}}\right) = 2.11 \text{ atm}$

c. $T_2 = 306 \text{ K} \times \left(\frac{7.00 \text{ atm}}{1.35 \text{ atm}}\right) \times \left(\frac{0.973 \text{ L}}{15.2 \text{ L}}\right) = 102 \text{ K}$

$102 \text{ K} - 273 = -171^\circ \text{C}$

d. $V_2 = 15{,}200 \text{ mL} \times \left(\frac{1.35 \text{ atm}}{6.70 \text{ atm}}\right) \times \left(\frac{370 \text{ K}}{306 \text{ K}}\right) = 3.70 \times 10^3 \text{ mL}$

7.19 $T = \frac{(5.23 \text{ atm})(5.23 \text{ L})}{(5.23 \text{ moles})\left(0.0821 \frac{\text{atm L}}{\text{mole K}}\right)} = 63.7 \text{ K}$

$63.7 \text{ K} - 273 = -209^\circ \text{C}$

7.21 $V = \frac{(0.100 \text{ mole})\left(0.0821 \frac{\text{atm L}}{\text{mole K}}\right)(273 \text{ K})}{(2.00 \text{ atm})} = 1.12 \text{ L}$

7.23 a. $V = \dfrac{(0.250 \text{ mole})\left(0.0821 \dfrac{\text{atm L}}{\text{mole K}}\right)(300 \text{ K})}{(1.50 \text{ atm})} = 4.11 \text{ L}$

b. $P = \dfrac{(0.250 \text{ mole})\left(0.0821 \dfrac{\text{atm L}}{\text{mole K}}\right)(308 \text{ K})}{(2.00 \text{ L})} = 3.16 \text{ atm}$

c. $T = \dfrac{(1.20 \text{ atm})(3.00 \text{ L})}{(0.250 \text{ mole})\left(0.0821 \dfrac{\text{atm L}}{\text{mole K}}\right)} = 175 \text{ K}$

$175 \text{ K} - 273 = -98° \text{ C}$

d. $V = \dfrac{(0.250 \text{ mole})\left(0.0821 \dfrac{\text{atm L}}{\text{mole K}}\right)(398 \text{ K})}{(0.500 \text{ atm})} = 16.3 \text{ L} = 16{,}300 \text{ mL}$

7.25 (1.50 – 0.75 – 0.33) atm = 0.42 atm

7.27 (623 – 125 – 175 – 225) mm Hg = 98 mm Hg

7.29 a. endothermic b. endothermic c. exothermic

7.31 a. no b. yes c. yes

7.33 a. boiling point b. vapor pressure c. boiling d. boiling point

7.35 a. different intermolecular force strengths result in different vapor pressures
b. vapor pressure becomes equal to atmospheric pressure at a lower temperature
c. evaporation is a cooling process because of the loss of the most energetic molecules
d. low-heat and high-heat boiling water have the same temperature and the same heat content

7.37 molecule must be polar

7.39 boiling point increases as intermolecular force strength increases

7.41 a. London b. hydrogen bonding c. dipole-dipole d. London

7.43 a. no b. yes c. yes d. no

7.45 four

7.47 a. $P_2 = 1.25 \text{ atm Hg} \times \left(\frac{575 \text{ mL}}{825 \text{ mL}}\right) = 0.871 \text{ atm}$

b. $T_2 = 398 \text{ K} \times \left(\frac{825 \text{ mL}}{575 \text{ mL}}\right) = 571 \text{ K} = 298^\circ \text{ C}$

c. $T_2 = 398 \text{ K} \times \left(\frac{50 \text{ atm}}{25 \text{ atm}}\right) \times \left(\frac{825 \text{ mL}}{575 \text{ mL}}\right) = 1142 \text{ K} = 869^\circ \text{ C}$

7.48 a. $T_2 = 297 \text{ K} \times \left(\frac{2P_1}{P_1}\right) \times \left(\frac{2V_1}{V_1}\right) = 1188 \text{ K} = 915^\circ \text{ C}$

b. $T_2 = 297 \text{ K} \times \left(\frac{\frac{1}{2}P_1}{P_1}\right) \times \left(\frac{\frac{1}{2}V_1}{V_1}\right) = 74.2 \text{ K} = -199^\circ \text{ C}$

c. $T_2 = 297 \text{ K} \times \left(\frac{2P_1}{P_1}\right) \times \left(\frac{\frac{1}{2}V_1}{V_1}\right) = 297 \text{ K} = 24^\circ \text{ C}$

d. $T_2 = 297 \text{ K} \times \left(\frac{\frac{1}{2}P_1}{P_1}\right) \times \left(\frac{3V_1}{V_1}\right) = 446 \text{ K} = 172^\circ \text{ C}$

7.49 a. $P = \left(\frac{(0.72 \text{ mole})\left(0.0821 \frac{\text{atm-L}}{\text{mole-K}}\right)(313 \text{ K})}{4.00 \text{ L}}\right) = 4.6 \text{ atm}$

b. $P = \left(\frac{(4.5 \text{ moles})\left(0.0821 \frac{\text{atm-L}}{\text{mole-K}}\right)(313 \text{ K})}{4.00 \text{ L}}\right) = 29 \text{ atm}$

c. $P = \left(\frac{\left(0.72 \text{ g} \times \frac{1 \text{ mole}}{32 \text{ g}}\right)\left(0.0821 \frac{\text{atm-L}}{\text{mole-K}}\right)(313 \text{ K})}{4.00 \text{ L}}\right) = 0.14 \text{ atm}$

d. $P = \left(\frac{\left(4.5 \text{ g} \times \frac{1 \text{ mole}}{32 \text{ g}}\right)\left(0.0821 \frac{\text{atm-L}}{\text{mole-K}}\right)(313 \text{ K})}{4.00 \text{ L}}\right) = 0.90 \text{ atm}$

7.50 $$\frac{PV}{RT} = \left(\frac{(1.00 \text{ atm})(2.00 \text{ L})}{\left(0.0821 \dfrac{\text{atm-L}}{\text{mole-K}}\right)(273 \text{ K})}\right) \times \left(\frac{6.02 \times 10^{23} \text{ molecules } H_2S}{1 \text{ mole } H_2S}\right)$$

$$= 5.37 \times 10^{22} \text{ molecules } H_2S$$

7.51 $$V = \left(\frac{(1.00 \text{ mole})\left(0.0821 \dfrac{\text{atm-L}}{\text{mole-K}}\right)(296 \text{ K})}{(0.983 \text{ atm})}\right) = 24.7 \text{ L}$$

7.52 $$n_1 = \left(\frac{(1.12 \text{ atm})(1.00 \text{ L})}{\left(0.0821 \dfrac{\text{atm-L}}{\text{mole-K}}\right)(299 \text{ K})}\right) = 0.0456 \text{ mole } N_2$$

$$n_2 = \left(\frac{(0.924 \text{ atm})(1.00 \text{ L})}{\left(0.0821 \dfrac{\text{atm-L}}{\text{mole-K}}\right)(297 \text{ K})}\right) = 0.0379 \text{ mole } N_2$$

$$n_1 - n_2 = (0.0456 - 0.0379) \text{ mole} = 0.0077 \text{ mole } N_2$$

$$0.0077 \text{ mole } N_2 \times \left(\frac{28 \text{ g } N_2}{1 \text{ mole } N_2}\right) = 0.22 \text{ g } N_2$$

7.53 a. PBr_3, lower vapor pressure b. PI_3, higher vapor pressure c. PI_3, higher vapor pressure

7.54 a. Br_2, larger mass b. H_2O, hydrogen bonding c. CO, dipole-dipole d. C_3H_8, larger size

7.55 a. 1, 3, 4, 6 b. 2, 5 c. 3, 6 d. 1, 2, 4, 5

7.56 a. 1, 2, 4 b. 2, 3 c. 2, 4, 5, 6 d. 1, 2, 3

7.57 a. 1, 2, 3 b. 5 c. 2, 3 d. 1, 2, 4, 6

7.58 a. all choices b. 3, 4, 5, 6 c. 4, 5 d. (4, 5)

Solutions **Chapter 8**

Solutions to Exercises

8.1 a. water b. ethyl alcohol

8.3 see Table 8.1

8.5 a. saturated b. unsaturated c. unsaturated d. saturated

8.7 a. dilute b. concentrated c. dilute d. concentrated

8.9 a. slightly soluble b. very soluble c. slightly soluble d. slightly soluble

8.11 a. all b. all c. CaS, $Ca(OH)_2$, $CaCl_2$ d. $NiSO_4$, $Ni(C_2H_3O_2)_2$

8.13 a. $\frac{6.50 \text{ g}}{91.5 \text{ g}} \times 100 = 7.10\% \text{ (m/m)}$

b. $\frac{2.31 \text{ g}}{37.3 \text{ g}} \times 100 = 6.19\% \text{ (m/m)}$

c. $\frac{12.5 \text{ g}}{138 \text{ g}} \times 100 = 9.06\% \text{ (m/m)}$

d. $\frac{0.0032 \text{ g}}{1.2 \text{ g}} \times 100 = 0.27\% \text{ (m/m)}$

8.15 a. $275 \text{ g } H_2O \times \left(\frac{1.30 \text{ g glucose}}{98.70 \text{ g } H_2O}\right) = 3.62 \text{ g glucose}$

b. $275 \text{ g } H_2O \times \left(\frac{5.00 \text{ g glucose}}{95.00 \text{ g } H_2O}\right) = 14.5 \text{ g glucose}$

c. $275 \text{ g } H_2O \times \left(\frac{20.0 \text{ g glucose}}{80.0 \text{ g } H_2O}\right) = 68.8 \text{ g glucose}$

d. $275 \text{ g } H_2O \times \left(\frac{31.0 \text{ g glucose}}{69.0 \text{ g } H_2O}\right) = 124 \text{ g glucose}$

8.17 $32.00 \text{ g solution} \times \left(\frac{2.000 \text{ g } K_2SO_4}{100.0 \text{ g solution}}\right) = 0.6400 \text{ g } K_2SO_4$

8.19 $20.0 \text{ g NaOH} \times \left(\frac{93.25 \text{ g } H_2O}{6.75 \text{ g NaOH}}\right) = 276 \text{ g } H_2O$

8.21 a. $\frac{20.0 \text{ mL}}{475 \text{ mL}} \times 100 = 4.21\% \text{ (v/v)}$

b. $\frac{4.00 \text{ mL}}{87.0 \text{ mL}} \times 100 = 4.60\% \text{ (v/v)}$

8.23 $\frac{22 \text{ mL}}{125 \text{ mL}} \times 100 = 18\% \text{ (v/v)}$

8.25 a. $\frac{5.0 \text{ g}}{250 \text{ mL}} \times 100 = 2.0\% \text{ (m/v)}$

b. $\frac{85 \text{ g}}{580 \text{ mL}} \times 100 = 15\% \text{ (m/v)}$

8.27 $25.0 \text{ mL solution} \times \left(\frac{2.00 \text{ g } Na_2CO_3}{100.0 \text{ mL solution}}\right) = 0.500 \text{ g } Na_2CO_3$

8.29 $50.0 \text{ mL solution} \times \left(\frac{7.50 \text{ g NaCl}}{100.0 \text{ mL solution}}\right) = 3.75 \text{ g NaCl}$

8.31 a. $\frac{3.0 \text{ moles}}{0.50 \text{ L}} = 6.0 \text{ M}$

b. $12.5 \text{ g } C_{12}H_{22}O_{11} \times \left(\frac{1 \text{ mole } C_{12}H_{22}O_{11}}{342.34 \text{ g } C_{12}H_{22}O_{11}}\right) = 0.0365 \text{ mole } C_{12}H_{22}O_{11}$

$\frac{0.0365 \text{ mole}}{0.0800 \text{ L}} = 0.456 \text{ M}$

c. $25.0 \text{ g NaCl} \times \left(\frac{1 \text{ mole NaCl}}{58.44 \text{ g NaCl}}\right) = 0.428 \text{ mole NaCl}$

$\frac{0.428 \text{ mole}}{1.250 \text{ L}} = 0.342 \text{ M}$

d. $\frac{0.00125 \text{ mole}}{0.00250 \text{ L}} = 0.500 \text{ M}$

8.33 a. $2.50 \text{ L solution} \times \left(\frac{3.00 \text{ moles HCl}}{1.00 \text{ L solution}}\right) \times \left(\frac{36.46 \text{ g HCl}}{1 \text{ mole HCl}}\right) = 273 \text{ g HCl}$

b. $0.0100 \text{ L solution} \times \left(\frac{0.500 \text{ mole KCl}}{1.00 \text{ L solution}}\right) \times \left(\frac{74.55 \text{ g KCl}}{1 \text{ mole KCl}}\right) = 0.373 \text{ g KCl}$

c. $0.875 \text{ L solution} \times \left(\frac{1.83 \text{ moles NaNO}_3}{1.00 \text{ L solution}}\right) \times \left(\frac{85.00 \text{ g NaNO}_3}{1 \text{ mole NaNO}_3}\right) = 136 \text{ g NaNO}_3$

d. $0.075 \text{ L solution} \times \left(\frac{12.0 \text{ moles H}_2\text{SO}_4}{1.00 \text{ L solution}}\right) \times \left(\frac{98.09 \text{ g H}_2\text{SO}_4}{1 \text{ mole H}_2\text{SO}_4}\right) = 88 \text{ g H}_2\text{SO}_4$

8.35 a. $1.00 \text{ g NaCl} \times \left(\frac{1 \text{ mole NaCl}}{58.44 \text{ g NaCl}}\right) \times \left(\frac{1000 \text{ mL solution}}{0.200 \text{ mole NaCl}}\right) = 85.6 \text{ mL solution}$

b. $2.00 \text{ g C}_6\text{H}_{12}\text{O}_6 \times \left(\frac{1 \text{ mole C}_6\text{H}_{12}\text{O}_6}{180.18 \text{ g C}_6\text{H}_{12}\text{O}_6}\right) \times \left(\frac{1000 \text{ mL solution}}{4.20 \text{ moles C}_6\text{H}_{12}\text{O}_6}\right) = 2.64 \text{ mL solution}$

c. $3.67 \text{ moles AgNO}_3 \times \left(\frac{1000 \text{ mL solution}}{0.400 \text{ mole AgNO}_3}\right) = 9180 \text{ mL solution}$

d. $0.0021 \text{ mole C}_{12}\text{H}_{22}\text{O}_{11} \times \left(\frac{1000 \text{ mL solution}}{8.7 \text{ moles C}_{12}\text{H}_{22}\text{O}_{11}}\right) = 0.24 \text{ mL solution}$

8.37 a. $0.220 \text{ M} \times \left(\frac{25.0 \text{ mL}}{30.0 \text{ mL}}\right) = 0.183 \text{ M}$

b. $0.220 \text{ M} \times \left(\frac{25.0 \text{ mL}}{75.0 \text{ mL}}\right) = 0.0733 \text{ M}$

c. $0.220 \text{ M} \times \left(\frac{25.0 \text{ mL}}{457 \text{ mL}}\right) = 0.0120 \text{ M}$

d. $0.220 \text{ M} \times \left(\frac{25.0 \text{ mL}}{2000 \text{ mL}}\right) = 0.00275 \text{ M}$

8.39 a. $50.0 \text{ mL} \times \left(\dfrac{3.00 \text{ M}}{0.100 \text{ M}}\right) = 1500 \text{ mL}$

$1500 \text{ mL} - 50.0 \text{ mL} = 1450 \text{ mL}$

b. $2.00 \text{ mL} \times \left(\dfrac{1.00 \text{ M}}{0.100 \text{ M}}\right) = 20.0 \text{ mL}$

$20.0 \text{ mL} - 2.00 \text{ mL} = 18.0 \text{ mL}$

c. $1450 \text{ mL} \times \left(\dfrac{6.00 \text{ M}}{0.100 \text{ M}}\right) = 87000 \text{ mL}$

$87000 \text{ mL} - 1450 \text{ mL} = 85600 \text{ mL}$

d. $75.0 \text{ mL} \times \left(\dfrac{0.110 \text{ M}}{0.100 \text{ M}}\right) = 82.5 \text{ mL}$

$82.5 \text{ mL} - 75.0 \text{ mL} = 7.5 \text{ mL}$

8.41 a. $5.0 \text{ M} \times \left(\dfrac{30.0 \text{ mL}}{50.0 \text{ mL}}\right) = 3.0 \text{ M}$

b. $5.0 \text{ M} \times \left(\dfrac{30.0 \text{ mL}}{50.0 \text{ mL}}\right) = 3.0 \text{ M}$

c. $7.5 \text{ M} \times \left(\dfrac{30.0 \text{ mL}}{50.0 \text{ mL}}\right) = 4.5 \text{ M}$

d. $2.0 \text{ M} \times \left(\dfrac{60.0 \text{ mL}}{80.0 \text{ mL}}\right) = 1.5 \text{ M}$

8.43 The presence of solute molecules decreases the ability of solvent molecules to escape

8.45 KCl produces ions, so more solute particles are present

8.47 a. same b. greater than c. less than d. greater than

8.49 2 to 1

8.51 a. swell b. remain the same c. swell d. shrink

8.53 a. hypotonic b. isotonic c. hypotonic d. hypertonic

8.55 a. K^+ and Cl^- leave the bag
b. K^+, Cl^- and glucose leave the bag

8.57 a. like (both soluble) b. unlike c. unlike d. like (both insoluble)

8.58 a. $134 \text{ g solution} \times \left(\frac{3.00 \text{ g } KNO_3}{100 \text{ g solution}}\right) = 4.02 \text{ g } KNO_3$

b. $75.02 \text{ g solution} \times \left(\frac{9.735 \text{ g NaOH}}{100.0 \text{ g solution}}\right) = 7.303 \text{ g NaOH}$

c. $1576 \text{ g solution} \times \left(\frac{0.800 \text{ g HI}}{100 \text{ g solution}}\right) = 12.6 \text{ g HI}$

d. $1.23 \text{ g solution} \times \left(\frac{12.0 \text{ g } NH_4Cl}{100 \text{ g solution}}\right) = 0.148 \text{ g } NH_4Cl$

8.59 $3.50 \text{ qt solution} \times \left(\frac{2.000 \text{ qt } H_2O}{100 \text{ qt solution}}\right) = 0.0700 \text{ qt } H_2O$

8.60 a. $60.0 \text{ g } NaNO_3 \times \left(\frac{1 \text{ mole } NaNO_3}{85.0 \text{ g } NaNO_3}\right) \times \left(\frac{1.0 \text{ L solution}}{0.10 \text{ mole } NaNO_3}\right) = 7.1 \text{ L solution}$

b. $60.0 \text{ g } HNO_3 \times \left(\frac{1 \text{ mole } HNO_3}{63.0 \text{ g } HNO_3}\right) \times \left(\frac{1.0 \text{ L solution}}{0.10 \text{ mole } HNO_3}\right) = 9.5 \text{ L solution}$

c. $60.0 \text{ g KOH} \times \left(\frac{1 \text{ mole KOH}}{56.1 \text{ g KOH}}\right) \times \left(\frac{1.0 \text{ L solution}}{0.10 \text{ mole KOH}}\right) = 11 \text{ L solution}$

d. $60.0 \text{ g LiCl} \times \left(\frac{1 \text{ mole LiCl}}{42.4 \text{ g LiCl}}\right) \times \left(\frac{1.0 \text{ L solution}}{0.10 \text{ mole LiCl}}\right) = 14 \text{ L solution}$

8.61 a. $0.400 \text{ M} \times \left(\frac{2212 \text{ mL}}{1875 \text{ mL}}\right) = 0.472 \text{ M}$

b. $0.400 \text{ M} \times \left(\frac{2212 \text{ mL}}{1250 \text{ mL}}\right) = 0.708 \text{ M}$

c. $0.400 \text{ M} \times \left(\frac{2212 \text{ mL}}{853 \text{ mL}}\right) = 1.04 \text{ M}$

d. $0.400 \text{ M} \times \left(\frac{2212 \text{ mL}}{553 \text{ mL}}\right) = 1.60 \text{ M}$

8.62 a. $\% \text{ (m/v)} = \left(\dfrac{3.75 \text{ g}}{10.0 \text{ mL solution}} \right) \times 100 = 37.5\%$

b. $$\left(\dfrac{3.75 \text{ g CsCl} \times \left(\dfrac{1 \text{ mole CsCl}}{168 \text{ g CsCl}} \right)}{10.0 \text{ mL solution} \times \left(\dfrac{10^{-3} \text{ L solution}}{1 \text{ mL solution}} \right)} \right) = 2.23 \text{ M}$$

8.63 a. 4.00 M; both solutions are 4.00 M so the mixture is 4.00 M

b. (4.00 M x 352 mL) + (2.00 M x 225 mL) = (y M)(577 mL)

y = 3.22 M

8.64 a. $$\text{OsM}_{\text{NaCl}} = \dfrac{8.00 \text{ g NaCl} \times \left(\dfrac{1 \text{ mole NaCl}}{58.5 \text{ g NaCl}} \right) \times \left(\dfrac{2 \text{ ions}}{1 \text{ NaCl}} \right)}{0.375 \text{ L solution}} = 0.729$$

$$\text{OsM}_{\text{NaBr}} = \dfrac{4.00 \text{ g NaBr} \times \left(\dfrac{1 \text{ mole NaBr}}{103 \text{ g NaBr}} \right) \times \left(\dfrac{2 \text{ ions}}{1 \text{ NaBr}} \right)}{0.155 \text{ L solution}} = 0.501$$

NaCl has the greater osmotic pressure

a. $$\text{OsM}_{\text{NaCl}} = \dfrac{6.00 \text{ g NaCl} \times \left(\dfrac{1 \text{ mole NaCl}}{58.5 \text{ g NaCl}} \right) \times \left(\dfrac{2 \text{ ions}}{1 \text{ NaCl}} \right)}{0.375 \text{ L solution}} = 0.547$$

$$\text{OsM}_{\text{MgCl}_2} = \dfrac{6.00 \text{ g MgCl}_2 \times \left(\dfrac{1 \text{ mole MgCl}_2}{95.2 \text{ g MgCl}_2} \right) \times \left(\dfrac{3 \text{ ions}}{1 \text{ MgCl}_2} \right)}{0.225 \text{ L solution}} = 0.840$$

$MgCl_2$ has the greater osmotic pressure

8.65 a. 1, 2, 4 b. 3, 5, 6 c. 1 d. (3, 6)

8.66 a. all choices b. all choices c. 2, 3, 5 d. (1, 4), (2, 3)

8.67 a. 2, 3 b. 4, 5 c. 4, 5 d. 1, 6

8.68 a. 1 b. 2, 3, 4, 5, 6 c. 4, 6 d. 3, 5

Chemical Reactions   *Chapter 9*

Solutions to Exercises

9.1 a. single replacement b. decomposition c. double replacement d. combination

9.3 a. +2 b. +6 c. 0 d. +5

9.5 a. +3 b. +4 c. +6 d. +6 e. +6 f. +6 g. +6 h. +5

9.7 a. +3 P, −1 F b. +1 Na, −2 O, +1 H c. +1 Na, +6 S, −2 O d. +4 C, −2 O

9.9 a. redox b. nonredox c. redox d. redox

9.11 a. H_2 oxidized, N_2 reduced b. KI oxidized, Cl_2 reduced c. Fe oxidized, Sb_2O_3 reduced
d. H_2SO_3 oxidized, HNO_3 reduced

9.13 a. N_2 oxidizing agent, H_2 reducing agent
b. Cl_2 oxidizing agent, KI reducing agent
c. Sb_2O_3 oxidizing agent, Fe reducing agent
d. HNO_3 oxidizing agent, H_2SO_3 reducing agent

9.15 Collision theory is a set of statements that give the conditions that must be met before a chemical reaction will take place. The statements of collision theory are:
1. Reactant particles must collide with one another before any reaction can occur.
2. Colliding particles must possess a certain minimum total amount of energy, called the activation energy, if the collision is to be effective.
3. Colliding particles must come together in the proper orientation unless the particles involved are single atoms or small, symmetrical molecules.

9.17 a. exothermic b. endothermic c. endothermic d. exothermic

9.19

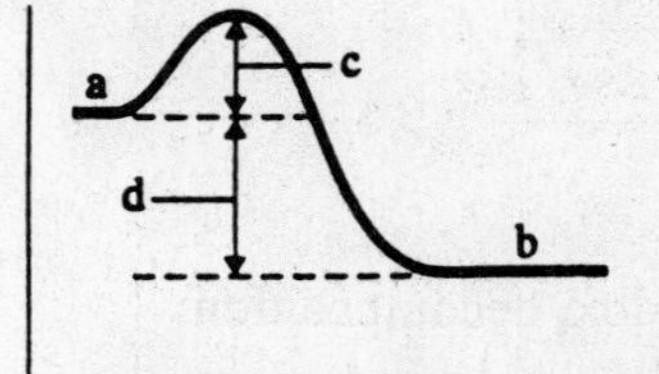

9.21 a. as temperature increases so does the number of collisions per second
b. a catalyst lowers the activation energy

9.23 the concentration of O_2 has increased from 21% to 100%

9.25

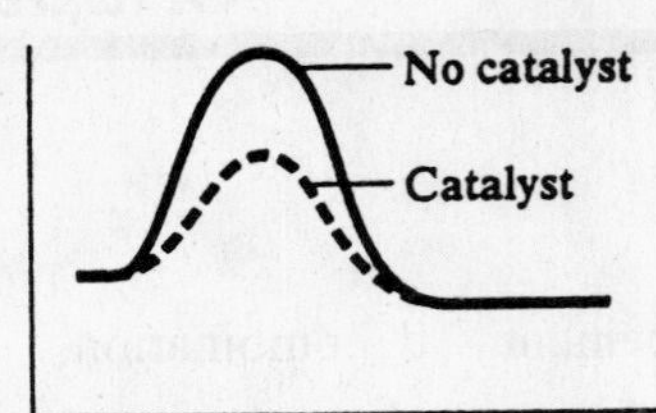

9.27 a. 1 b. 3 c. 4 d. 3

9.29 rate of forward reaction is equal to rate of reverse reaction

9.31

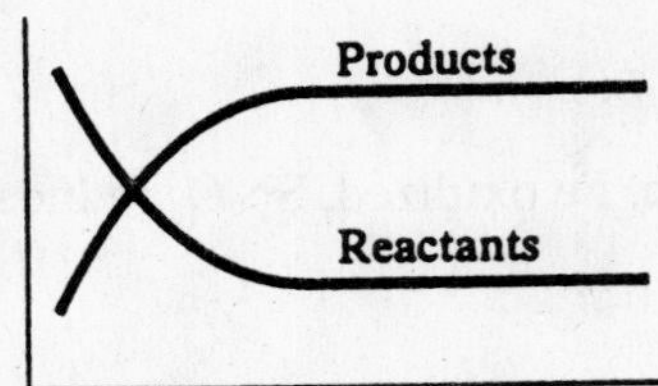

9.33 a. $K_{eq} = \dfrac{[NO_2]^2}{[N_2O_4]}$ b. $K_{eq} = \dfrac{[Cl_2][CO]}{[COCl_2]}$ c. $K_{eq} = \dfrac{[H_2S]^2[CH_4]}{[H_2]^4[CS_2]}$ d. $K_{eq} = \dfrac{[SO_3]^2}{[O_2][SO_2]^2}$

9.35 a. $K_{eq} = [SO_3]$ b. $K_{eq} = \dfrac{1}{[Cl_2]}$ c. $K_{eq} = \dfrac{[NaCl]^2}{[Na_2SO_4][BaCl_2]}$ d. $K_{eq} = [O_2]$

9.37 $K_{eq} = \dfrac{[0.0032]^2}{[0.213]} = 4.8 \times 10^{-5}$

9.39 a. more products than reactants b. essentially all reactants c. significant amounts of both reactants and products d. significant amounts of both reactants and products

9.41 a. right b. left c. left d. right

9.43 a. left b. left c. left d. left

9.45 a. shifts left b. no effect c. shifts right d. no effect

9.47 a. redox, single replacement b. redox, combination c. redox, decomposition d. nonredox, double replacement

9.48 a. redox b. redox c. redox d. can't tell

9.49 a. gain b. reduction c. decrease d. increase

9.50 a. decrease b. increase c. increase d. decrease

9.51 a. no b. no c. yes d. no

9.52 $CH_{4\,(g)} + 2H_2S_{(g)} \rightarrow CS_{2\,(g)} + 4H_{2\,(g)}$

9.53 a. yes b. yes c. no d. yes

9.54 a. no effect b. right c. right d. right

9.55 a. 1, 2, 4 b. 1, 2, 3, 4 c. 4 d. 3

9.56 a. 1, 2, 3, 4 b. 5 c. 6 d. 5

9.57 a. 1, 2, 3, 5, 6 b. 1, 2, 5 c. 3 d. 1

9.58 a. 4 b. 6 c. 1, 3, 6 d. 2, 5

Solutions to Exercises

10.1 a. $HI \xrightarrow{H_2O} H^+ + I^-$ b. $HClO \xrightarrow{H_2O} H^+ + ClO^-$

c. $LiOH \xrightarrow{H_2O} Li^+ + OH^-$ d. $CsOH \xrightarrow{H_2O} Cs^+ + OH^-$

10.3 a. acid b. base c. acid d. acid b. $HClO_4 + NH_3 \rightarrow NH_4^+ + ClO_4^-$

10.5 a. $HClO + H_2O \rightarrow H_3O^+ + ClO^-$

c. $H_3O^+ + OH^- \rightarrow H_2O + H_2O$ d. $H_3O^+ + NH_2^- \rightarrow H_2O + NH_3$

10.7 a. HSO_3^- b. HCN c. $C_2O_4^{2-}$ d. $H_2PO_4^-$

10.9 a. $HS^- + H_2O \rightarrow H_3O^+ + S^{2-}$; $HS^- + H_2O \rightarrow H_2S + OH^-$

b. $HPO_4^{2-} + H_2O \rightarrow H_3O^+ + PO_4^{3-}$; $HPO_4^{2-} + H_2O \rightarrow H_2PO_4^- + OH^-$

c. $NH_3 + H_2O \rightarrow H_3O^+ + NH_2^-$; $NH_3 + H_2O \rightarrow NH_4^+ + OH^-$

d. $OH^- + H_2O \rightarrow H_3O^+ + O^{2-}$; $OH^- + H_2O \rightarrow H_2O + OH^-$

10.11 a. monoprotic b. diprotic c. monoprotic d. diprotic

10.13 $H_3C_6H_5O_7 + H_2O \rightarrow H_3O^+ + H_2C_6H_5O_7^-$;

$H_2C_6H_5O_7^- + H_2O \rightarrow H_3O^+ + HC_6H_5O_7^{2-}$;

$HC_6H_5O_7^{2-} + H_2O \rightarrow H_3O^+ + C_6H_5O_7^{3-}$

10.15 to show that it is a monoprotic acid

10.17 monoprotic; only one H atom is involved in a polar bond

10.19 a. strong b. weak c. weak d. strong

10.21 a. $K_a = \frac{[H^+][F^-]}{[HF]}$ b. $K_a = \frac{[H^+][C_2H_3O_2^-]}{[HC_2H_3O_2]}$

10.23 a. $K_b = \frac{[NH_4^+][OH^-]}{[NH_3]}$ b. $K_b = \frac{[C_6H_5NH_3^+][OH^-]}{[C_6H_5NH_2]}$

10.25 a. H_3PO_4 b. HF c. H_2CO_3 d. HNO_2

10.27 $[H_3O^+] = [A^-] = (0.12)(0.00300\ M) = 0.00036\ M$

$[HA] = (0.00300 - 0.00036)\ M = 0.00264\ M$

$$K_a = \frac{[0.00036][0.00036]}{[0.00264]} = 4.9 \times 10^{-5}$$

10.29 a. acid b. salt c. salt d. base e. salt f. base g. acid h. acid

10.31 a. $Ba(NO_3)_2 \xrightarrow{H_2O} Ba^{2+} + 2NO_3^-$ b. $Na_2SO_4 \xrightarrow{H_2O} 2Na^+ + SO_4^{2-}$

c. $CaBr_2 \xrightarrow{H_2O} Ca^{2+} + 2Br^-$ d. $K_2CO_3 \xrightarrow{H_2O} 2K^+ + CO_3^{2-}$

10.33 a. $HCl + NaOH \rightarrow NaCl + H_2O$ b. $HNO_3 + KOH \rightarrow KNO_3 + H_2O$

c. $H_2SO_4 + 2LiOH \rightarrow Li_2SO_4 + 2H_2O$ d. $2H_3PO_4 + 3Ba(OH)_2 \rightarrow Ba_3(PO_4)_2 + 6H_2O$

10.35 a. $H_2SO_4 + 2LiOH \rightarrow Li_2SO_4 + 2H_2O$ b. $HCl + NaOH \rightarrow NaCl + H_2O$

c. $HNO_3 + KOH \rightarrow KNO_3 + H_2O$ d. $2H_3PO_4 + 3Ba(OH)_2 \rightarrow Ba_3(PO_4)_2 + 6H_2O$

10.37 a. $\frac{1.00 \times 10^{-14}\ M}{3.00 \times 10^{-3}\ M} = 3.3 \times 10^{-12}\ M$

b. $\frac{1.00 \times 10^{-14}\ M}{6.7 \times 10^{-6}\ M} = 1.5 \times 10^{-9}\ M$

c. $\frac{1.00 \times 10^{-14}\ M}{9.1 \times 10^{-8}\ M} = 1.1 \times 10^{-7}\ M$

d. $\frac{1.00 \times 10^{-14}\ M}{1.2 \times 10^{-11}\ M} = 8.3 \times 10^{-4}\ M$

10.39 a. acidic b. basic c. basic d. acidic

10.41 a. 4.00 b. 11.00 c. $[H_3O^+] = 1.0 \times 10^{-11}$; 11.00 d. $[H_3O^+] = 1.0 \times 10^{-7}$; 7.00

10.43 a. 7.68 b. 7.40 c. $[H_3O^+] = 1.4 \times 10^{-4}$; 3.85 d. $[H_3O^+] = 1.4 \times 10^{-12}$; 11.85

10.45 a. 1×10^{-2} M b. 1×10^{-6} M c. 1×10^{-8} M d. 1×10^{-10} M

10.47 a. 2.1×10^{-4} M b. 8.1×10^{-6} M c. 4.5×10^{-8} M d. 3.5×10^{-13} M

10.49 a. 3.14 b. 6.37 c. 7.21 d. 1.82

10.51 acid B

10.53 a. strong acid – strong base salt b. weak acid – strong base salt
c. strong acid – weak base salt d. strong acid – strong base salt

10.55 a. none b. $C_2H_3O_2^-$ c. NH_4^+ d. none

10.57 a. neutral b. basic c. acidic d. neutral

10.59 a. no b. yes c. no d. yes

10.61 a. $F^- + H_3O^+ \rightarrow HF + H_2O$ b. $H_2CO_3 + OH^- \rightarrow HCO_3^- + H_2O$

c. $CO_3^{2-} + H_3O^+ \rightarrow HCO_3^- + H_2O$ d. $H_3PO_4 + OH^- \rightarrow H_2PO_4^- + H_2O$

10.63 $\mathrm{pH} = 6.72 + \log \left[\frac{0.500\ \mathrm{M}}{0.230\ \mathrm{M}}\right] = 7.06$

10.65 $\mathrm{pH} = -\log [6.8 \times 10^{-6}] + \log \left[\frac{0.150\ \mathrm{M}}{0.150\ \mathrm{M}}\right] = 5.17$

10.67 a. weak b. strong c. strong d. strong

10.69 a. $5.00 \text{ mL } HNO_3 \times \left(\dfrac{0.250 \text{ mole } HNO_3}{1000 \text{ mL } HNO_3}\right) \times \left(\dfrac{1 \text{ mole NaOH}}{1 \text{ mole } HNO_3}\right) = 0.00125 \text{ mole NaOH}$

$\dfrac{0.00125 \text{ mole NaOH}}{0.0250 \text{ L solution}} = 0.0500 \text{ M}$

b. $20.00 \text{ mL } H_2SO_4 \times \left(\dfrac{0.500 \text{ mole } H_2SO_4}{1000 \text{ mL } H_2SO_4}\right) \times \left(\dfrac{2 \text{ moles NaOH}}{1 \text{ mole } H_2SO_4}\right) = 0.0200 \text{ mole NaOH}$

$\dfrac{0.0200 \text{ mole NaOH}}{0.0250 \text{ L solution}} = 0.800 \text{ M}$

c. $23.76 \text{ mL HCl} \times \left(\dfrac{1.00 \text{ mole HCl}}{1000 \text{ mL HCl}}\right) \times \left(\dfrac{1 \text{ mole NaOH}}{1 \text{ mole HCl}}\right) = 0.0238 \text{ mole NaOH}$

$\dfrac{0.0238 \text{ mole NaOH}}{0.0250 \text{ L solution}} = 0.952 \text{ M}$

d. $10.00 \text{ mL } H_3PO_4 \times \left(\dfrac{0.100 \text{ mole } H_3PO_4}{1000 \text{ mL } H_3PO_4}\right) \times \left(\dfrac{3 \text{ moles NaOH}}{1 \text{ mole } H_3PO_4}\right) = 0.00300 \text{ mole NaOH}$

$\dfrac{0.00300 \text{ mole NaOH}}{0.0250 \text{ L solution}} = 0.120 \text{ M}$

10.71 a. yes b. no c. no d. yes

10.72 a. no b. yes, both strong c. no d. yes, both weak

10.73 a. B b. A

10.74 7.0; NaCl does not affect the pH and the equal amounts of NaOH and HCl of the same concentration neutralize each other to produce more NaCl.

10.75 HCl (strong acid), HCN (weak acid), KCl (salt that does not hydrolyze), NaOH (strong base)

10.76 HCN/CN^-

10.77 CN^- ion undergoes hydrolysis to a greater extent than NH_4^+ ion resulting in a basic solution; NH_4^+ ion and $C_2H_3O_2^-$ ion hydrolyze to an equal extent resulting in a neutral solution.

10.78 $[H^+] = 1.0 \times 10^{-10}$

$$[OH^-] = \frac{1.0 \times 10^{-14}}{1.0 \times 10^{-10}} = 1.0 \times 10^{-4}$$

$$0.875 \text{ L solution} \times \left(\frac{1.0 \times 10^{-4} \text{ mole NaOH}}{1 \text{ L solution}}\right) \times \left(\frac{40.0 \text{ g NaOH}}{1 \text{ mole NaOH}}\right) = 0.0035 \text{ g NaOH}$$

10.79 a. 2, 3, 5, 6 b. 1, 2, 4, 5 c. 2, 5 d. (2, 5), (2, 6)

10.80 a. 1 b. 2, 4, 6 c. 2, 6 d. 1, 2, 3, 5

10.81 a. 3, 6 b. 4, 5 c. (1, 2), (3, 6), (4, 5) d. (1, 2), (3, 6), (4, 5)

10.82 a. 1, 4 b. 2, 5 c. 2, 5 d. 1, 4

Nuclear Chemistry — Chapter 11

Solutions to Exercises

11.1 a. $^{10}_{4}Be$, Be–10 b. $^{25}_{11}Na$, Na–25 c. $^{96}_{41}Nb$, Nb–96 d. $^{257}_{103}Lr$, Lr–257

11.3 a. $^{14}_{7}N$ b. $^{197}_{79}Au$ c. tin–121 d. boron–10

11.5 a. $^{4}_{2}\alpha$ b. $^{0}_{-1}\beta$ c. $^{0}_{0}\gamma$

11.7 2 protons and 2 neutrons

11.9 a. $^{200}_{84}Po \rightarrow {}^{4}_{2}\alpha + {}^{196}_{82}Pb$

b. $^{240}_{96}Cm \rightarrow {}^{4}_{2}\alpha + {}^{236}_{94}Pu$

c. $^{244}_{96}Cm \rightarrow {}^{4}_{2}\alpha + {}^{240}_{94}Pu$

d. $^{238}_{92}U \rightarrow {}^{4}_{2}\alpha + {}^{234}_{90}Th$

11.11 a. $^{10}_{4}Be \rightarrow {}^{0}_{-1}\beta + {}^{10}_{5}B$

b. $^{14}_{6}C \rightarrow {}^{0}_{-1}\beta + {}^{14}_{7}N$

c. $^{21}_{9}F \rightarrow {}^{0}_{-1}\beta + {}^{21}_{10}Ne$

d. $^{25}_{11}Na \rightarrow {}^{0}_{-1}\beta + {}^{25}_{12}Mg$

11.13 $A \rightarrow A - 4$, $Z \rightarrow Z - 2$

11.15 a. $^{0}_{-1}\beta$ b. $^{28}_{12}Mg$ c. $^{4}_{2}\alpha$ d. $^{200}_{80}Hg$

11.17 a. $^{4}_{2}\alpha$ b. $^{0}_{-1}\beta$

11.19 a. $\frac{1}{2^2} = \frac{1}{4}$ b. $\frac{1}{2^6} = \frac{1}{64}$ c. $\frac{1}{2^3} = \frac{1}{8}$ d. $\frac{1}{2^6} = \frac{1}{64}$

11.21 a. $\dfrac{1}{16} = \dfrac{1}{2^4}$; $\dfrac{5.4 \text{ days}}{4} = 1.4 \text{ days}$

b. $\dfrac{1}{64} = \dfrac{1}{2^6}$; $\dfrac{5.4 \text{ days}}{6} = 0.90 \text{ day}$

c. $\dfrac{1}{256} = \dfrac{1}{2^8}$; $\dfrac{5.4 \text{ days}}{8} = 0.68 \text{ day}$

d. $\dfrac{1}{1024} = \dfrac{1}{2^{10}}$; $\dfrac{5.4 \text{ days}}{10} = 0.54 \text{ day}$

11.23 $\dfrac{60.0 \text{ hr}}{15.0 \text{ hr}} = 4$; $\dfrac{1}{2^4} = \dfrac{1}{16}$; $\dfrac{1}{16} \times 4.00 \text{ g} = 0.250 \text{ g}$

11.25 2000

11.27 92, uranium

11.29 a. ${}^{4}_{2}\alpha$ b. ${}^{25}_{12}\text{Mg}$ c. ${}^{4}_{2}\alpha$ d. ${}^{1}_{1}\text{p}$

11.31 stable; termination of a decay series requires a stable isotope

11.33 ${}^{232}_{90}\text{Th} \rightarrow {}^{4}_{2}\alpha + {}^{228}_{88}\text{Ra}$

${}^{228}_{88}\text{Ra} \rightarrow {}^{0}_{-1}\beta + {}^{228}_{89}\text{Ac}$

${}^{228}_{89}\text{Ac} \rightarrow {}^{0}_{-1}\beta + {}^{228}_{90}\text{Th}$

${}^{228}_{90}\text{Th} \rightarrow {}^{4}_{2}\alpha + {}^{224}_{88}\text{Ra}$

11.35 alpha is stopped; beta and gamma go through

11.37 alpha, 0.1 the speed of light; beta, up to 0.9 the speed of light; gamma, the speed of light

11.39 a. no detectable effects b. nausea, fatigue, lowered blood cell count

11.41 19% human-made, 81% natural sources

11.43 to monitor the extent of radiation exposure

11.45 so the radiation can be detected externally (outside the body)

11.47 a. bone tumors b. circulatory problems c. iron metabolism d. intercellular space problems

11.49 they usually are α or β emitters instead of γ emitters

11.51 a. fusion b. fusion c. both d. fission

11.53 a. fusion b. fission c. neither d. neither

11.55 a. $^{206}_{80}Hg \rightarrow ^{0}_{-1}\beta + ^{206}_{81}Tl$

b. $^{109}_{46}Pd \rightarrow ^{0}_{-1}\beta + ^{109}_{47}Ag$

c. $^{245}_{96}Cm \rightarrow ^{4}_{2}\alpha + ^{241}_{94}Pu$

d. $^{249}_{100}Fm \rightarrow ^{4}_{2}\alpha + ^{245}_{98}Cf$

11.56 a. $\frac{1}{8}$ left $= \frac{1}{2^3}$

$$3 \text{ half-lives} \times \frac{18 \text{ hrs}}{1 \text{ half-life}} = 54 \text{ hrs}$$

b. $\frac{1}{32}$ left $= \frac{1}{2^5}$

$$5 \text{ half-lives} \times \frac{18 \text{ hrs}}{1 \text{ half-life}} = 90 \text{ hrs}$$

c. $\frac{1}{64}$ left $= \frac{1}{2^6}$

$$6 \text{ half-lives} \times \frac{18 \text{ hrs}}{1 \text{ half-life}} = 108 \text{ hrs}$$

d. $\frac{1}{128}$ left $= \frac{1}{2^7}$

$$7 \text{ half-lives} \times \frac{18 \text{ hrs}}{1 \text{ half-life}} = 126 \text{ hrs}$$

11.57 a. $^{243}_{94}Pu + ^{4}_{2}\alpha \rightarrow ^{246}_{96}Cm + ^{1}_{0}n$

b. $^{246}_{96}Cm + ^{12}_{6}C \rightarrow ^{254}_{102}No + 4^{1}_{0}n$

c. $^{27}_{13}Al + ^{4}_{2}\alpha \rightarrow ^{1}_{0}n + ^{30}_{15}P$

d. $^{23}_{11}Na + ^{2}_{1}H \rightarrow ^{21}_{10}Ne + ^{4}_{2}\alpha$

11.58 $^{142}_{60}Nd + ^{1}_{0}n \rightarrow ^{143}_{61}Pm + ^{0}_{-1}\beta$

11.59 fourteen elements

11.60 four neutrons

11.61 A = 0 (negligible amount), B = 0 (negligible amount), C = 63 atoms, D = 937 atoms

11.62 ^{228}Ac, ^{228}Th, ^{224}Ac

11.63 a. 3, 4 b. 2, 3, 6 c. 3, 5 d. 1, 5

11.64 a. 1, 6 b. 2, 5 c. 4 d. 3

11.65 a. 1 b. 2, 5 c. 3, 6 d. 4

11.66 a. 1 b. 2, 4 c. 5, 6 d. 3

Saturated Hydrocarbons — Chapter 12

Solutions to Exercises

12.1 a. saturated b. unsaturated c. unsaturated d. unsaturated

12.3 a. 18 b. 4 c. 13 d. 22

12.5 a. $CH_3-CH_2-CH_2-CH_3$

b. $CH_3-CH_2-CH(CH_3)-CH_2-CH_3$

c. $CH_3-CH_2-CH(CH_3)-CH_2-CH(CH_3)-CH_3$

d. $CH_3-CH_2-CH(CH_2-CH_3)-CH_2-CH_3$

12.7 a. $CH_3-CH(CH_3)-CH_2-CH_3$

b. $CH_3-CH(CH_3)-CH(CH_3)-CH(CH_3)-CH_2-CH_3$

c. $CH_3-CH_2-CH_2-CH_2-CH_2-CH_3$

d. $CH_3-C(CH_3)_2-CH_2-CH_3$

12.9 a. $H-C-C-C-C-C-H$ (each C bearing two H)

b. $H-C-C-C-C-C-C-C-C-H$ (each C bearing two H)

c. $CH_3-(CH_2)_8-CH_3$ d. C_6H_{14}

12.11 a. different compounds that are not structural isomers
b. different compounds that are structural isomers
c. different conformations of the same molecule
d. different compounds that are structural isomers

12.13 a. 7 b. 8 c. 8 d. 7

12.15 a. 2-methylpentane
b. 2,4,5-trimethylheptane
c. 3-ethyl-2,3-dimethylpentane
d. 3-ethyl-2,4-dimethylhexane
e. decane
f. 4-propylheptane

12.17 horizontal chain, because it has more substituents

12.19

a.
```
CH3-CH-CH2-CH3
    |
    CH3
```

b.
```
CH3-CH2-CH-CH-CH2-CH3
        |  |
        CH3 CH3
```

c.
```
            CH3
            |
CH3-CH2-C-CH2-CH3
            |
            CH2
            |
            CH3
```

d.
```
CH3-CH-CH-CH-CH-CH2-CH3
    |  |  |  |
    CH3 CH3 CH3 CH3
```

e.
```
CH3-CH2-CH-CH2-CH-CH2-CH2-CH
        |      |
        CH2    CH2
        |      |
        CH3    CH3
```

f.
```
CH3-CH2-CH2-CH-CH2-CH2-CH2-CH2-CH3
            |
            CH2
            |
            CH2
            |
            CH3
```

12.21 a. 1, 1 b. 2, 2 c. 2, 2 d. 4, 4 e. 2, 2 f. 1, 1

12.23
a. carbon chain numbered from wrong end; 2-methylpentane
b. not based on longest carbon chain; 2,2-dimethylbutane
c. carbon chain numbered from wrong end; 2,2,3-trimethylbutane
d. not based on longest carbon chain; 3,3-dimethylhexane
e. carbon chain numbered from wrong end, substituents not alphabetical; 3-ethyl-4-methylhexane
f. like alkyl groups listed separately; 2,4-dimethylhexane

12.25 a. 3, 2, 1, 0 b. 5, 2, 3, 0 c. 5, 2, 1, 1 d. 5, 2, 3, 0 e. 2, 8, 0, 0 f. 3, 6, 1, 0

12.27 a. isopropyl b. isobutyl c. isopropyl d. sec-butyl

12.29

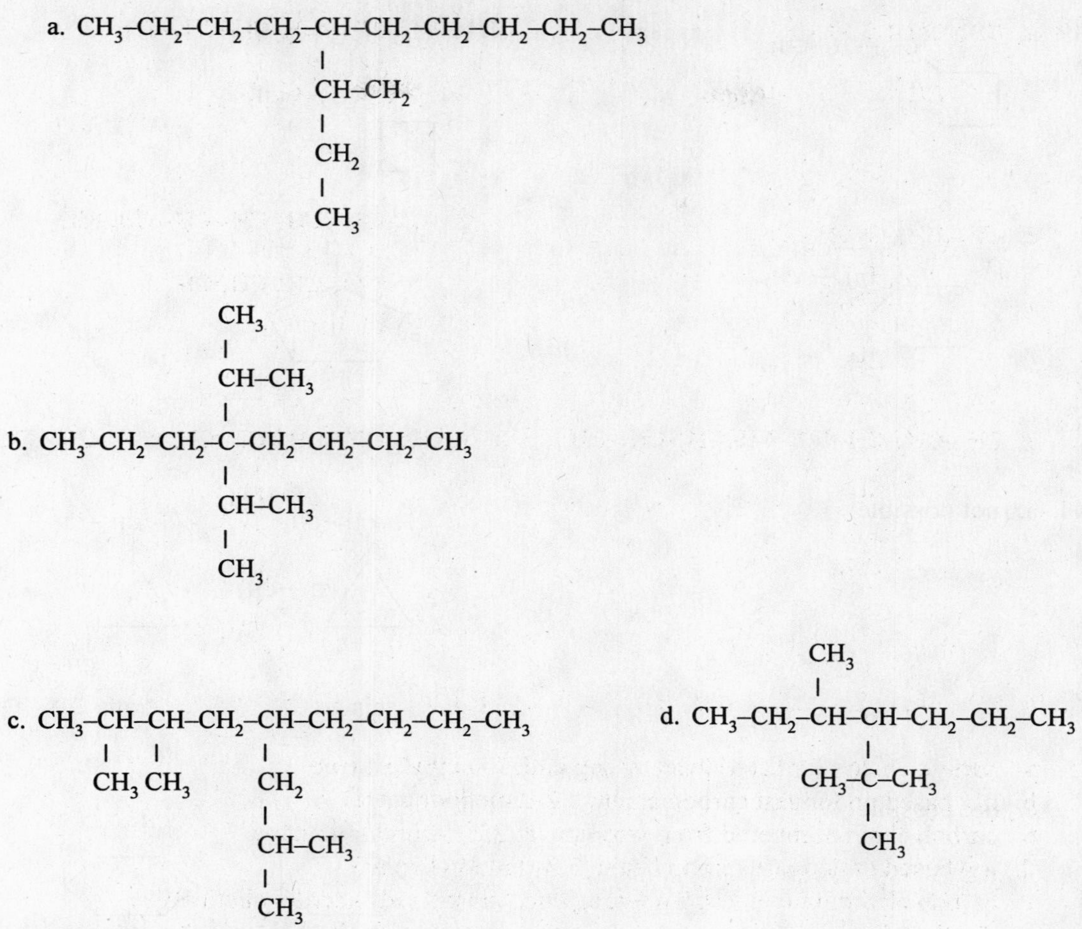

a.
```
CH3–CH2–CH2–CH2–CH–CH2–CH2–CH2–CH2–CH3
                 |
                 CH–CH2
                 |
                 CH2
                 |
                 CH3
```

b.
```
               CH3
               |
               CH–CH3
               |
CH3–CH2–CH2–C–CH2–CH2–CH2–CH3
               |
               CH–CH3
               |
               CH3
```

c.
```
CH3–CH–CH–CH2–CH–CH2–CH2–CH2–CH3
    |   |      |
  CH3 CH3      CH2
               |
               CH–CH3
               |
               CH3
```

d.
```
                 CH3
                 |
CH3–CH2–CH–CH–CH2–CH2–CH3
              |
          CH3–C–CH3
              |
              CH3
```

12.31 a. 16 b. 6 c. 5 d. 15

12.33 a. C_6H_{12} b. C_6H_{12} c. C_4H_8 d. C_7H_{14}

12.35 a. cyclohexane
b. 1,2-dimethylcyclobutane
c. methylcyclopropane
d. 1,2-dimethylcyclopentane

12.37 a. must locate methyl groups with numbers
b. wrong numbering system for ring
c. no number needed
d. wrong numbering system for ring

12.39 a. (cyclobutane with $CH_2-CH_2-CH_3$) b. (cyclobutane with $CH(CH_3)-CH_3$)

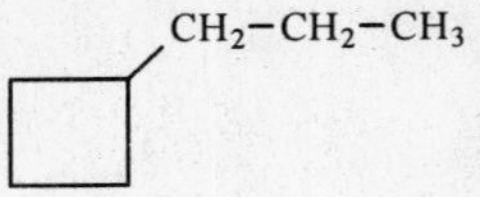

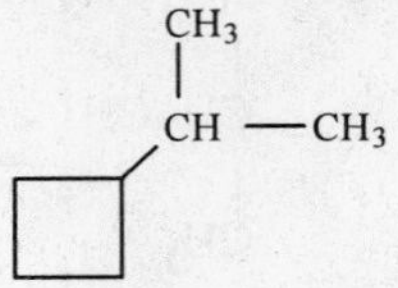

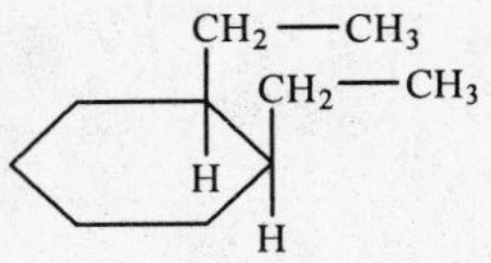

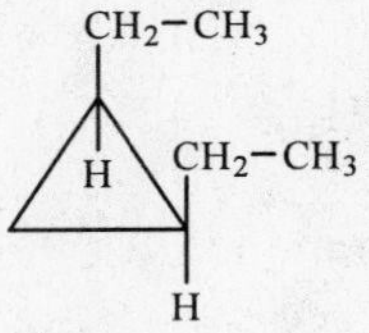

c. (cyclohexane with CH_2-CH_3, CH_2-CH_3, H, H) d. (cyclopentane with CH_2-CH_3, H, H, $CH_2-CH_2-CH_3$)

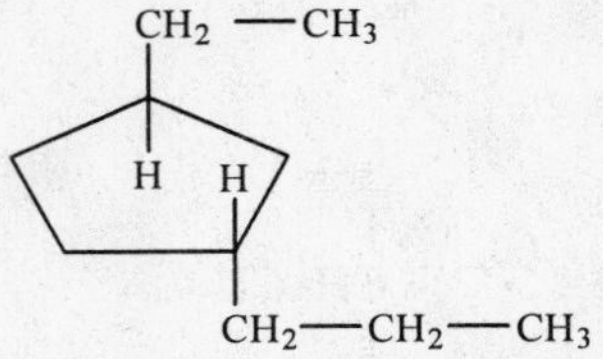

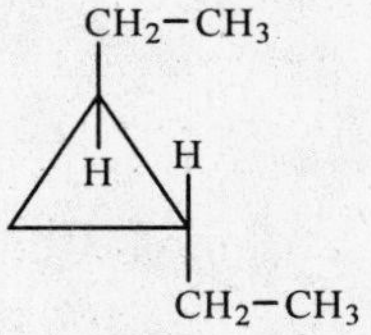

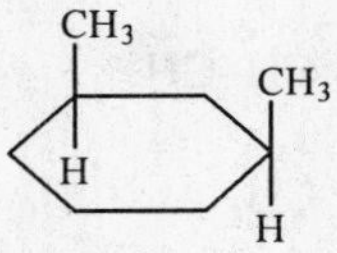

12.41 a. not possible b. (cyclopropane with CH_2-CH_3, H, CH_2-CH_3, H) cis (cyclopropane with CH_2-CH_3, H, H, CH_2-CH_3) trans

c. not possible d. (cyclohexane with CH_3, H, CH_3, H) cis (cyclohexane with CH_3, H, H, CH_3) trans

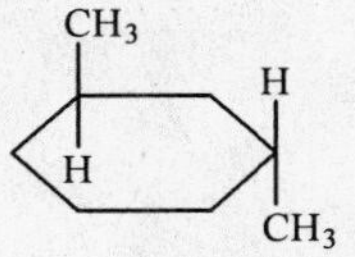

12.43 boiling point

12.45 a. octane b. cyclopentane c. pentane d. cyclopentane

12.47 a. different states b. same state c. same state d. same state

12.49 a. CO_2 and H_2O b. CO_2 and H_2O c. CO_2 and H_2O d. CO_2 and H_2O

12.51 CH_3Br, CH_2Br_2, $CHBr_3$, CBr_4

12.53 a. $CH_3–CH_2$
$\quad\quad\quad\quad$ |
$\quad\quad\quad\quad$ Cl

b. $CH_2–CH_2–CH_2–CH_3$, $CH_3–CH–CH_2–CH_3$
$\quad$ | $\quad\quad\quad\quad\quad\quad\quad\quad\quad$ |
$\quad$ Cl $\quad\quad\quad\quad\quad\quad\quad\quad$ Cl

c. $Cl–CH_2–CH–CH_3$ (with CH_3 on the CH), $CH_3–C–CH_3$ (with Cl above and CH_3 below the C)

d. Cl (on a cyclopentane ring)

12.55 a. iodomethane, methyl iodide
b. 1-chloropropane, propyl chloride
c. 2-fluorobutane, sec-butyl fluoride
d. chlorocyclobutane, cyclobutyl chloride

12.57 a. H–C–Cl (with Cl above and Cl below the C)
b. Cl–C–C–Cl (with F F above and F F below)
c. $CH_3–CH–Br$ (with CH_3 below the CH)
d.

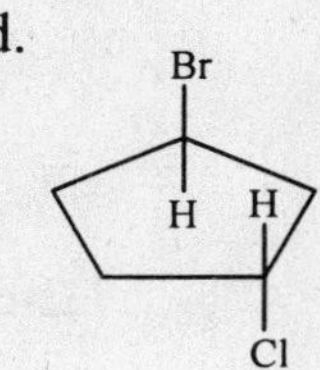

12.59 a. no b. yes c. no d. no

12.60 a. F, H, H, F (on a cyclohexane ring)
b. Cl, CH_3, H, H (on a cyclobutane ring)
c. $CH_3 – C – Br$ (with CH_3 above and CH_3 below the C)
d. $CH_3 – CH – CH_2 – I$ (with CH_3 below the CH)

12.61 4-tert-butyl-5-ethyl-2,2,6-trimethyloctane

12.62 a. $C_{18}H_{38}$ b. C_7H_{14} c. $C_7H_{14}F_2$ d. $C_6H_{10}Br_2$

12.63 a. 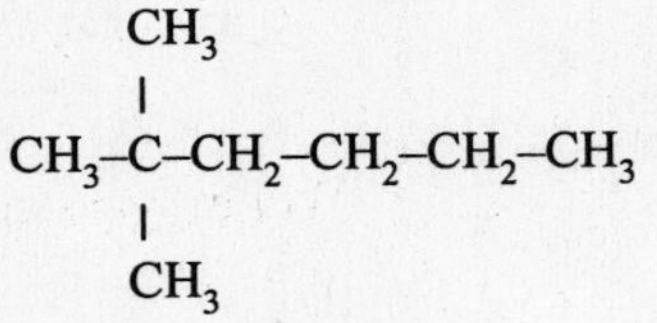
$$
\begin{array}{c}
CH_3 \\
| \\
CH_3\text{–}C\text{–}CH_2\text{–}CH_2\text{–}CH_2\text{–}CH_3 \\
| \\
CH_3
\end{array}
$$

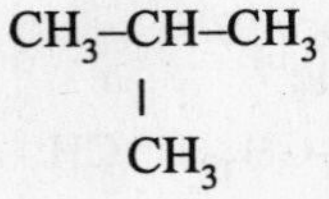

b.
$$
\begin{array}{c}
CH_3\text{–}CH\text{–}CH_3 \\
| \\
CH_3
\end{array}
$$

c.
$$
\begin{array}{c}
Cl \\
| \\
Cl-C-H \\
| \\
H
\end{array}
$$

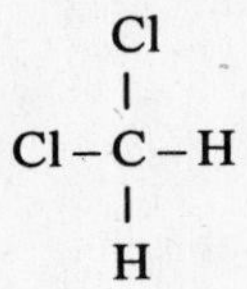

d. 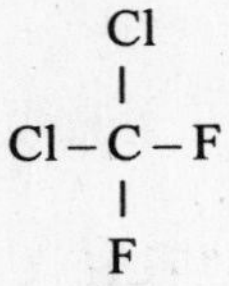
$$
\begin{array}{c}
Cl \\
| \\
Cl-C-F \\
| \\
F
\end{array}
$$

12.64 a. alkane b. halogenated cycloalkane c. halogenated alkane d. cycloalkane

12.65 a.

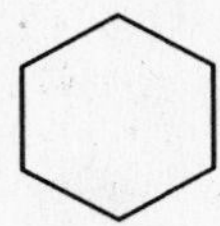

cyclohexane

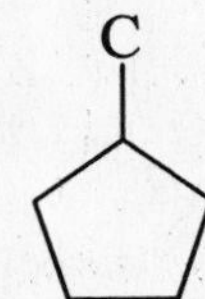

methylcyclopentane

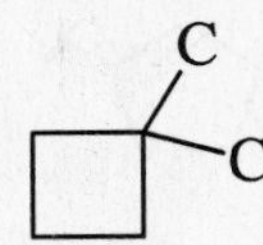

1,1-dimethylcyclobutane

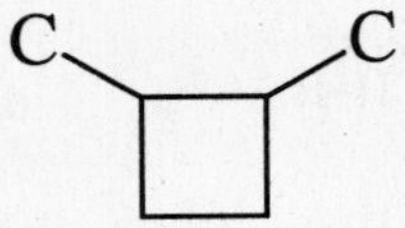

1,2-dimethylcyclobutane

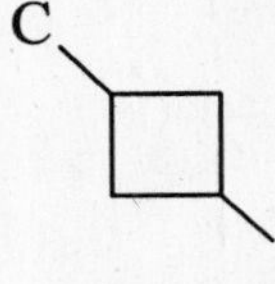

1,3-dimethylcyclobutane

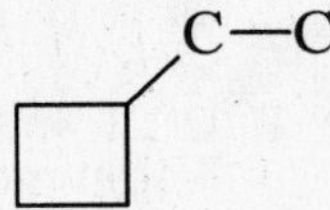

ethylcyclobutane

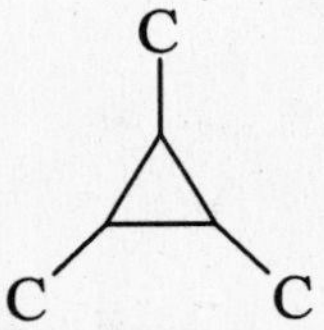

1,2,3-trimethylcyclopropane

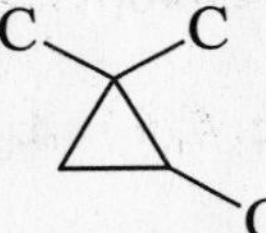

1,1,2-trimethylcyclopropane

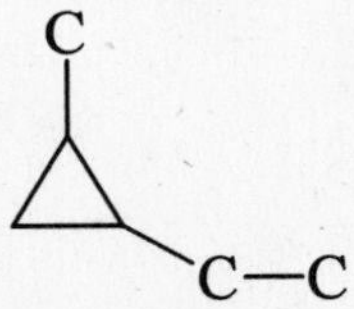

1-ethyl-2-methylcyclopropane

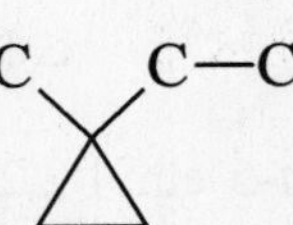

1-ethyl-1-methylcyclopropane

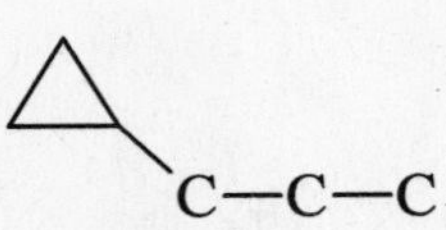

propylcyclopropane

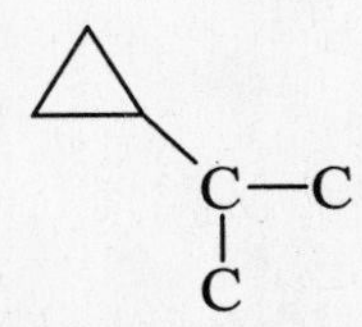

isopropylcyclopropane

b.

```
C–C–C–C–C–C        C–C–C–C–C        C–C–C–C–C
                     |                  |
                     C                  C
  hexane          2-methylpentane    3-methylpentane

                         C
                         |
C–C–C–C              C–C–C–C
  | |                    |
  C C                    C
2.3-dimethylbutane   2,2-dimethylbutane
```

c.

```
  Br                Br
  |                 |
  C–C–C           C–C–C            C–C–C
  |                 |              |  |
  Br                Br             Br Br
1,1-dibromopropane  2,2-dibromopropane  1,2-dibromopropane

  C–C–C
  |   |
  Br  Br
1,3-dibromopropane
```

12.66 a. 1,2-diethylcyclohexane
b. 3-methylhexane
c. 2,3-dimethyl-3-propylnonane
d. 1-isopropyl-3,5-dipropylcyclohexane

12.67 a. 1, 2, 5 b. 6 c. 4, 6 d. 1, 2

12.68 a. 2, 3 b. 5 c. 6 d. 1, 3, 4

12.69 a. all choices b. 3 c. (1, 4), (2, 5) d. 3, 6

12.70 a. 1, 3, 6 b. 2, 4, 5 c. 2, 3, 6 d. 2

Unsaturated Hydrocarbons **Chapter 13**

Solutions to Exercises

13.1 a. unsaturated, alkene
b. saturated
c. unsaturated, alkene
d. unsaturated, diene
e. unsaturated, triene
f. unsaturated, diene

13.3 a. C_4H_{10} b. C_5H_{10} c. C_5H_8 d. C_7H_{10}

13.4 a. C_6H_{10} b. C_6H_8 c. C_5H_{10} d. C_8H_8

13.5 a. C_nH_{2n-2} b. C_nH_{2n-2} c. C_nH_{2n-2} d. C_nH_{2n-6}

13.7 a. 2-butene
b. 2,4-dimethyl-2-pentene
c. cyclohexene
d. 1,3-cyclopentadiene
e. 2-ethyl-1-pentene
f. 2,4,6-octatriene

13.9 a. 2-pentene
b. pentane
c. 2,3,3-trimethyl-1-butene
d. 2-methyl-1,4-pentadiene
e. 1,3,5-hexatriene
f. 2,3-pentadiene

13.11 a. $CH_2{=}CH{-}CH{-}CH_2{-}CH_3$ (with CH_3 on the third carbon)

b. CH_3 (ring structure)

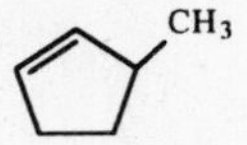

c. $CH_2{=}CH{-}CH{=}CH_2$

d. $CH_2{=}CH{-}CH{-}CH{=}CH_2$ (with $CH_2{-}CH_3$ on the third carbon)

e. $CH_3{-}CH{=}CH{-}CH{-}CH_2{-}CH_2{-}CH_3$ (with $CH_2{-}CH_2{-}CH_3$ on the fourth carbon)

f. $CH_2{-}CH_3$ (ring structure) $CH_2{-}CH_3$

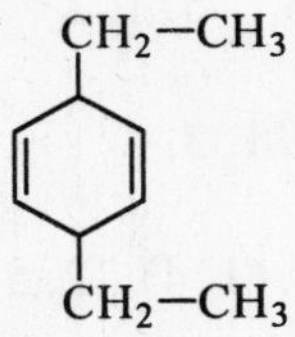

13.13 a. 3-methyl-3-hexene
b. 2,3-dimethyl-2-hexene
c. 1,3-cyclopentadiene
d. 4,5-dimethylcyclohexene

13.15 σ-bond: orbital overlap along internuclear axis
π-bond: orbital overlap above and below (but not on) internuclear axis

13.17 a. 4 σ, 0 π b. 3 σ, 1 π c. 3 σ, 2 π d. 1 σ, 2 π e. 5 σ, 0 π f. 10 σ, 2 π

13.19 a. 16 σ, 0 π b. 11 σ, 1 π c. 9 σ, 2 π d. 3 σ, 2 π e. 15 σ, 1 π f. 26 σ, 2 π

13.21

```
C=C–C–C–C–C        C–C=C–C–C–C        C–C–C=C–C–C
 1-hexene           2-hexene           3-hexene

C=C–C–C–C          C=C–C–C–C          C=C–C–C–C
  |                    |                    |
  C                    C                    C
2-methyl-1-pentene  3-methyl-1-pentene  4-methyl-1-pentene

C–C=C–C–C          C–C=C–C–C          C–C=C–C–C
  |                    |                    |
  C                    C                    C
2-methyl-2-pentene  3-methyl-2-pentene  4-methyl-2-pentene

                        C
                        |
C=C–C–C             C=C–C–C            C–C=C–C
  | |                   |                | |
  C C                   C                C C
2,3-dimethyl-1-butene  3,3-dimethyl-1-butene  2,3-dimethyl-2-butene

C=C–C–C
  |
  C
  |
  C
2-ethyl-1-butene
```

13.23 a. no b. no c. no d.

```
CH3–CH2       CH2–CH3        CH3–CH2        H
       \     /                      \      /
        C=C                          C=C
       /   \                        /   \
      H     H                      H     CH2–CH3
        cis                           trans
```

e.

```
             CH3
             |
CH3         CH–CH3        CH3        H
   \       /                 \      /
    C=C                       C=C
   /   \                     /   \
  H     H                   H     CH–CH3
                                  |
                                  CH3
     cis                      trans
```

f.

```
  CH3                CH3
   |   CH3            |    H
   H   H              H    CH3
    cis                trans
```

13.25 a. cis-2-pentene
b. trans-1-bromo-2-iodoethene
c. tetrafluoroethene
d. 2-methyl-2-butene

13.27 a. 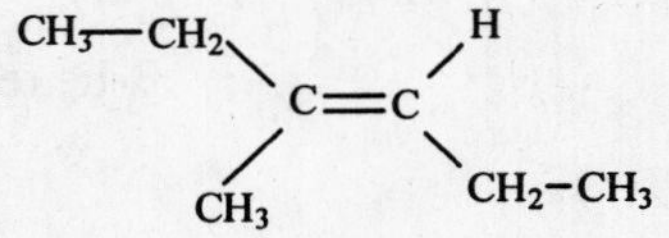 $CH_3{-}CH_2(CH_3)C{=}C(H)CH_2{-}CH_3$

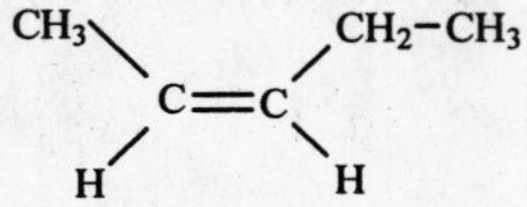

b. $CH_3(H)C{=}C(H)CH_2{-}CH_3$

c. $CH_3(H)C{=}C(H)CH_2{-}CH(CH_3){-}CH_2{-}CH_3$

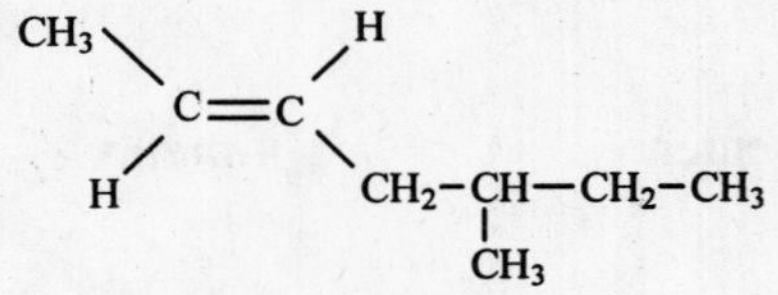

d. 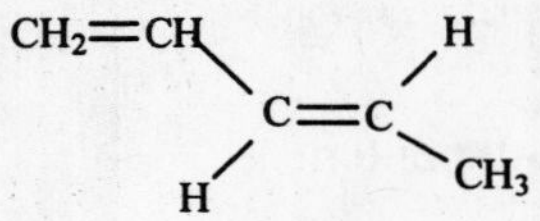 $CH_2{=}CH(H)C{=}C(H)CH_3$

13.29 a. yes b. no c. yes d. no

13.31 a. $CH_2{=}CH_2 + Cl_2 \rightarrow CH_2(Cl){-}CH_2(Cl)$

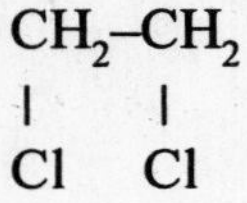

b. $CH_2{=}CH_2 + HCl \rightarrow CH_3{-}CH_2(Cl)$

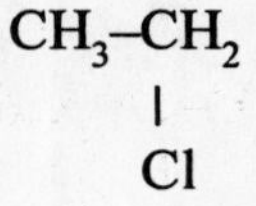

c. $CH_2{=}CH_2 + H_2 \xrightarrow{Ni} CH_3{-}CH_3$

d. $CH_2{=}CH_2 + HBr \rightarrow CH_3{-}CH_2(Br)$

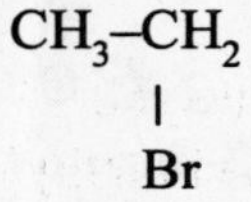

13.33 a. $CH_2{=}CH{-}CH_3 + Cl_2 \xrightarrow{H_2SO_4} CH_2(Cl){-}CH(Cl){-}CH_3$

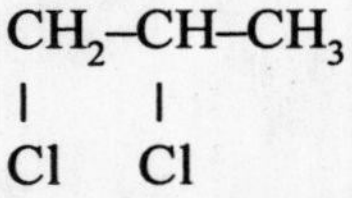

b. $CH_2{=}CH{-}CH_3 + HCl \rightarrow CH_3{-}CH(Cl){-}CH_3$

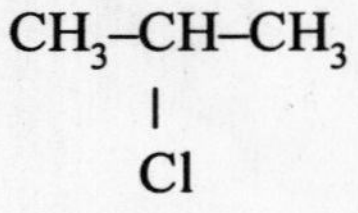

c. $CH_2{=}CH{-}CH_3 + H_2 \xrightarrow{Ni} CH_3{-}CH_2{-}CH_3$

d. $CH_2{=}CH{-}CH_3 + HBr \rightarrow CH_3{-}CH(Br){-}CH_3$

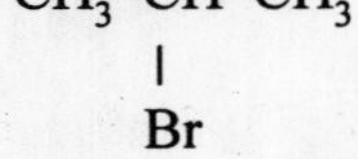

13.35 a. 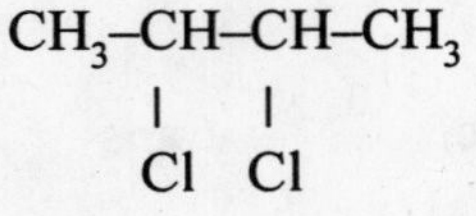 $CH_3{-}CH(Cl){-}CH(Cl){-}CH_3$

b. 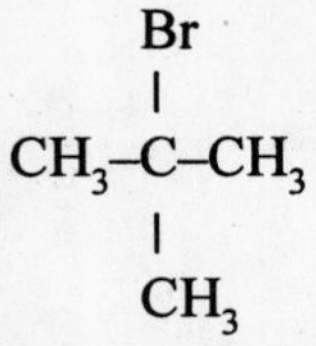 $CH_3{-}C(Br)(CH_3){-}CH_3$

c. $CH_3{-}CH_2{-}CH(Cl){-}CH_3$

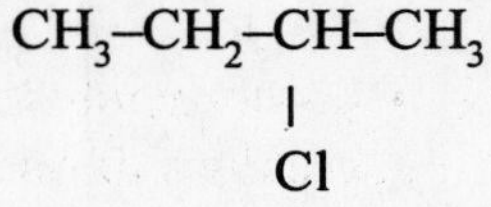

d.

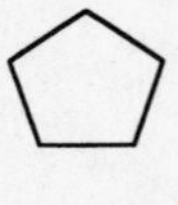

e.

f. 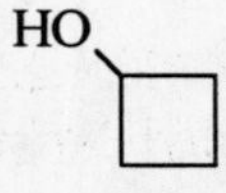 HO

13.37 a. Br_2 b. H_2 + Ni catalyst c. HCl d. $H_2O + H_2SO_4$ catalyst

13.39 a. 2 b. 2 c. 2 d. 3

13.41

a.
```
F   F
|   |
C = C
|   |
F   F
```

b.
```
H            H
|            |
C = C – C = C
|   |   |    |
H   Cl  H    H
```

c.
```
H   H
|   |
C = C
|   |
H   Cl
```

d.
```
H   H
|   |
C = C
|
H
(C₆H₅ ring on second C)
```

13.43

a. $-CH_2-CH_2-CH_2-CH_2-CH_2-CH_2-$

b.
```
–CH2–CH–CH2–CH–CH2–CH–
     |       |       |
     Cl      Cl      Cl
```

c.
```
–CH–CH–CH–CH–CH–CH–
  |   |   |   |   |   |
  Cl  Cl  Cl  Cl  Cl  Cl
```

d.
```
–CH2–CH–CH2–CH–CH2–CH–
     |       |       |
     Cl      Cl      Cl
```

13.45 a. 1-hexyne
b. 4-methyl-2-pentyne
c. 2,2-dimethyl-3-heptyne
d. 2-butyne
e. 3-methyl-1,4-hexadiyne
f. 4-methyl-2-hexyne

13.47 a. CH_3-CH_3

b.
```
       Br Br
       |  |
CH3–C–CH
       |  |
       Br Br
```

c.
```
       Br
       |
CH3–C–CH3
       |
       Br
```

d.
```
CH2=CH
     |
     Cl
```

e.

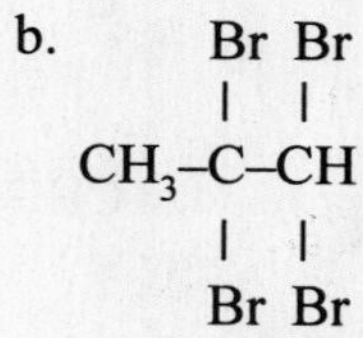

f.
```
CH3–CH2–C=CH2
         |
         Br
```

13.49 a. 1,3-dibromobenzene
b. 2-nitroaniline
c. 1-chloro-4-nitrobenzene
d. 3-chlorotoluene
e. 1-bromo-2-ethylbenzene
f. 4-bromophenol

13.51 a. m-dibromobenzene
b. o-nitiroaniline
c. p-chloronitrobenzene
d. m-chlorotoluene
e. o-bromoethylbenzene
f. p-bromophenol

13.53 a. 2,4-dibromo-1-chlorobenzene
b. 3,5-dinitrotoluene
c. 1-bromo-3-chloro-2-fluorobenzene
d. 1,4-dibromo-2,5-dichlorobenzene

13.55 a. 2-phenylbutane
b. 3-phenyl-1-butene
c. 3-methyl-1-phenylbutane
d. 2,4-diphenylpentane

13.57 a. CH_2-CH_3, CH_2-CH_3 b. CH_3, CH_3 c. CH_3, CH_2-CH_3

d. COOH, Cl e. NH_2 f. CH_3, $CH_3-CH_2-C-CH_2-CH_3$

13.59 a. substitution b. addition c. substitution d. addition

13.61 a. Br_2 b. HNO_3 c. CH_3, $CH-CH_3$ d. CH_3-CH_2-Br

13.63 a. $CH_3-C{\equiv}C-CH_2-CH-CH_3$ with CH_3 on CH

b. $CH_2-CH{=}CH-CH_3$ with Cl on CH_2

c. $CH_3-C{\equiv}C-CH_2-CH-CH-CH_3$ with CH_3 CH_3 on the two CH

d. $CH_2{=}CH-CH-CH_2-CH_2-CH_3$ with $CH-CH_3$ on CH, and CH_3 on that CH

e. $CH_2{=}CH-CH_2-CH_2-CH_2-CH{=}CH_2$

f. $CH{\equiv}C-CH-C{\equiv}CH$ with CH_3 on CH

13.64 a. two b. one c. two d. one e. two f. four

13.65 a. $CH_2{=}CH{-}C_6H_5$ (CH₂ =CH — benzene ring)

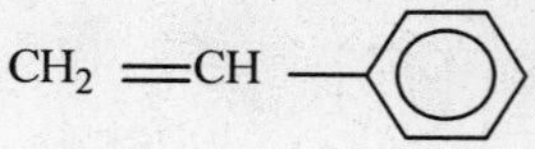

b. $CH_2{=}CH{-}CH_2{-}Cl$

c. $CH_3{-}CH_2{-}CH_2{-}C{\equiv}CH$

d. $CH_3{-}CH_2{-}CH_2{-}C{\equiv}C{-}CH_2{-}CH_2{-}CH_3$

e. benzene ring with COOH and meta $CH_2{-}CH_3$

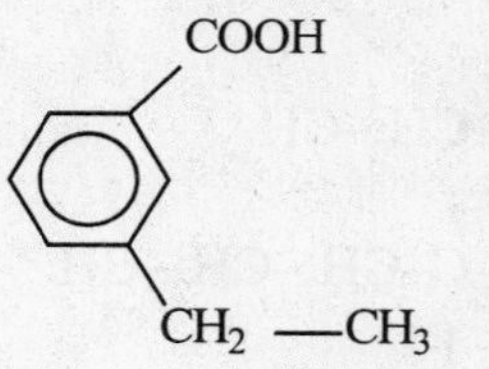

f. benzene ring with CH_3 and meta phenyl ring

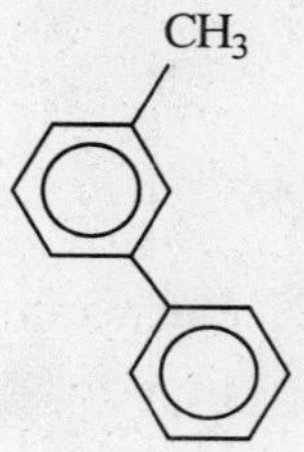

13.66 It would require a carbon atom that formed five bonds.

13.67 Each carbon atom in 1,2-dichlorobenzene has only one substituent.

13.68 $CH_2{=}CH{-}CH_2{-}CH_2{-}CH_3$ $CH_3{-}CH{=}CH{-}CH_2{-}CH_3$ (cis-trans forms) $CH_2{=}C(CH_3){-}CH_2{-}CH_3$

$CH_2{=}CH{-}CH(CH_3){-}CH_3$ $CH_3{-}C(CH_3){=}CH{-}CH_3$ cyclopentane

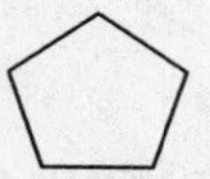

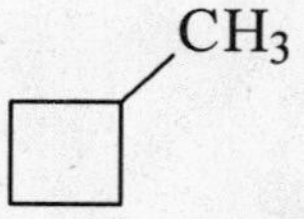

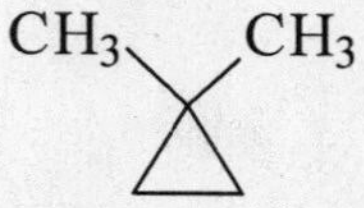

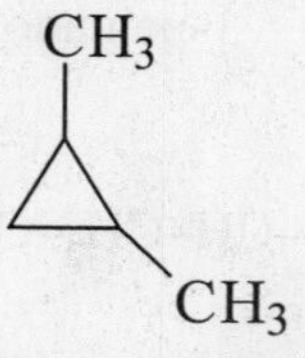

methylcyclobutane (CH₃ on square) 1,1-dimethylcyclopropane (CH₃, CH₃ on triangle) 1,2-dimethylcyclopropane (CH₃, CH₃ on triangle) (cis-trans forms)

13.69 1,2,3-trimethylbenzene; 1,2,4-trimethylbenzene; 1,3,5-trimethylbenzene; 2-ethyltoluene; 3-ethyltoluene; 4-ethyltoluene; propylbenzene; isopropylbenzene

13.70 a. 3 b. 3 c. 11 d. 3

13.71 a. 1, 5, 6 b. 2, 3 c. 1, 5, 6 d. 2, 6

13.72 a. 4, 6 b. 1, 2, 3, 5 c. 1, 6 d. 1, 2, 5, 6

13.73 a. 1, 2, 3 b. 4, 5, 6 c. 2, 3, 4, 6 d. 1, 5

13.74 a. 4 b. 1, 2, 3, 5, 6 c. 5 d. 1, 2, 4, 5, 6

Alcohols, Phenols, Ethers And Thiols — Chapter 14

Solutions to Exercises

14.1 a. alcohol b. ether c. phenol d. alcohol e. ether f. ether

14.3 a. 2-pentanol b. ethanol c. 3-methyl-2-butanol d. 2-ethyl-1-pentanol
e. 2-butanol f. 3,3-dimethyl-1-butanol

14.5 a. $CH_3-CH_2-CH-CH_2-CH_3$
|
OH

b.
CH_2-CH_3
|
$CH_3-CH_2-C-CH_2-CH_2-CH_3$
|
OH

c. $CH_2-CH-CH_3$
| |
OH CH_3

d. $CH_3-CH-CH_2-CH-CH_3$
| |
OH CH_3

e.
OH
|
CH_3-C-CH_3
(attached to a benzene ring)

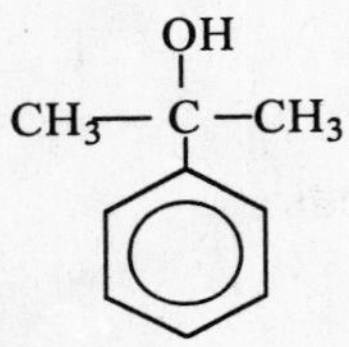

f.
OH
(cyclobutane ring)
CH_3

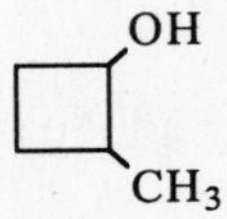

4.7 a. $CH_3-CH_2-CH_2-CH_2-CH_2-OH$

1-pentanol

b. $CH_3-CH_2-CH_2-OH$

1-propanol

c. $CH_3-CH-CH_2-OH$
|
CH_3

2-methyl-1-propanol

d. $CH_3-CH_2-CH-OH$
|
CH_3

2-butanol

14.9 a. 1,2-propanediol
c. 1,3-pentanediol

b. 1,4-pentanediol
d. 3-methyl-1,2,4-butanetriol

14.11 a. cyclohexanol
c. cis-2-methylcyclohexanol

b. trans-3-chlorocyclohexanol
d. 1-methylcyclobutanol

14.13 a. 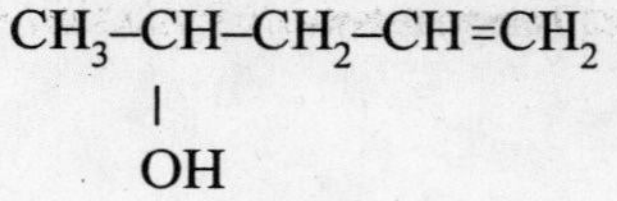 CH_3–CH–CH_2–CH=CH_2 (OH on C-2)

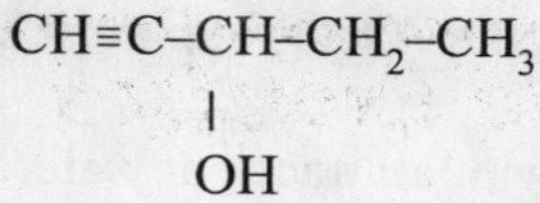

b. CH≡C–CH–CH_2–CH_3 (OH on C-3)

c. CH_3–CH–C=CH_2 (OH on C-2; CH_3 on C-3)

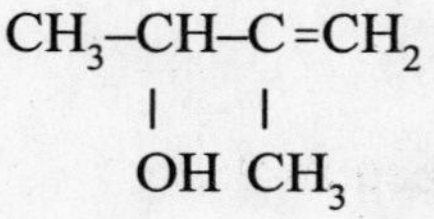

d. 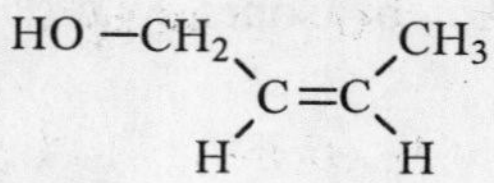 HO–CH_2(H)C=C(H)CH_3

14.15 a. 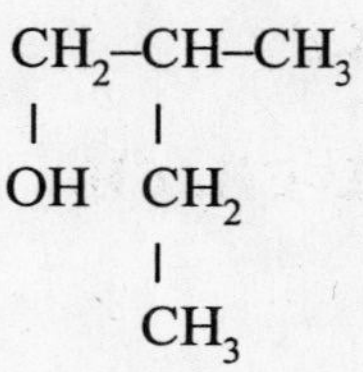 CH_2–CH–CH_3 (OH on C-1; CH_2–CH_3 on C-2)

2-methyl-1-butanol

b. 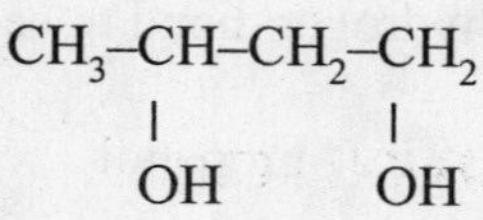 CH_3–CH–CH_2–CH_2 (OH on C-2 and C-4)

1,3-butanediol

c. 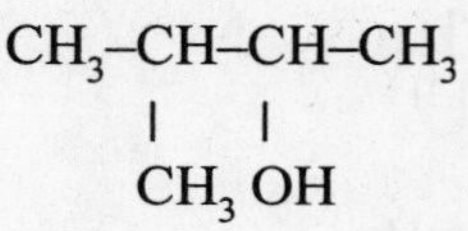 CH_3–CH–CH–CH_3 (CH_3 on C-2; OH on C-3)

3-methyl-2-butanol

d. 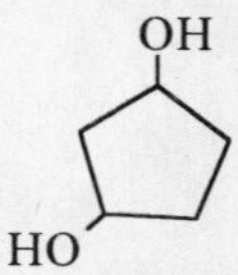 OH, HO

1,3-cyclopentanediol

14.17 a. 3-ethylphenol
b. 2-chlorophenol
c. o-cresol
d. hydroquinone
e. 2-bromophenol
e. 2-bromo-3-ethylphenol

14.19 a. 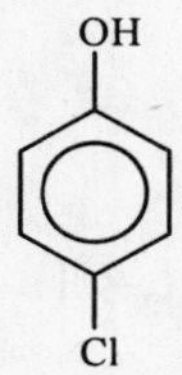 OH, Cl

b. 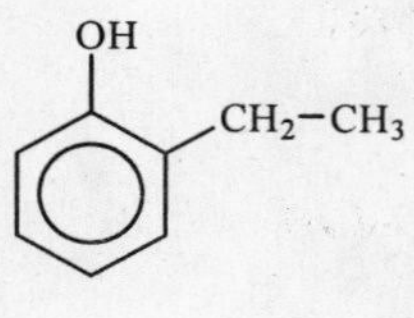 OH, CH_2–CH_3

c. 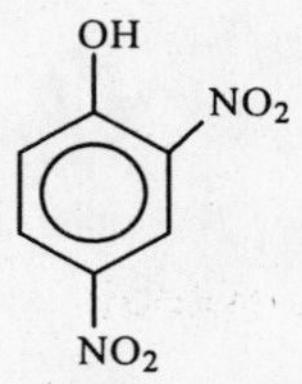 OH, NO_2, NO_2

d. 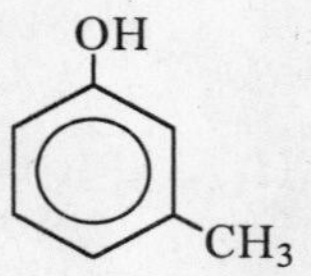 OH, CH_3

e. 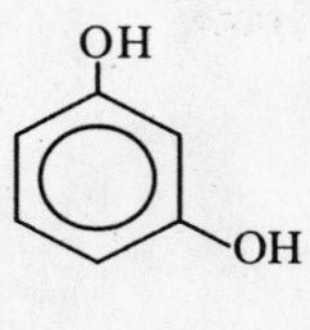 OH, OH

f. 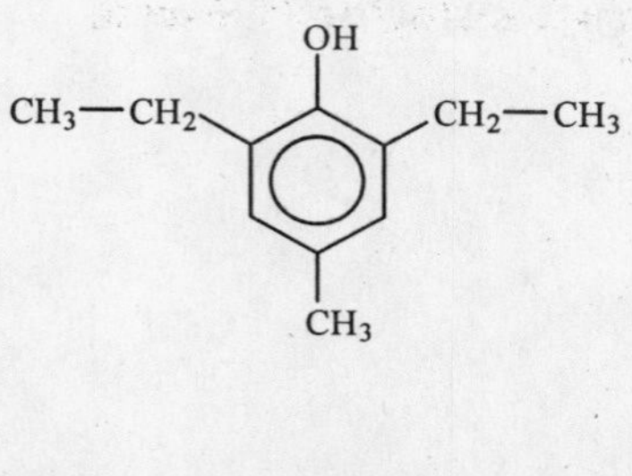 OH, CH_3–CH_2, CH_2–CH_3, CH_3

14.21 In a phenol the –OH group must be directly attached to the benzene ring.

14.23 a. ethanol with all traces of water removed
b. ethanol
c. 70% solution of isopropyl alcohol
d. ethanol

14.25 a. glycerol b. ethanol c. methanol d. methanol

14.27 alcohols can hydrogen-bond to each other; alkanes cannot

14.29 a. 1-heptanol b. 1-propanol c. 1,2-ethanediol

14.31 a. 1-butanol b. 1-pentanol c. 1,2-butanediol

14.33 a. 3 b. 3 c. 3 d. 3

14.35 a. $CH_3–CH_2–OH$ (OH on the CH_2)

b. $CH_3–CH_2–CH_2–OH$ (OH on the terminal CH_2)

c. $CH_3–CH_2–C(OH)(CH_3)–CH_3$

d. $CH_3–CH_2–CH(OH)–CH_2–CH_3$

14.37 a. $CH_2=CH–CH_3$

b. $CH_3–C(CH_3)=CH_2$

c. $CH_2=CH–CH_3$ (with CH_3 drawn below the CH)

d. $CH_3–CH_2–CH_2–O–CH_2–CH_2–CH_3$

14.39 a. $CH_3–CH(OH)–CH(CH_3)–CH_3$

b. $CH_3–CH_2–CH_2–OH$ (OH on the terminal CH_2)

c. $CH_3–CH_2–OH$

d. $CH_3–CH(CH_3)–CH_2–OH$

14.41 a. $CH_3-CH_2-CH(OH)-CH_3$

b. $CH_3-CH_2-CH_2-OH$

c. $CH_3-CH_2-CH_2-OH$

d. 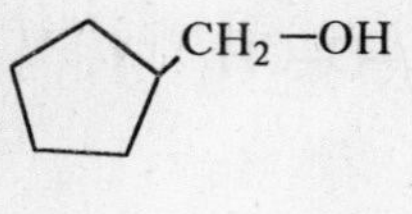 cyclopentyl$-CH_2-OH$

14.43 a. $CH_3-CH_2-CH_2-Cl$

b. 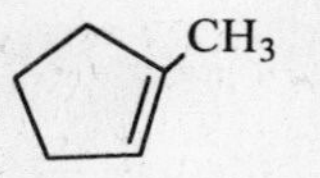 1-methylcyclopentene (CH_3 on ring double bond)

c. $CH_3-C(=O)-CH_2-CH_3$

d. $CH_3-CH_2-CH(Cl)-CH_2-CH_3$

e. $CH_3-CH_2-O-CH_2-CH_3$

f. $CH_2(Br)-CH_2(Br)$

14.45 $\left(-CH(OH)-CH(OH)- \right)_n$

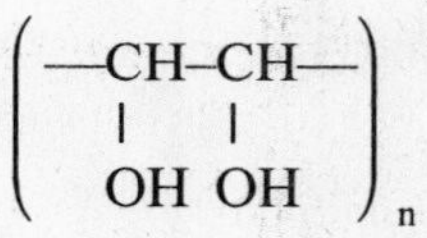

14.47 An antiseptic kills microorganisms on living tissue; a disinfectant kills microorganisms on inanimate objects

14.49

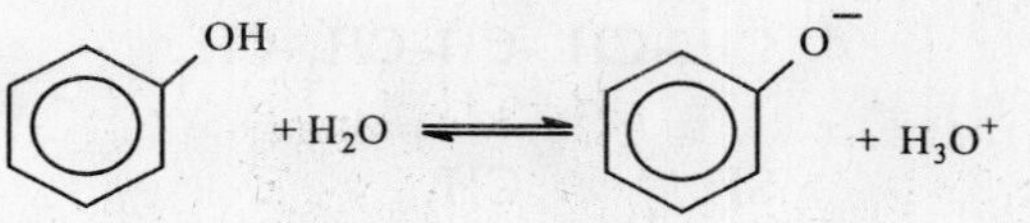

$$C_6H_5OH + H_2O \rightleftharpoons C_6H_5O^- + H_3O^+$$

14.51 a. 1-methoxypropane
b. 1-ethoxypropane
c. 2-methoxypropane
d. methoxybenzene
e. cyclohexoxycyclohexane
f. ethoxybenzene

14.53 a. methyl propyl ether
b. ethyl propyl ether
c. isopropyl methyl ether
d. methyl phenyl ether
e. dicyclohexyl ether
f. cyclobutyl ethyl ether

14.55 a. $CH_3-CH(CH_3)-O-CH_2-CH_2-CH_3$

b. $CH_3-CH_2-O-C_6H_5$

c. $CH_3-CH(CH_3)-CH_2-O-CH_3$

d. $CH_3-CH(O-CH_2-CH_3)-CH_2-CH_2-CH_3$

e. cyclobutyl $-O-CH_2-CH_3$

f. $CH_3-O-CH_2-C(CH_3)_2-CH_3$

14.57 There is no hydrogen bonding in the ether; there is hydrogen bonding in the alcohol

14.59 Flammability and peroxide formation

14.61 There are no hydrogen atoms bonded to oxygen atoms

14.63 a. noncyclic ether b. noncyclic ether c. cyclic ether d. cyclic ether
e. noncyclic ether f. nonether

14.65 R–S–H versus R–O–H

14.67 a. CH_3-SH

b. $CH_3-CH(SH)-CH_3$

c. $CH_2(SH)-CH_2-CH_2-CH_3$

d. $CH_2(SH)-CH_2-CH(CH_3)-CH_2-CH_3$

e. cyclopentyl $-SH$

f. $CH_2(SH)-CH_2(SH)$

14.69 Alcohol oxidation produces aldehydes and ketones; thiol oxidation produces disulfides

14.71 a. methylthioethane b. 2-methylthiopropane c. methylthiocyclohexane
d. cyclohexylthiocyclohexane e. 1-methylthio-1-propene f. 2-methylthiobutane

14.73 a. secondary b. secondary c. primary d. primary e. tertiary f. secondary

14.74 a. primary, primary b. primary, secondary c. secondary, secondary
d. primary, secondary, secondary

14.75 a. 2-hexanol b. 3-pentanol c. 3-phenoxy-1-propene d. 2-methyl-1-propanol
e. 2-methyl-2-propanol f. ethoxyethane

14.76

```
CH2–CH2–CH2–CH2–CH3      CH3–CH–CH2–CH2–CH3      CH3–CH2–CH–CH2–CH3
|                            |                              |
OH                           OH                             OH

                                                            CH3
                                                            |
CH2–CH–CH2–CH3           CH2–CH2–CH–CH3             CH3–C–CH2–CH3
|   |                    |       |                      |
OH  CH3                  OH      CH3                    OH

                            CH3
                            |
CH3–CH–CH–CH3            CH2–C–CH3                  CH3–CH2–CH2–CH2–O–CH3
    |   |                |   |
    OH  CH3              OH  CH3

                                                            CH3
                                                            |
CH3–CH–CH2–O–CH3         CH3–CH2–CH–O–CH3           CH3–C–O–CH3
    |                            |                      |
    CH3                          CH3                    CH3

CH3–CH2–CH2–O–CH2–CH3    CH3–CH–O–CH2–CH3
                             |
                             CH3
```

14.77 1-pentanol

14.78 $CH_3–O–CH_3$, $CH_3–CH_2–CH_2–O–CH_2–CH_2–CH_3$, and $CH_3–O–CH_2–CH_2–CH_3$

14.79 a. disulfide b. thiol, thioalcohol c. alcohol d. peroxide e. alcohol, thiol, thioalcohol
f. ether, sulfide, thioether

14.80 a. 1,2-ethanedithiol b. 3-methoxy-1-propanol c. 1-propanol d. 1,2-dimethoxyethane
e. methylthioethane f. 1-ethylthio-2-methoxyethane

14.81 a. 3, 4 b. 6 c. 3, 4 d. 1, 2, 5

14.82 a. 1, 2, 3, 4, 5 b. all c. all d. (1, 5), (2, 3)

14.83 a. 6 b. 1, 3, 5, 6 c. 2, 4 d. (2, 6), (3, 4)

14.84 a. 1 b. 6 c. 1 d. 3

Aldehydes And Ketones — Chapter 15

Solutions to Exercises

15.1 a. yes b. no c. yes d. yes e. no f. no

15.3 similarity: both have σ and π components
difference: C=O is polar, C=C is nonpolar

15.5 a. neither b. aldehyde c. ketone d. neither e. aldehyde f. aldehyde

15.7 $\text{H–}\overset{\text{O}}{\overset{\|}{\text{C}}}\text{–H}$, $\text{CH}_3\text{–}\overset{\text{O}}{\overset{\|}{\text{C}}}\text{–H}$; $\text{CH}_3\text{–}\overset{\text{O}}{\overset{\|}{\text{C}}}\text{–CH}_3$, $\text{CH}_3\text{–}\overset{\text{O}}{\overset{\|}{\text{C}}}\text{–CH}_2\text{–CH}_3$

15.9 a. neither b. aldehyde c. neither d. ketone e. ketone f. aldehyde

15.11 a. butanal b. 2-methylbutanal c. 4-methylheptanal d. 3-phenylpropanal
e. propanal f. 3,3-dimethylbutanal

15.13 a. $\text{CH}_3\text{–CH}_2\text{–}\underset{\underset{\text{CH}_3}{|}}{\text{CH}}\text{–CH}_2\text{–}\overset{\text{O}}{\overset{\|}{\text{C}}}\text{–H}$

b. $\text{CH}_3\text{–CH}_2\text{–CH}_2\text{–CH}_2\text{–}\underset{\underset{\underset{\text{CH}_3}{|}}{\text{CH}_2}}{\underset{|}{\text{CH}}}\text{–}\overset{\text{O}}{\overset{\|}{\text{C}}}\text{–H}$

c. $\text{CH}_3\text{–CH}_2\text{–CH}_2\text{–}\underset{\underset{\text{CH}_3}{|}}{\text{CH}}\text{ – }\underset{\underset{\text{CH}_3}{|}}{\text{CH}}\text{–CH}_2\text{–}\overset{\text{O}}{\overset{\|}{\text{C}}}\text{–H}$

d. $\text{CH}_3\text{–}\overset{\text{Cl}}{\overset{|}{\underset{\underset{\text{Cl}}{|}}{\text{C}}}}\text{–}\overset{\text{O}}{\overset{\|}{\text{C}}}\text{–H}$

e. $\text{CH}_3\text{–CH}_2\text{–}\underset{\underset{\text{CH}_3}{|}}{\text{CH}}\text{ – }\underset{\underset{\text{CH}_3}{|}}{\text{CH}}\text{–CH}_2\text{–}\underset{\underset{\text{CH}_3}{|}}{\text{CH}}\text{–}\overset{\text{O}}{\overset{\|}{\text{C}}}\text{–H}$

f. $\text{CH}_3\text{–CH}_2\text{–CH}_2\text{–CH}_2\text{–}\underset{\underset{\text{OH}}{|}}{\text{CH}}\text{–CH}_2\text{–}\underset{\underset{\text{CH}_3}{|}}{\text{CH}}\text{–}\overset{\text{O}}{\overset{\|}{\text{C}}}\text{–H}$

15.15 a. $H{-}C({=}O){-}H$ b. $CH_3{-}CH_2{-}C({=}O){-}H$

c. $CH_2(Cl){-}C({=}O){-}H$ d. $CH_3{-}CH_2{-}CH(Cl){-}C({=}O){-}H$

e. $C({=}O){-}H$, CH_3 f. $C({=}O){-}H$, CH_3, CH_3

15.17 a. propionaldehyde b. 2-methylpropionaldehyde
c. butyraldehyde d. dichloroacetaldehyde
e. 2-chlorobenzaldehyde f. 3-chloro-4-hydroxybenzaldehyde

15.19 a. 2-butanone b. 2,4,5-trimethyl-3-hexanone
c. 6-methyl-3-heptanone d. 2-octanone
e. 1,5-dichloro-3-pentanone f. 1,1-dichloro-2-butanone

15.21 a. cyclohexanone b. 3-methylcyclohexanone
c. 2-methylcyclohexanone d. 3-chlorocyclopentanone

15.23 a. $CH_3{-}C({=}O){-}CH(CH_3){-}CH_2{-}CH_3$ b. $CH_3{-}CH_2{-}C({=}O){-}CH_2{-}CH_2{-}CH_3$

c. O d. $CH_3{-}CH(CH_3){-}C({=}O){-}CH(CH_3){-}CH_3$

e. $CH_2(Cl){-}C({=}O){-}CH_3$ f. $CH_2(Cl){-}C({=}O){-}CH_2(Cl)$

15.25 a. $CH_3-CH_2-\overset{O}{\overset{\|}{C}}-CH_2-CH_3$

b. $CH_3-\overset{O}{\overset{\|}{C}}-CH_3$

c. $CH_3-\underset{CH_3}{\underset{|}{CH}}-\overset{O}{\overset{\|}{C}}-CH_2-CH_2-CH_3$

d. $\underset{Cl}{\underset{|}{CH_2}}-\overset{O}{\overset{\|}{C}}-CH_3$

e. 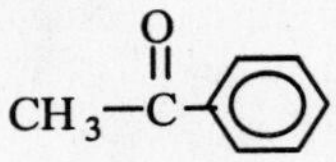 $CH_3-\overset{O}{\overset{\|}{C}}-C_6H_5$

f. 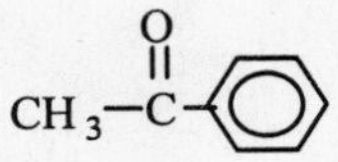 $CH_3-\overset{O}{\overset{\|}{C}}-C_6H_5$

15.27 dipole-dipole attractions between molecules raise the boiling point

15.29 2

15.31 ethanal, because it has a short carbon chain

15.33 a. $CH_3-CH_2-CH_2-CH_2-\overset{O}{\overset{\|}{C}}-H$

b. $CH_3-CH_2-\overset{O}{\overset{\|}{C}}-CH_3$

c. $CH_3-\underset{CH_3}{\underset{|}{\overset{CH_3}{\overset{|}{C}}}}-CH_2-\overset{O}{\overset{\|}{C}}-H$

d. $CH_3-CH_2-\overset{O}{\overset{\|}{C}}-CH_2-CH_3$

e. 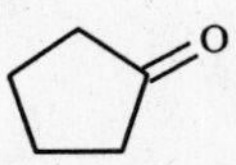 cyclopentanone (ring with =O)

f. 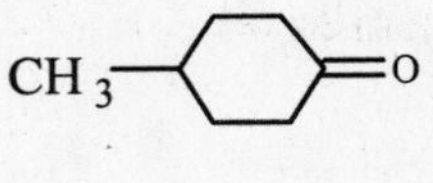 CH_3–cyclohexane ring=O

15.35 a. CH_3-CH_2-OH

b. $CH_3-CH_2-\underset{OH}{\underset{|}{CH}}-CH_2-CH_3$

c. 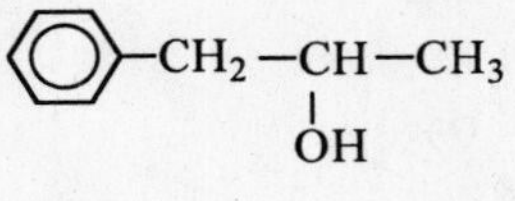 $C_6H_5-CH_2-\underset{OH}{\underset{|}{CH}}-CH_3$

d. CH_3-CH_2-OH

e. $CH_3-\underset{OH}{\underset{|}{CH}}-CH_3$

f. $CH_3-CH_2-CH_2-CH_2-\underset{\underset{CH_3}{\underset{|}{CH_2}}}{\underset{|}{CH}}-CH_2-OH$

15.37 a. $CH_3-\overset{\overset{O}{\|}}{C}-OH$ b. $CH_3-CH_2-CH_2-CH_2-\overset{\overset{O}{\|}}{C}-OH$

c. $H-\overset{\overset{O}{\|}}{C}-OH$ d. $CH_3-CH_2-\underset{\underset{Cl}{|}}{CH}-\underset{\underset{Cl}{|}}{CH}-CH_2-\overset{\overset{O}{\|}}{C}-OH$

15.39 appearance of a silver mirror

15.41 Cu^{2+} ion

15.43 a. no b. yes c. yes d. no

15.45 a. $CH_3-CH_2-CH_2-\underset{\underset{OH}{|}}{CH_2}$ b. $CH_3-CH_2-\underset{\underset{OH}{|}}{CH}-CH_2-CH_3$

c. $CH_3-\underset{\underset{CH_3}{|}}{CH}-CH_2-\underset{\underset{OH}{|}}{CH_2}$ d. $CH_3-\underset{\underset{CH_3}{|}}{CH}-\underset{\underset{OH}{|}}{CH}-CH_2-CH_2-CH_3$

15.47 R–O– and H–

15.48 H–

15.49 a. neither b. hemiacetal c. neither d. hemiketal e. hemiacetal f. neither

15.50 a. neither b. hemiketal c. neither d. neither e. neither f. hemiketal

15.51 a. $CH_3-\overset{\overset{OH}{|}}{CH}-O-CH_2-CH_3$ b. $CH_3-\overset{\overset{OH}{|}}{\underset{\underset{O-CH_3}{|}}{C}}-CH_2-CH_2-CH_3$

c. $CH_3-CH_2-CH_2-\overset{\overset{OH}{|}}{\underset{\underset{O-CH_2-CH_3}{|}}{CH}}$ d. $CH_3-\overset{\overset{OH}{|}}{\underset{\underset{O-\underset{\underset{CH_3}{|}}{CH}-CH_3}{|}}{CH}}-CH_3$

15.53 a. $CH_3-(CH_2)_2-CH(OH)-O-CH_2-CH_3$

b. $CH_3-CH_2-C(=O)-H$

c. $CH_3-CH_2-CH(OH)(O-CH_3)-CH_3$

d. (ring structure: six-membered ring containing O, with CH_2OH, OH, and HO substituents)

15.55 a. acetal b. ketal c. neither d. ketal

15.57 a. CH_3-OH

b. $CH_3-CH(OH)-O-CH_3$

c. $CH_3-CH(O-CH(CH_3)-CH_3)-O-CH_3$

d. $CH_3-CH(OH)-O-CH_3$, CH_3-OH

15.59 a. $CH_3-C(=O)-H$, 2 CH_3-OH

b. $CH_3-C(=O)-CH_3$, 2 CH_3-OH

c. $CH_3-CH_2-C(=O)-CH_2-CH_3$, CH_3-OH , CH_3-CH_2-OH

d. $CH_3-CH_2-CH_2-CH_2-C(=O)-H$, 2 CH_3-OH

15.61 a number is not needed because there is only one possible location for the double bond.

15.62 a. a ketone carbonyl group cannot be on a terminal carbon atom
b. it requires a carbon atom with five bonds
c. it requires a carbon atom with five bonds
d. it requires a carbon atom with five bonds

15.63 **heptanal**

15.64 **three**

15.65

$$\begin{array}{l} \qquad\qquad\qquad\qquad\qquad\qquad\quad O \\ \qquad\qquad\qquad\qquad\qquad\qquad\quad \| \\ CH_3\text{–}CH_2\text{–}CH_2\text{–}CH_2\text{–}CH_2\text{–}C\text{–}H \end{array}$$

hexanal

$$\begin{array}{l} \qquad\qquad\qquad\quad CH_3 \;\; O \\ \qquad\qquad\qquad\quad\;\; | \qquad \| \\ CH_3\text{–}CH_2\text{–}CH_2\text{–}CH - C\text{–}H \end{array}$$

2-methylpentanal

$$\begin{array}{l} \qquad\quad CH_3 \qquad\quad O \\ \qquad\quad\;\; | \qquad\qquad \| \\ CH_3\text{–}CH_2\text{–}CH\text{–}CH_2 - C\text{–}H \end{array}$$

3-methylpentanal

$$\begin{array}{l} \qquad\quad CH_3 \;\; O \\ \qquad\quad\;\; | \qquad \| \\ CH_3\text{–}CH_2\text{–}C - C\text{–}H \\ \qquad\quad\;\; | \\ \qquad\quad CH_3 \end{array}$$

2,2-dimethylbutanal

$$\begin{array}{l} \quad CH_3 \qquad O \\ \quad\;\; | \qquad\quad \| \\ CH_3\text{–}C\text{–}CH_2\text{–}C\text{–}H \\ \quad\;\; | \\ \quad CH_3 \end{array}$$

3,3-dimethylbutanal

$$\begin{array}{l} \quad CH_3 \;\; CH_3 \quad O \\ \quad\;\; | \qquad\; | \qquad\; \| \\ CH_3\text{–}CH - CH - C\text{–}H \end{array}$$

2,3-dimethylbutanal

$$\begin{array}{l} \quad\;\; O \\ \quad\;\; \| \\ CH_3\text{–}C\text{–}CH_2\text{–}CH_2\text{–}CH_2\text{–}CH_3 \end{array}$$

2-hexanone

$$\begin{array}{l} \qquad\qquad\; O \\ \qquad\qquad\; \| \\ CH_3\text{–}CH_2\text{–}C\text{–}CH_2\text{–}CH_2\text{–}CH_3 \end{array}$$

3-hexanone

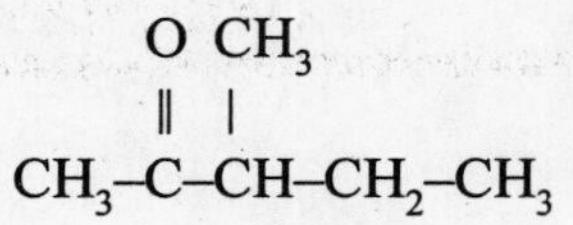

$$\begin{array}{l} \quad\;\; O \;\; CH_3 \\ \quad\;\; \| \quad\; | \\ CH_3-C-CH-CH_2-CH_3 \end{array}$$ 3-methyl-2-pentanone

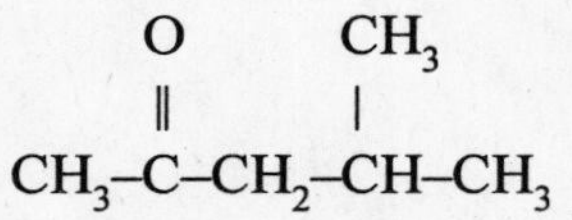

$$\begin{array}{l} \quad\;\; O \qquad\quad CH_3 \\ \quad\;\; \| \qquad\qquad | \\ CH_3-C-CH_2-CH-CH_3 \end{array}$$ 4-methyl-2-pentanone

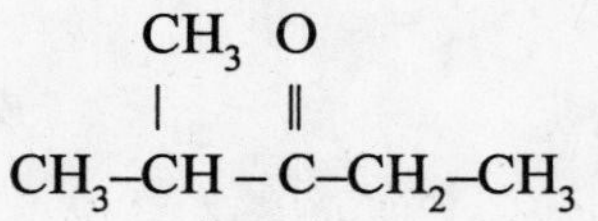

$$\begin{array}{l} \quad\; CH_3 \;\; O \\ \quad\;\;\; | \qquad \| \\ CH_3-CH-C-CH_2-CH_3 \end{array}$$ 2-methyl-3-pentanone

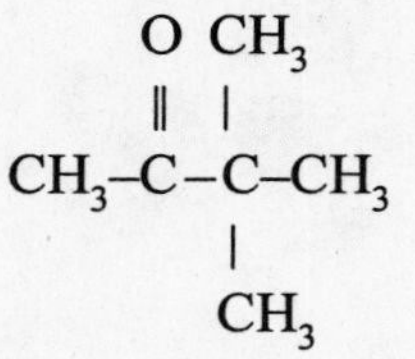

$$\begin{array}{l} \quad\;\; O \;\; CH_3 \\ \quad\;\; \| \quad | \\ CH_3-C-C-CH_3 \\ \qquad\quad\;\;\; | \\ \qquad\quad\; CH_3 \end{array}$$ 3,3-dimethyl-2-butanone

15.66

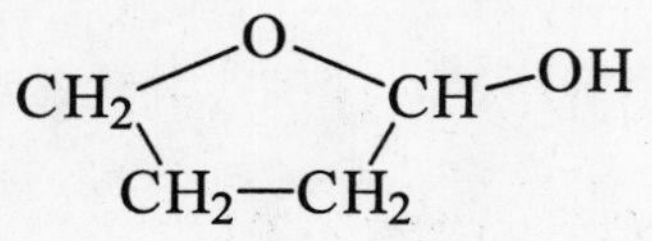

$$\begin{array}{c} \quad O \\ CH_2 \diagup \quad \diagdown CH-OH \\ \diagdown \qquad\quad \diagup \\ CH_2-CH_2 \end{array}$$

15.67 a. ketone, alkene b. aldehyde, alcohol, ether c. ketone, alkyne d. aldehyde, ketone

15.68 a. 2-hexanone b. propanal c. pentanedial d. 2-hexenal

15.69 a. 2, 4, 6 b. 1, 3, 5 c. 1, 2, 3, 4 d. 6

15.70 a. 1, 2, 3 b. all c. 1, 3, 6 d. 4, 5, 6

15.71 a. 1, 4 b. 2, 3, 5, 6 c. 3, 6 d. 2, 5

15.72 a. 3, 6 b. 1, 4 c. 1 d. 2, 5

Carboxylic Acids And Esters — Chapter 16

Solutions to Exercises

16.1 a. yes b. no c. yes d. yes e. no f. yes

16.3 a. butanoic acid
b. heptanoic acid
c. 2,3-dimethylpentanoic acid
d. 4-bromopentanoic acid
e. 3-methylpentanoic acid
f. chloroethanoic acid

16.5 a. $CH_3{-}CH_2{-}CH(CH_2{-}CH_3){-}C(=O){-}OH$

b. $CH_3{-}CH(CH_3){-}CH_2{-}CH_2{-}CH(CH_3){-}C(=O){-}OH$

c. $CH_3{-}CH(CH_3){-}C(=O){-}OH$

d. $CHCl_2{-}C(=O){-}OH$ (Cl–CH(Cl)–C(=O)–OH)

e. $CH_3{-}CH_2{-}CH_2{-}CH(Cl){-}CH_2{-}CH(Br){-}CH_2{-}C(=O){-}OH$

f. $CH_3{-}CH(CH_3){-}CH(CH_3){-}C(=O){-}OH$

16.7 a. butanedioic acid
b. propanedioic acid
c. 3-methylpentanedioic acid
d. 2-chlorobenzoic acid
e. m-toluic acid
f. 2-bromo-4-chlorobenzoic acid

16.9 a. $CH_3\text{–}CH_2\text{–}C(CH_3)_2\text{–}C(=O)\text{–}OH$

b. (benzene ring) $C(=O)\text{–}OH$, Br

c. (benzene ring) $C(=O)\text{–}OH$, Cl, Cl

d. (benzene ring) $C(=O)\text{–}OH$, H_3C

e. $HO\text{–}C(=O)\text{–}C(CH_3)_2\text{–}CH_2\text{–}CH_2\text{–}C(=O)\text{–}OH$

f. $HO\text{–}C(=O)\text{–}C(CH_3)_2\text{–}CH_2\text{–}C(=O)\text{–}OH$

16.11 a. $CH_3\text{–}CH_2\text{–}CH_2\text{–}CH_2\text{–}C(=O)\text{–}OH$

b. $CH_3\text{–}CH_2\text{–}C(=O)\text{–}OH$

c. $CH_3\text{–}C(=O)\text{–}OH$

d. $CH_3\text{–}CH_2\text{–}CH(Cl)\text{–}C(=O)\text{–}OH$

e. $CH_3\text{–}CH_2\text{–}CH_2\text{–}CH(Br)\text{–}CH_2\text{–}C(=O)\text{–}OH$

f. $CH_3\text{–}CH(Cl)\text{–}CH_2\text{–}CH(CH_3)\text{–}C(=O)\text{–}OH$

16.13 a. $HO{-}C({=}O){-}CH_2{-}C({=}O){-}OH$

b. $HO{-}C({=}O){-}CH_2{-}CH_2{-}C({=}O){-}OH$

c. $HO{-}C({=}O){-}(CH_2)_4{-}C({=}O){-}OH$

d. $HO{-}C({=}O){-}(CH_2)_2{-}CH(Br){-}(CH_2)_2{-}C({=}O){-}OH$

e. $HO{-}C({=}O){-}CH(CH_3){-}(CH_2)_2{-}C({=}O){-}OH$

f. $HO{-}C({=}O){-}C(Cl)(Br){-}CH_2{-}C({=}O){-}OH$

16.15 a. 3 b. 1 c. 2 d. 1

16.17 a. carbon-carbon double bond
b. hydroxyl group
c. keto group
d. two hydroxyl groups

16.19 a. propenoic acid
b. 2-hydroxypropanoic acid
c. 2-oxobutanedioic acid
d. 2,3-dihydroxypropanoic acid

16.21 a. $CH_3{-}CH_2{-}C({=}O){-}CH_2{-}C({=}O){-}OH$

b. $CH_3{-}CH_2{-}C(OH) - C({=}O){-}OH$

c $CH_3(H)C{=}C(H){-}CH_2{-}CH_2{-}C({=}O){-}OH$

d. $HO{-}C({=}O){-}CH(OH){-}CH(OH){-}CH_2{-}C({=}O){-}OH$

16.23 a. 2 b. 5

16.25 a. solid b. solid c. liquid d. solid

16.27 a. $CH_3-\overset{O}{\overset{\|}{C}}-OH$ b. $CH_3-\overset{O}{\overset{\|}{C}}-OH$

c. $CH_3-CH_2-\underset{CH_3}{\underset{|}{CH}}-CH_2-\overset{O}{\overset{\|}{C}}-OH$ d. $C_6H_5-\overset{O}{\overset{\|}{C}}-OH$

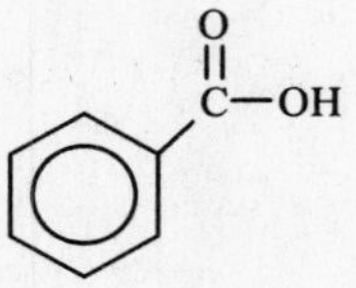

16.29 a. 1 b. 3 c. 2 d. 2

16.31 a. –1 b. –3 c. –2 d. –2

16.33 a. pentanoate ion b. citrate ion c. succinate ion d. oxalate ion

16.35 a. $CH_3-\overset{O}{\overset{\|}{C}}-OH + H_2O \rightarrow H_3O^+ + CH_3-\overset{O}{\overset{\|}{C}}-O^-$

b. $HO-\overset{O}{\overset{\|}{C}}-CH_2-\underset{\underset{OH}{\underset{|}{C=O}}}{\underset{|}{\overset{OH}{\overset{|}{C}}}}-CH_2-\overset{O}{\overset{\|}{C}}-OH + 3H_2O \rightarrow 3H_3O^+ + {}^-O-\overset{O}{\overset{\|}{C}}-CH_2-\underset{\underset{O^-}{\underset{|}{C=O}}}{\underset{|}{\overset{OH}{\overset{|}{C}}}}-CH_2-\overset{O}{\overset{\|}{C}}-O^-$

c. $CH_3-\overset{O}{\overset{\|}{C}}-OH + H_2O \rightarrow H_3O^+ + CH_3-\overset{O}{\overset{\|}{C}}-O^-$

d. $CH_3-CH_2-\underset{CH_3}{\underset{|}{CH}}-\overset{O}{\overset{\|}{C}}-OH + H_2O \rightarrow H_3O^+ + CH_3-CH_2-\underset{CH_3}{\underset{|}{CH}}-\overset{O}{\overset{\|}{C}}-O^-$

16.37 a. potassium ethanoate b. calcium propanoate
c. potassium butanedioate d. sodium pentanoate

16.39 a. $CH_3-\overset{O}{\overset{\|}{C}}-OH + KOH \rightarrow CH_3-\overset{O}{\overset{\|}{C}}-O^- K^+ + H_2O$

b. $2\ CH_3-CH_2-\overset{O}{\overset{\|}{C}}-OH + Ca(OH)_2 \rightarrow \left(CH_3-CH_2-\overset{O}{\overset{\|}{C}}-O^-\right)_2 Ca^{2+} + 2H_2O$

c. $HO-\overset{O}{\overset{\|}{C}}-CH_2-CH_2-\overset{O}{\overset{\|}{C}}-OH + 2KOH \rightarrow K^+\ {}^-O-\overset{O}{\overset{\|}{C}}-CH_2-CH_2-\overset{O}{\overset{\|}{C}}-O^- K^+ + 2H_2O$

d. $CH_3-CH_2-CH_2-CH_2-\overset{O}{\overset{\|}{C}}-OH + NaOH \rightarrow CH_3-CH_2-CH_2-CH_2-\overset{O}{\overset{\|}{C}}-O^- Na^+ + H_2O$

16.41 a. $CH_3-CH_2-CH_2-\overset{O}{\overset{\|}{C}}-O^- Na^+ + HCl \rightarrow CH_3-CH_2-CH_2-\overset{O}{\overset{\|}{C}}-OH + NaCl$

b. $K^+\ {}^-O-\overset{O}{\overset{\|}{C}}-\overset{O}{\overset{\|}{C}}-O^- K^+ + 2HCl \rightarrow HO-\overset{O}{\overset{\|}{C}}-\overset{O}{\overset{\|}{C}}-OH + 2KCl$

c. $\left({}^-O-\overset{O}{\overset{\|}{C}}-CH_2-\overset{O}{\overset{\|}{C}}-O^-\right) Ca^{2+} + 2HCl \rightarrow HO-\overset{O}{\overset{\|}{C}}-CH_2-\overset{O}{\overset{\|}{C}}-OH + CaCl_2$

d.

$C_6H_5-\overset{O}{\overset{\|}{C}}-O^- Na^+ + HCl \longrightarrow C_6H_5-\overset{O}{\overset{\|}{C}}-OH + NaCl$

16.43 a. yes b. yes c. no d. yes e. yes f. no

16.45 a. $CH_3-CH_2-\overset{O}{\overset{\|}{C}}-O-CH_3$

b. $CH_3-\overset{O}{\overset{\|}{C}}-O-CH_2-CH_2-CH_3$

c. $CH_3-CH_2-\underset{CH_3}{\underset{|}{CH}}-\overset{O}{\overset{\|}{C}}-O-\underset{CH_3}{\underset{|}{CH}}-CH_3$

d. $CH_3-CH_2-CH_2-CH_2-\overset{O}{\overset{\|}{C}}-O-\underset{CH_3}{\underset{|}{CH}}-CH_2-CH_3$

16.47 a. $CH_3-CH_2-\overset{O}{\overset{\|}{C}}-OH$, CH_3-CH_2-OH

b. $CH_3-CH_2-CH_2-\overset{O}{\overset{\|}{C}}-OH$, CH_3-OH

c. $CH_3-CH_2-CH_2-\overset{O}{\overset{\|}{C}}-OH$, CH_3-OH

d. $CH_3-\overset{O}{\overset{\|}{C}}-OH$, (phenol: benzene ring with OH)

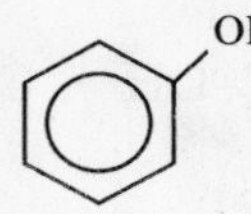

e. 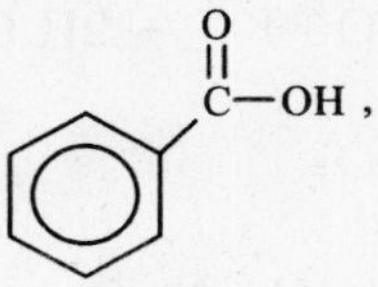 (benzene ring)$-\overset{O}{\overset{\|}{C}}-OH$, CH_3-OH

f. $CH_3-\underset{Cl}{\underset{|}{CH}}-\overset{O}{\overset{\|}{C}}-OH$, CH_3-CH_2-OH

16.49 a. methyl propanoate
b. methyl methanoate
b. methyl ethanoate
d. propyl ethanoate
e. isopropyl propanoate
f. ethyl benzoate

16.51 a. methyl propionate
b. methyl formate
c. methyl acetate
d. propyl acetate
e. isopropyl propionate
f. ethyl benzoate

16.53 a. $H-\overset{O}{\overset{\|}{C}}-O-CH_3$

b. $CH_3-\overset{O}{\overset{\|}{C}}-O-CH_2-CH_2-CH_3$

c. $CH_3-(CH_2)_8-\overset{O}{\overset{\|}{C}}-O-(CH_2)_7-CH_3$

d. 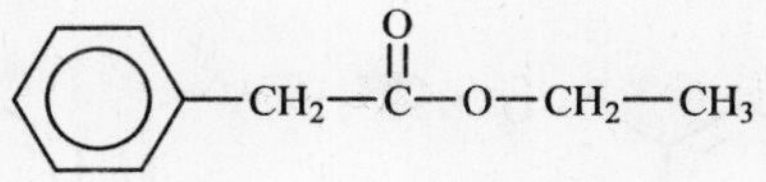 (benzene ring)$-CH_2-\overset{O}{\overset{\|}{C}}-O-CH_2-CH_3$

e. $CH_3-\overset{O}{\overset{\|}{C}}-O-\underset{CH_3}{\underset{|}{CH}}-CH_3$

f. $CH_3-\overset{O}{\overset{\|}{C}}-O-CH_2-\underset{Br}{\underset{|}{CH}}-CH_3$

16.55 a. ethyl ethanoate
b. methyl ethanoate
c. ethyl butanoate
d. propyl 2-hydroxypropanoate
e. pentyl pentanoate
f. 1-methylpropyl hexanoate

16.57 no oxygen-hydrogen bonds are present within an ester

16.59 ester molecules cannot hydrogen-bond to each other; acid molecules can hydrogen-bond to each other

16.61 a. $CH_3-CH_2-\overset{O}{\overset{\|}{C}}-OH$, CH_3-CH_2-OH

b. $CH_3-\overset{O}{\overset{\|}{C}}-OH$, CH_3-CH_2-OH

c. $CH_3-\underset{CH_3}{\underset{|}{CH}}-\overset{O}{\overset{\|}{C}}-OH$, C_6H_5-OH

d. $CH_3-CH_2-CH_2-\overset{O}{\overset{\|}{C}}-OH$, CH_3-OH

e. $H-\overset{O}{\overset{\|}{C}}-OH$, CH_3-CH_2-OH

f. $C_6H_5-\overset{O}{\overset{\|}{C}}-OH$, $CH_3-\underset{CH_3}{\underset{|}{CH}}-OH$

16.63 a. $CH_3-CH_2-\overset{O}{\overset{\|}{C}}-O^-\ Na^+$, CH_3-CH_2-OH

b. $CH_3-\overset{O}{\overset{\|}{C}}-O^-\ Na^+$, CH_3-CH_2-OH

c. $CH_3-\underset{CH_3}{\underset{|}{CH}}-\overset{O}{\overset{\|}{C}}-O^-\ Na^+$, C_6H_5-OH

d. $CH_3-CH_2-CH_2-\overset{O}{\overset{\|}{C}}-O^-\ Na^+$, CH_3-OH

e. $H-\overset{O}{\overset{\|}{C}}-O^-\ Na^+$, CH_3-CH_2-OH

f. $C_6H_5-\overset{O}{\overset{\|}{C}}-O^-\ Na^+$, $CH_3-\underset{CH_3}{\underset{|}{CH}}-OH$

16.65 a. $CH_3{-}CH(CH_3){-}C({=}O){-}OH$, $CH_3{-}CH_2{-}OH$

b. $CH_3{-}CH(CH_3){-}C({=}O){-}O^- \; Na^+$, $CH_3{-}CH_2{-}OH$

c. $H{-}C({=}O){-}OH$, $CH_3{-}CH_2{-}CH_2{-}CH_2{-}OH$

d. $CH_3{-}C({=}O){-}O^- \; Na^+$, $CH_3{-}CH(CH_3){-}CH(CH_3){-}CH_2{-}OH$

16.67 a. $CH_3{-}C({=}O){-}S{-}CH_2{-}CH_3$

b. $CH_3{-}(CH_2)_8{-}C({=}O){-}S{-}CH_3$

c. $C_6H_5{-}C({=}O){-}S{-}CH(CH_3){-}CH_3$

d. $H{-}C({=}O){-}S{-}CH_2{-}CH_2{-}CH_3$

16.69 $-C({=}O){-}C({=}O){-}O{-}(CH_2)_3{-}O{-}C({=}O){-}C({=}O){-}O{-}(CH_2)_3{-}O{-}$

16.71 $HO{-}C({=}O){-}(CH_2)_2{-}C({=}O){-}OH$, $HO{-}(CH_2)_3{-}OH$

16.73 a. $HO{-}P({=}O)(OH){-}O{-}CH_3$

b. $HO{-}P({=}O)(O{-}CH_3){-}O{-}CH_3$

c. $O{=}N{-}O{-}CH_3$ (printed as O–N–O–CH$_3$ with O above N)

d. $O{=}N({=}O)$… printed as: O–N(=O)–O–CH_2–CH_2–O–N(=O)–O

16.75 H_3PO_4 is a triprotic acid and H_2SO_4 is a diprotic acid

16.77 a. 2, 2 b. 7, 1 c. 7, 1 d. 6, 3 e. 3, 1 f. 2, 1

16.78 a. $HO{-}\overset{\overset{\large O}{\|}}{C}{-}CH_2{-}\overset{\overset{\large O}{\|}}{C}{-}OH$ malonic acid

b. $HO{-}\overset{\overset{\large O}{\|}}{C}{-}CH{=}CH{-}\overset{\overset{\large O}{\|}}{C}{-}OH$ maleic acid (cis isomer)

c. $HO{-}\overset{\overset{\large O}{\|}}{C}{-}\overset{\overset{\large OH}{|}}{CH}{-}CH_2{-}\overset{\overset{\large O}{\|}}{C}{-}OH$ malic acid

16.79 $C_nH_{2n-2}O_2$

16.80 $CH_3{-}CH_2{-}CH_2{-}CH_2{-}CH_2{-}\overset{\overset{\large O}{\|}}{C}{-}OH$ hexanoic acid

$CH_3{-}\underset{\underset{\large CH_3}{|}}{CH}{-}CH_2{-}CH_2{-}\overset{\overset{\large O}{\|}}{C}{-}OH$ 4-methylpentanoic acid

$CH_3{-}CH_2{-}\underset{\underset{\large CH_3}{|}}{CH}{-}CH_2{-}\overset{\overset{\large O}{\|}}{C}{-}OH$ 3-methylpentanoic acid

$CH_3{-}CH_2{-}CH_2{-}\underset{\underset{\large CH_3}{|}}{CH}{-}\overset{\overset{\large O}{\|}}{C}{-}OH$ 2-methylpentanoic acid

$CH_3{-}\underset{\underset{\large CH_3}{|}}{\overset{\overset{\large CH_3}{|}}{C}}{-}CH_2{-}\overset{\overset{\large O}{\|}}{C}{-}OH$ 3,3-dimethylbutanoic acid

$CH_3{-}CH_2{-}\underset{\underset{\large CH_3}{|}}{\overset{\overset{\large CH_3}{|}}{C}}-\overset{\overset{\large O}{\|}}{C}{-}OH$ 2,2-dimethylbutanoic acid

$CH_3{-}\underset{\underset{\large CH_3}{|}}{CH}-\underset{\underset{\large CH_3}{|}}{CH}{-}\overset{\overset{\large O}{\|}}{C}{-}OH$ 2,3-dimethylbutanoic acid

$CH_3{-}CH_2{-}\underset{\underset{\underset{\underset{\large CH_3}{|}}{\large CH_2}}{|}}{CH}{-}\overset{\overset{\large O}{\|}}{C}{-}OH$ 2-ethylbutanoic acid

16.81 methyl propanoate; ethyl ethanoate; propyl methanoate; 1-methylethyl methanoate

16.82 a. ethyl 2-methylpropanoate b. 2-methylbutanoic acid c. ethyl thiobutanoate
d. sodium propanoate

16.83 $CH_3{-}\overset{\displaystyle O}{\overset{\|}{C}}{-}O{-}CH_2{-}CH_3$

16.84 a. $CH_3{-}CH_2{-}\overset{\displaystyle O}{\overset{\|}{C}}{-}O^-\ Na^+$; $CH_3{-}OH$

b. $CH_3{-}CH_2{-}\overset{\displaystyle O}{\overset{\|}{C}}{-}S{-}CH_3$

c. $CH_3{-}\overset{\displaystyle O}{\overset{\|}{C}}{-}O^-\ Na^+$

d. $HO{-}\overset{\displaystyle O}{\overset{\|}{C}}{-}CH_2{-}CH_2{-}\underset{\underset{\displaystyle CH_3}{|}}{\overset{\overset{\displaystyle OH}{|}}{CH}}{-}CH_2$

16.85 a. 2, 3, 5, 6 b. 1, 4 c. 4, 6 d. (1, 3), (2, 4)

16.86 a. all choices b. 6 c. 2, 3, 4, 5, 6 d. 1, 2

16.87 a. 1, 4, 6 b. 2, 5 c. 2, 3 d. (1, 4), (3, 6)

16.88 a. 1, 4, 6 b. 1, 2, 4, 6 c. 1, 3, 4, 5, 6 d. 2, 3, 5

Amines And Amides

Chapter 17

Solutions to Exercises

17.1 a. yes b. yes c. no d. yes e. no f. yes

17.3 a. 1° b. 1° c. 2° d. 2° e. 1° f. 3°

17.5 a. 2° b. 3° c. 3° d. 1° e. 2° f. 2°

17.7 a. ethylmethylamine b. propylamine
c. diethylmethylamine d. diphenylamine
e. isopropylmethylamine f. diisopropylamine

17.9 a. 3-pentanamine b. 2-methyl-3-pentanamine
c. N-methyl-3-pentanamine d. 1,5-pentanediamine
e. 2,3-butanediamine f. N,N-dimethyl-1-butanamine

17.11 a. 2-bromoaniline b. N-isopropylaniline
c. N-ethyl-N-methylaniline d. N-methyl-N-phenylaniline
e. N-ethyl-N-methylaniline f. N-(1-chloroethyl)aniline

17.13 a. $CH_3-CH_2-NH_2$

b. $CH_3-CH(CH_3)-N(CH(CH_3)-CH_3)-CH(CH_3)-CH_3$

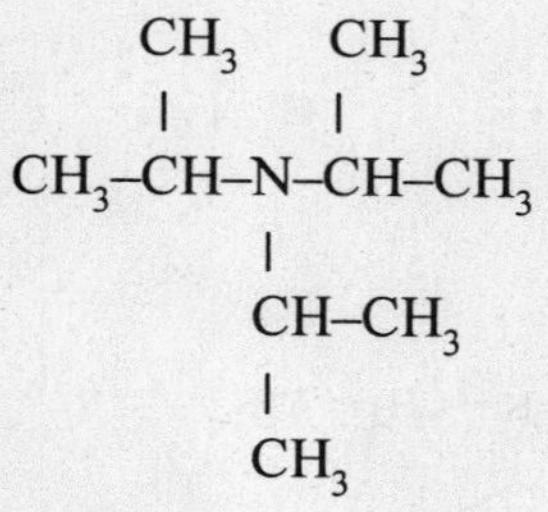

c. (benzene ring with NH_2 and adjacent CH_3)

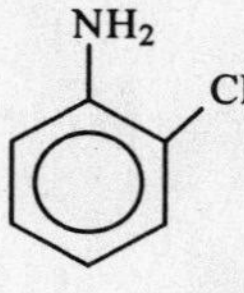

d. (benzene ring with $NH-CH_3$)

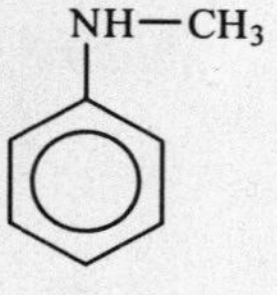

e. $CH_3-C(CH_3)(NH_2)-CH_2-CH_3$

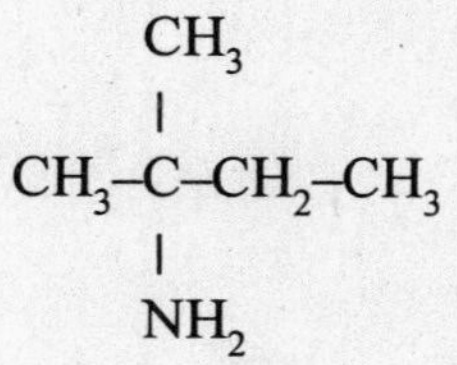

f. $H_2N-CH_2-CH_2-CH_2-CH_2-CH_2-CH_2-NH_2$

g. $CH_3-CH(NH_2)-C(=O)-CH_2-CH_3$

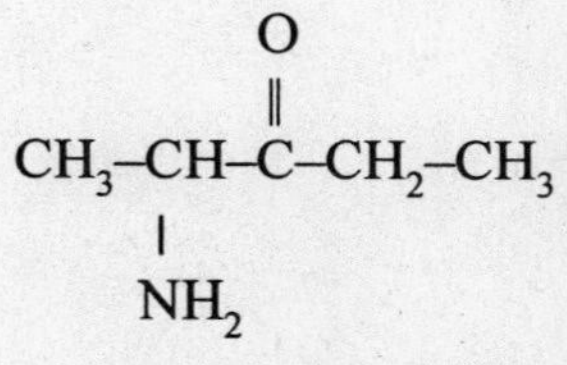

h. $CH_3-CH(NH_2)-C(=O)-OH$

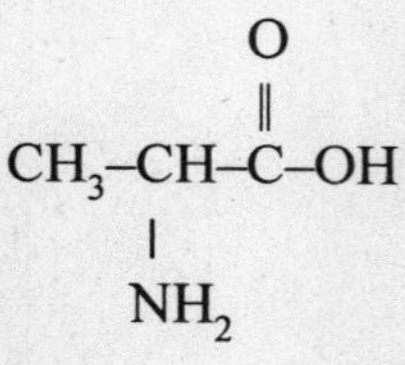

17.15 a. liquid b. gas c. gas d. liquid

17.17 a. 1 b. 3

17.19 hydrogen bonding is possible for the amine

17.21 a. $CH_3-CH_2-NH_2$; it has a shorter carbon chain

b. $H_2N-CH_2-CH_2-CH_2-NH_2$; it has two amino groups rather than one

17.23 a. $CH_3-CH_2-\overset{+}{N}H_3$

b. OH^-

c. $CH_3-\underset{\displaystyle CH_3}{\underset{|}{CH}}-NH-CH_3$

d. $CH_3-CH_2-\overset{+}{N}H_2-CH_2-CH_3 + OH^-$

17.25 a. dimethylammonium ion
b. triethylammonium ion
c. N,N-diethylanilinium ion
d. dimethylpropylammonium ion
e. propylammonium ion
f. N-isopropylanilinium ion

17.27 a. $CH_3-NH-CH_3$

b. $CH_3-CH_2-\underset{\displaystyle CH_2-CH_3}{\underset{|}{N}}-CH_2-CH_3$

c. $CH_3-CH_2-\underset{\displaystyle C_6H_5}{\underset{|}{N}}-CH_2-CH_3$

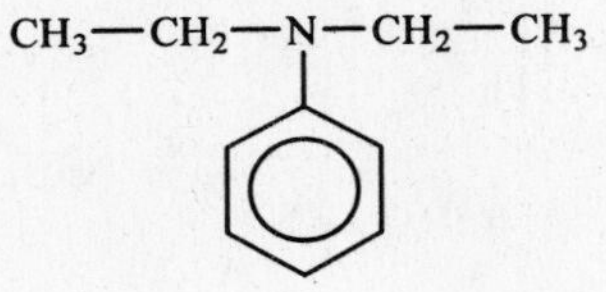

d. $CH_3-\underset{\displaystyle CH_3}{\underset{|}{N}}-CH_2-CH_2-CH_3$

e. $CH_3-CH_2-CH_2-NH_2$

f. 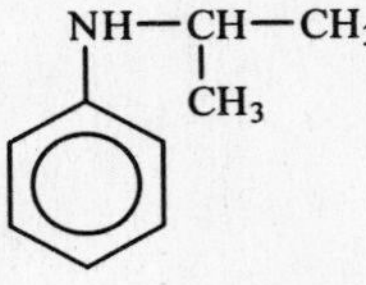 $C_6H_5-NH-\underset{\displaystyle CH_3}{\underset{|}{CH}}-CH_3$

17.29 a. $CH_3-CH_2-\overset{+}{N}H_3\ Cl^-$

b. $C_6H_5-\overset{+}{N}H_3\ Br^-$

c. $CH_3-\underset{\displaystyle CH_3}{\underset{|}{\overset{\displaystyle CH_3}{\overset{|}{C}}}}-NH_2$

d. HCl

17.31 a. $CH_3-CH-NH_2$
 |
 CH_3

b. $CH_3-\overset{+}{N}H_2-CH_3\ Cl^-$

c. 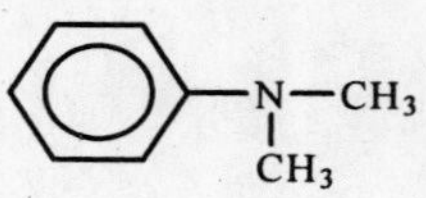 $C_6H_5-N-CH_3$, NaBr
 |
 CH_3

d. $CH_3-NH-CH_3$

17.33 a. propylammonium chloride
b. methylpropylammonium chloride
c. ethyldimethylammonium bromide
d. N,N-dimethylanilinium bromide

17.35 to increase the water solubility of the drug

17.37 ethylmethylamine hydrochloride

17.39 a. $CH_3-CH_2-CH_2-NH_2$, NaCl , H_2O

b. $CH_3-CH - N-CH_3$, NaBr , H_2O
 | |
 CH_3 CH_3

c. $CH_3-CH_2-NH-CH_2-CH_3$, NaCl , H_2O

d. CH_3
 |
CH_3-C-NH_2 , NaBr , H_2O
 |
 CH_3

17.41 ethylmethylamine, propyl chloride
ethylpropylamine, methyl chloride
methylpropylamine, ethyl chloride

17.43 a. CH_3
 |+
$CH_3-N-CH_2-CH_3\ Br^-$
 |
 CH_3

b. $CH_3-CH - N - CH-CH_3$
 | | |
 CH_3 CH_3 CH_3

c. CH_3
 |+
$CH_3-CH_2-CH_2-N-CH_2-CH_3\ Cl^-$
 |
 CH_3

d. $CH_3-CH_2-NH-CH_2-CH_3$

17.45 a. amine salt
b. quaternary ammonium salt
c. amine salt
d. quaternary ammonium salt

17.47 a. trimethylammonium bromide
b. tetramethylammonium chloride
c. ethylmethylammonium bromide
d. diethyldimethylammonium chloride

17.49 a. purine b. pyrrole c. imidazole d. indole

17.51 a. true b. false c. true d. false e. false f. false

17.53 a. yes b. yes c. no d. yes e. no f. yes

17.55 a. monosubstituted
b. disubstituted
c. unsubstituted
d. monosubstituted

17.57 a. 2° b. 3° c. 1° d. 2°

17.59 a. N-ethylethanamide
b. N,N-dimethylpropanamide
c. butanamide
d. N-methylmethanamide
e. 2-chloropropanamide
f. 2,N-dimethylpropanamide

17.61 a. N-ethylacetamide
b. N,N-dimethylpropionamide
c. butyramide
d. N-methylformamide
e. 2-chloropropionamide
f. 2,N-dimethylpropionamide

17.63 a. 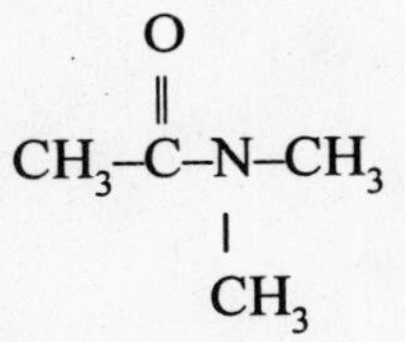 $CH_3-\overset{\overset{O}{\|}}{C}-\underset{\underset{CH_3}{|}}{N}-CH_3$

b. $CH_3-CH_2-\underset{\underset{CH_3}{|}}{CH}-\overset{\overset{O}{\|}}{C}-NH_2$

c. $CH_3-\underset{\underset{CH_3}{|}}{CH}-CH_2-\overset{\overset{O}{\|}}{C}-NH-CH_3$

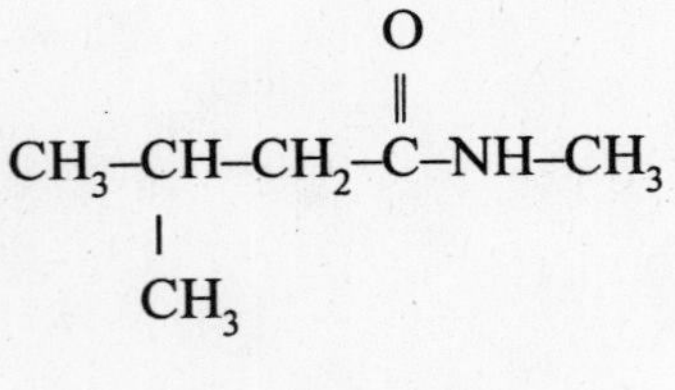

d. 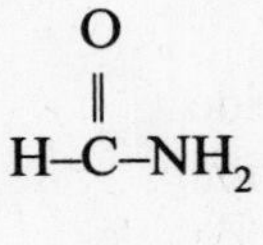 $H-\overset{\overset{O}{\|}}{C}-NH_2$

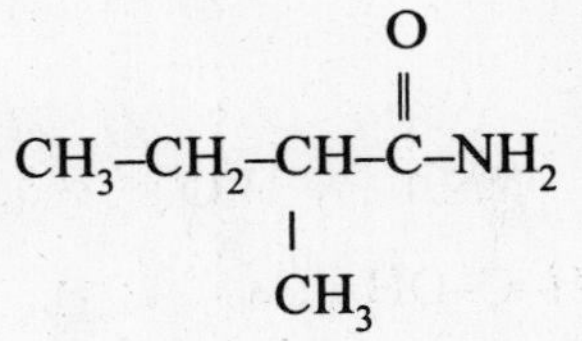

e. 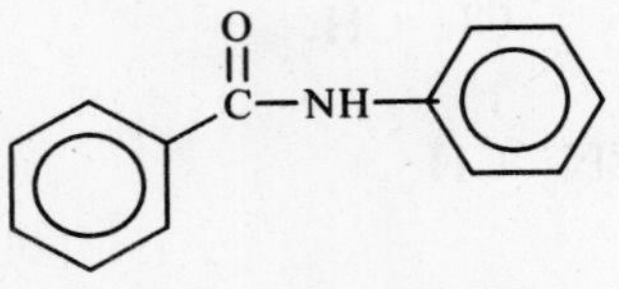 $\overset{\overset{O}{\|}}{C}-NH-$

f. 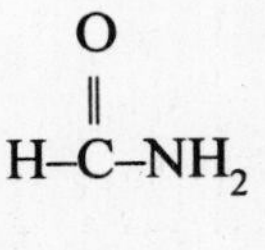 $H-\overset{\overset{O}{\|}}{C}-NH_2$

17.65 An electronegativity effect induced by the carbonyl oxygen atom makes the lone pair of electrons on the nitrogen atom unavailable

17.67 a. 5 b. 5

17.69 a. CH_3-NH_2

b. $CH_3-C(CH_3)_2-C(=O)-N(CH_3)-CH_3$

```
     CH3 O
      |  ||
b. CH3-C - C-N-CH3
      |     |
     CH3   CH3
```

c. NH_3

d. 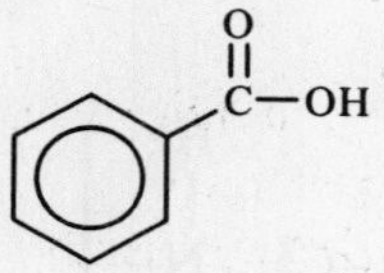 $C_6H_5-C(=O)-OH$ (benzoic acid)

17.71 a. $CH_3-C(=O)-OH$, $CH_3-NH-CH(CH_3)-CH_3$

b. $CH_3-CH_2-CH_2-CH_2-C(=O)-OH$, CH_3-NH_2

c. $CH_3-CH(CH_3)-C(=O)-OH$, CH_3-NH_2

d. $CH_3-CH(CH_3)-CH(CH_3)-C(=O)-OH$, CH_3-NH_2

17.73 a. $CH_3-CH_2-CH_2-C(=O)-OH$, CH_3-NH_2

b. $CH_3-CH_2-CH_2-C(=O)-OH$, $CH_3-\overset{+}{N}H_3\ Cl^-$

c. $CH_3-CH_2-CH_2-C(=O)-O^-\ Na^+$, CH_3-NH_2

d. 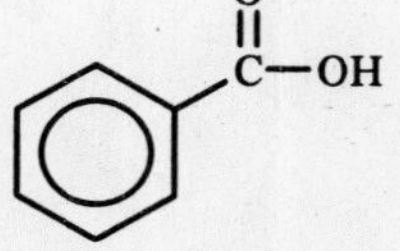 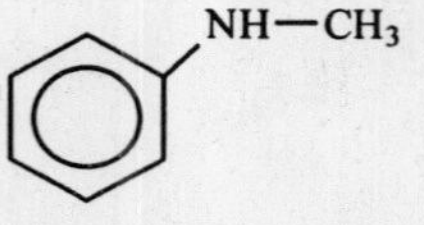 $C_6H_5-C(=O)-OH$, $C_6H_5-NH-CH_3$

17.75 diacid and diamine

17.77 $\left(-\overset{O}{\overset{\|}{C}}-CH_2-CH_2-\overset{O}{\overset{\|}{C}}-\overset{H}{\overset{|}{N}}-(CH_2)_4-\overset{H}{\overset{|}{N}}- \right)_n$

17.78 $\left(-\overset{O}{\overset{\|}{C}}-(CH_2)_4-\overset{O}{\overset{\|}{C}}-\overset{H}{\overset{|}{N}}-CH_2-CH_2-\overset{H}{\overset{|}{N}}- \right)_n$

17.79 a. $H-\overset{O}{\overset{\|}{C}}-NH_2$

b. $CH_3-\underset{NH_2}{\underset{|}{CH}}-CH_2-CH_2-CH_3$

c. $CH_3-CH_2-CH_2-\overset{CH_3}{\overset{|}{CH}}-\overset{O}{\overset{\|}{C}}-NH_2$

d. $CH_3-\overset{O}{\overset{\|}{C}}-NH-\overset{CH_3}{\overset{|}{CH}}-CH_3$

e. $CH_3-CH_2-\overset{+}{NH_2}-CH_2-CH_3\ Cl^-$

f. 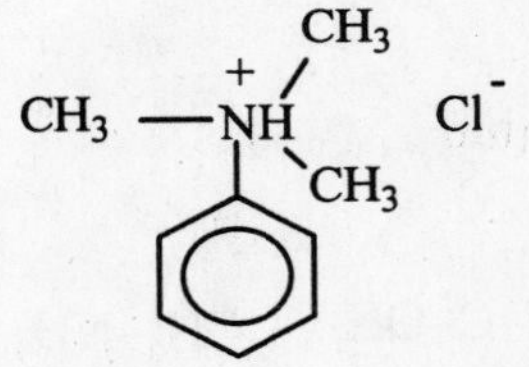 $CH_3-\overset{+}{N}H(CH_3)(C_6H_5)\ Cl^-$ (N bonded to CH_3, CH_3, CH_3 and a benzene ring)

17.80 a. $CH_3-CH_2-NH-CH_3$

b. $CH_3-CH_2-\overset{CH_3}{\overset{|}{CH}}-\overset{O}{\overset{\|}{C}}-\underset{CH_3}{\underset{|}{N}}-CH_3$

c. $CH_3-\underset{CH_3}{\underset{|}{\overset{CH_3}{\overset{|}{N^+}}}}-CH_3\ Cl^-$

d. $CH_3-CH_2-\overset{CH_3}{\overset{|}{CH}}-\overset{O}{\overset{\|}{C}}-O^-\ Na^+$

e. $CH_3-NH-CH_3$

f. $CH_3-CH_2-NH_3^+\ Cl^-$

17.81 $CH_3–CH_2–CH_2–CH_2–NH_2$

1-butanamine

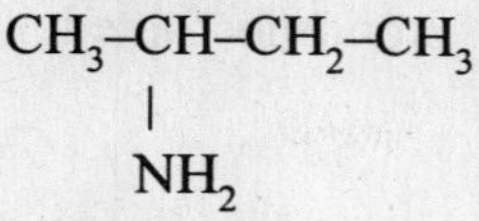

$CH_3–CH(NH_2)–CH_2–CH_3$

2-butanamine

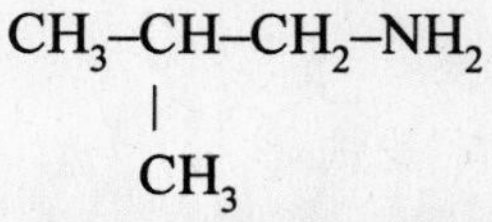

$CH_3–CH(CH_3)–CH_2–NH_2$

2-methyl-1-propanamine

$CH_3–C(CH_3)_2–NH_2$

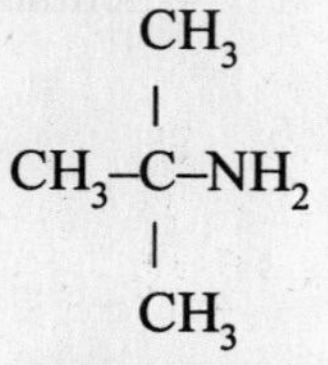

2-methyl-2-propanamine

$CH_3–CH_2–CH_2–NH–CH_3$

N-methyl-1-propanamine

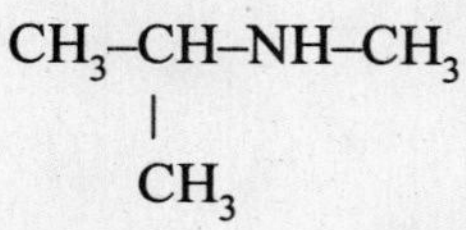

$CH_3–CH(CH_3)–NH–CH_3$

N-methyl-2-propanamine

$CH_3–CH_2–NH–CH_2–CH_3$

N-ethylethanamine

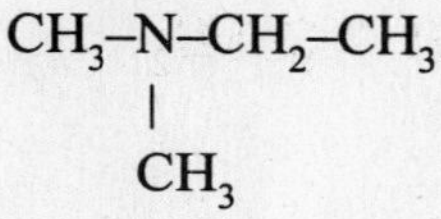

$CH_3–N(CH_3)–CH_2–CH_3$

N,N-dimethylethanamine

17.82 $CH_3–CH_2–C(=O)–NH_2$

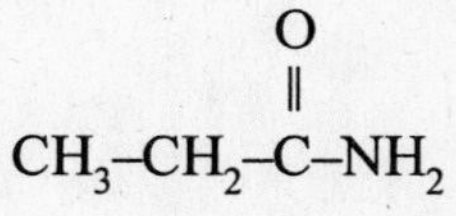

propanamide

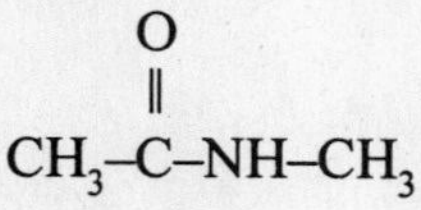

$CH_3–C(=O)–NH–CH_3$

N-methylethanamide

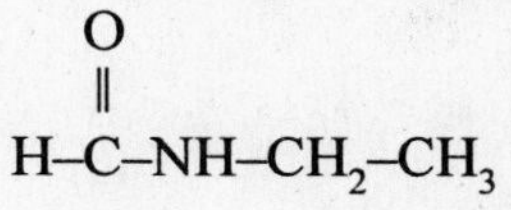

$H–C(=O)–NH–CH_2–CH_3$

N-ethylmethanamide

$H–C(=O)–N(CH_3)–CH_3$

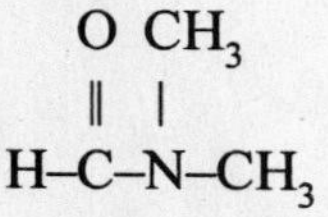

N,N-dimethylmethanamide

17.83 $CH_3–N^+(CH_3)_2–CH_2–CH_3\ Cl^-$

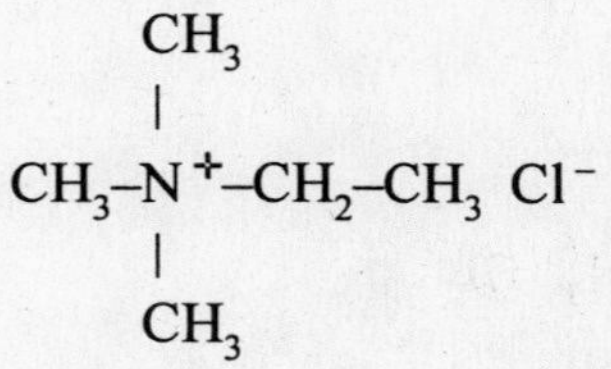

17.84 a. unsubstituted b. monosubstituted c. disubstituted d. disubstituted e. unsubstituted f. unsubstituted

17.85 a. amide b. amine c. amide d. amine e. amine f. amide

17.86 a. 1-butanamine b. 2-methyl-1-pentanamine c. N,2-dimethylpentanamine d. 3-methylpentanamide e. 1,4-pentandiamine f. 4-bromo-N-ethyl-N-methylpentanamide

17.87 a. 1, 2 b. 3 c. 4, 5 d. 6

17.88 a. 6 b. 3 c. 1 d. 5

17.89 a. 2, 3, 6 b. 6 c. 1, 4, 5 d. 2, 4

17.90 a. 2 b. 5 c. 3 d. 1, 2

Carbohydrates Chapter 18

Solutions to Exercises

18.1 Biochemistry is the study of the chemical substances found in living systems and the chemical interactions of these substances with each other.

18.3 proteins, lipids, carbohydrates, nucleic acids

18.5 $CO_2 + H_2O$ + solar energy $\xrightarrow[\text{plant enzymes}]{\text{chlorophyll}}$ carbohydrates + O_2

18.7 serve as structural elements, provide energy reserves

18.9 polyhydroxyaldehydes or polyhydroxyketones or compounds that yield such substances upon hydrolysis

18.11 a. one monosaccharide unit versus a few monosaccharide units
b. two monosaccharide units versus four monosaccharide units

18.13 superimposable objects have parts that coincide exactly at all points when the objects are laid upon each other

18.15 a. drill bit b. hand, foot, ear c. pop, peep

18.17 a. no b. no c. yes d. yes

18.19 a. CH_2–*CH–Br with Cl on CH_2 and Cl on *CH

b. CH_2–*C–*CH with Br on CH_2; Cl and Br on *C; Cl and Br on *CH

c. CH_2–*CH–*CH–*CH–C(=O)–H with OH on CH_2 and on each *CH

d. CH_2–*CH–*CH–*CH–*CH–CH_2 with OH on each carbon

18.21 a. zero b. two c. zero d. zero

18.23 structural isomers have different connectivity of atoms; stereoisomers have the same connectivity of atoms with different arrangements of atoms in space

18.25
a. H at top; Br— —Cl across; CH_3 at bottom
b. CH_3 at top; Br— —Cl across; H at bottom
c. CH_3 at top; Br— —H across; Cl at bottom
d. CH_3 at top; H— —Br across; Cl at bottom

18.27
a. CHO; H— —OH; HO— —H; HO— —H; CH_2OH
b. CH_2OH; C=O; HO— —H; H— —OH; HO— —H; CH_2OH
c. CHO; HO— —H; HO— —H; HO— —H; H— —OH; CH_2OH
d. CHO; HO— —H; H— —OH; HO— —H; H— —OH; CH_2OH

18.29 a. D-enantiomer b. D-enantiomer c. L-enantiomer d. L-enantiomer

18.31 a. diastereomers b. neither enantiomers nor diastereomers
c. enantiomers d. diastereomers

18.33 effect on plane-polarized light

18.35 a. same b. different c. same d. different

18.37 a. aldose b. ketose c. ketose d. ketose

18.39 a. aldohexose b. ketohexose c. ketotriose d. ketotetrose

18.41 a. D-galactose b. D-psicose c. dihydroxyacetone d. L-erythrulose

18.43 they differ at carbons 1 and 2; glucose has an aldehyde group and fructose has a ketone group

18.45 carbon 5

18.47 the difference involves the –OH group on the hemiacetal carbon atom; in the β-form it is on the same side of the ring as the $-CH_2OH$ group and in the α-form it is on the opposite side of the ring as the $-CH_2OH$ group

18.49 the fructose cyclization process involves carbons 2 and 5 and the ribose cyclization process involves carbons 1 and 4; both give five-membered rings

18.51 the cyclic and noncyclic forms interconvert; an equilibrium exists between forms

18.53 a. α-D-monosaccharide b. α-D-monosaccharide
c. β-D-monosaccharide d. α-D-monosaccharide

18.55 a. hemiacetal b. hemiacetal c. hemiacetal d. hemiketal

18.57 a.

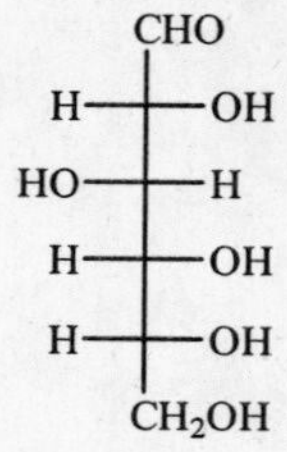

CHO
H—OH
HO—H
H—OH
H—OH
CH_2OH

b.

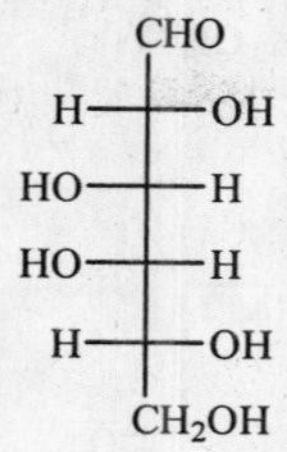

CHO
H—OH
HO—H
HO—H
H—OH
CH_2OH

c.

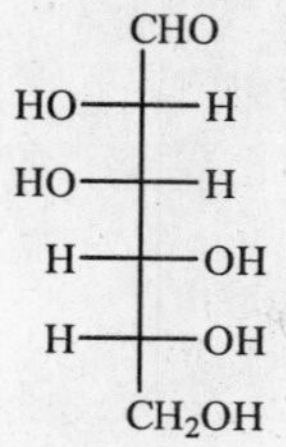

CHO
HO—H
HO—H
H—OH
H—OH
CH_2OH

d.

CH_2OH
C=O
H—OH
HO—H
H—OH
CH_2OH

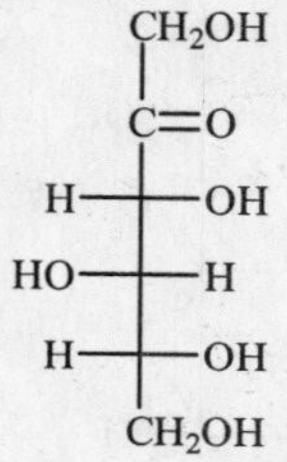

18.59 a. α-D-glucose b. α-D-galactose c. β-D-mannose d. α-D-sorbose

18.61 a.

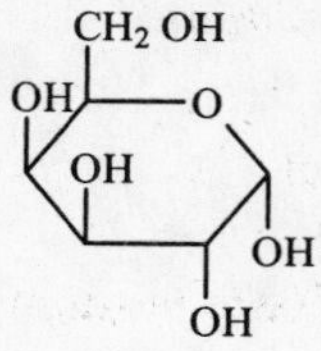

CH_2OH, O, OH, OH, OH, OH

b.

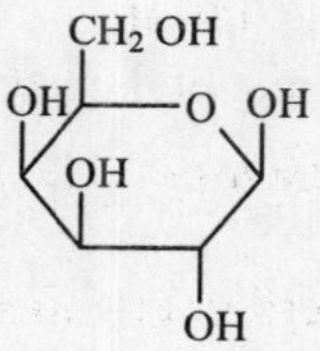

CH_2OH, O, OH, OH, OH, OH

c.

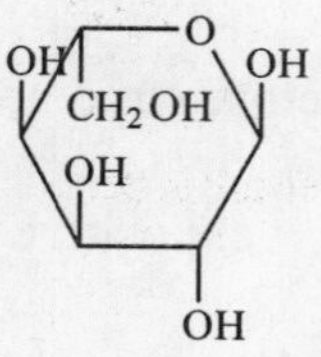

O, OH, OH, CH_2OH, OH, OH

d.

O, OH, CH_2OH, OH, OH, OH

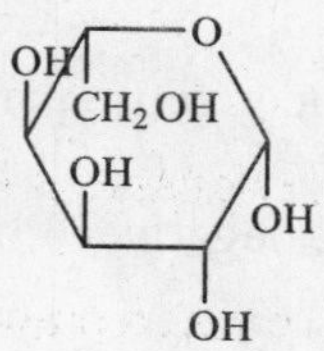

18.63 a. reducing sugar b. reducing sugar c. reducing sugar d. reducing sugar

18.65 the aldehyde group in glucose is oxidized to a carboxylic acid group; the Ag^+ ion in Tollens solution is reduced to Ag

18.67 a.

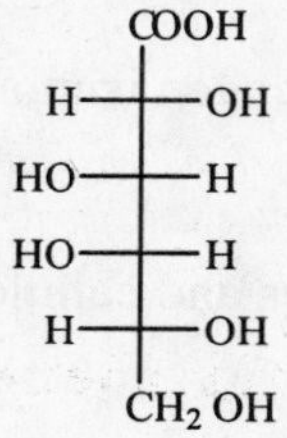

COOH
H—OH
HO—H
HO—H
H—OH
CH_2OH

b.

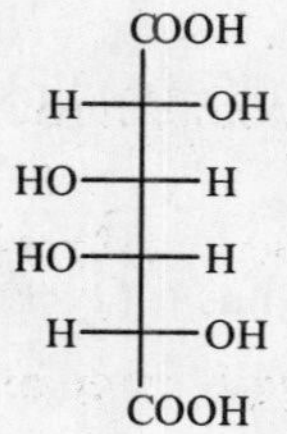

COOH
H—OH
HO—H
HO—H
H—OH
COOH

c.

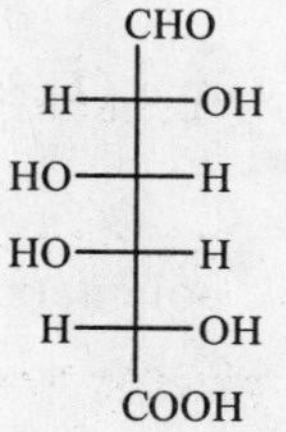

CHO
H—OH
HO—H
HO—H
H—OH
COOH

d.

CH_2OH
H—OH
HO—H
HO—H
H—OH
CH_2OH

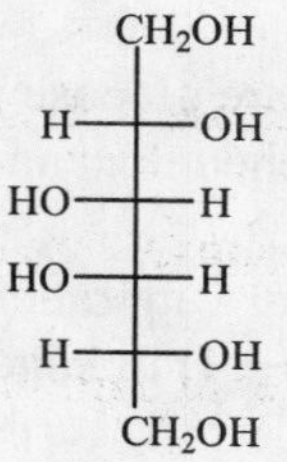

18.69 a. acetal b. acetal c. ketal d. acetal

18.71 a. alpha b. beta c. alpha d. beta

18.73 a. methyl alcohol b. ethyl alcohol c. ethyl alcohol d. methyl alcohol

18.75 a glycoside is an acetal or ketal formed from a cyclic form of a monosaccharide; a glucoside is a glycoside in which the monosaccharide is glucose

18.77 a.

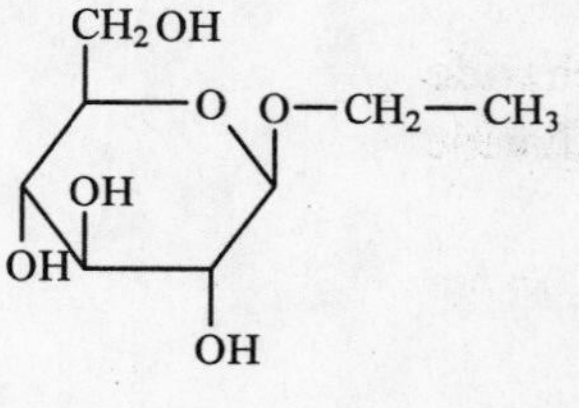

CH_2OH, O, $O-CH_2-CH_3$, OH, OH, OH

b.

CH_2OH, O, OH, OH, $O-CH_3$, OH

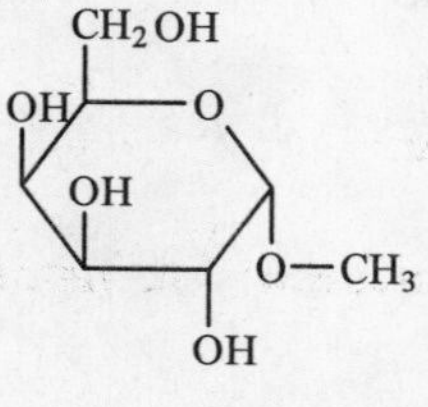

18.79 a.

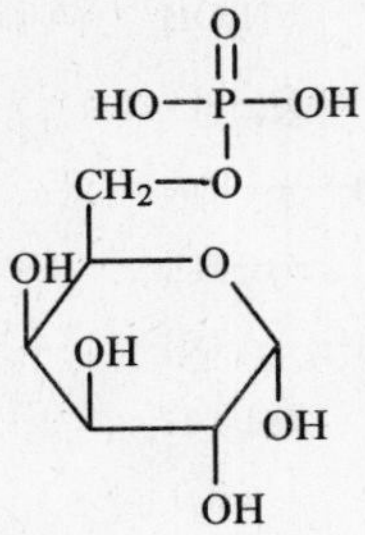

O
HO—P—OH
CH_2—O
OH O
OH
OH
OH

b.

CH_2OH
OH O
OH
OH
NH
C=O
CH_3

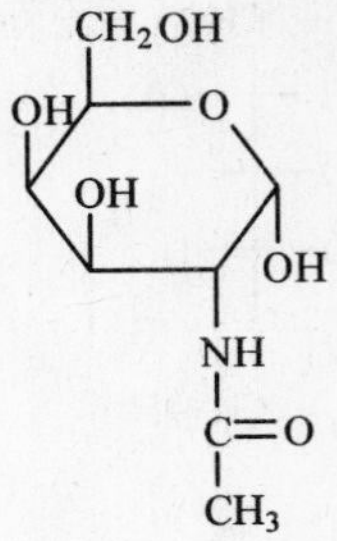

18.81 a. glucose and fructose b. glucose c. glucose and galactose d. glucose

18.83 the glucose part of the lactose structure has a hemiacetal carbon atom

18.85 a. negative b. positive c. positive d. positive

18.87 a. α (1 → 6) b. β (1 → 4) c. α (1 → 4) d. α (1 → 4)

18.89 a. alpha b. beta c. alpha d. beta

18.91 a. reducing sugar b. reducing sugar c. reducing sugar d. reducing sugar

18.93 a. glucose b. galactose and glucose c. glucose and altrose d. glucose

18.95 a. both are glucose polymers with α (1 → 4) and α (1 → 6) linkages; glycogen is more highly branched than amylopectin

b. both are unbranched glucose polymers; amylose has α (1 → 4) linkages and cellulose has β (1 → 4) linkages

18.97 a. amylopectin, glycogen
b. amylopectin, amylose, glycogen, cellulose
c. amylose, cellulose, chitin
d. cellulose, chitin

18.99 the human body possesses enzymes for α (1 → 4) linkages (starch) but not for β (1 → 4) linkages (cellulose)

18.101 a. chiral b. achiral c. chiral d. chiral

18.102 a. no b. no c. no d. yes

18.103 a. no b. no c. yes d. yes

18.104

CH_2OH	CH_2OH	CH_2OH	CH_2OH
C=O	C=O	C=O	C=O
H—OH	HO—H	H—OH	HO—H
H—H	HO—H	HO—H	H—OH
CH_2OH	CH_2OH	CH_2OH	CH_2OH

18.105 3-methyl hexane (hydrogen, methyl, ethyl and propyl groups attached to a carbon atom)

18.106 a. glucose, fructose b. glucose c. glucose d. glucose

18.107 a. homopolysaccharide b. homopolysaccharide c. heteropolysaccharide d. homopolysaccharide

18.108 a. strong oxidizing agent b. ethyl alcohol, H^+ ion c. water (H^+ ion or enzymes) d. enzymes

18.109 a. 4, 6 b. all choices c. 3, 4, 6 d. all choices

18.110 a. 2, 3, 5 b. 1, 3, 4 c. all choices d. 1, 2, 3, 4, 5

18.111 a. 6 b. 2, 3, 6 c. 1, 4, 5 d. 1, 2, 3, 4, 6

18.112 a. 2, 3, 5 b. (1, 4), (2, 5) c. (1, 6), (4, 6) d. all choices

Lipids

Chapter 19

Solutions to Exercises

19.1 lipids are organic compounds of biological origin that are water-insoluble but soluble in nonpolar solvents

19.3 a. insoluble b. soluble c. insoluble d. soluble

19.5 saponifiable lipids can be hydrolyzed under basic conditions while nonsaponifiable lipids cannot

19.7 a. long-chain b. medium-chain c. long-chain d. medium-chain

19.9 a. saturated b. polyunsaturated c. polyunsaturated d. monounsaturated

19.11 a. neither b. omega-3 c. omega-3 d. neither

19.13 a. nonessential b. essential c. nonessential d. nonessential

19.15 there are fewer attractions between fatty-acid carbon chains because of bends in the chains caused by the presence of double bonds

19.17 a. 18:1 acid b. 18:3 acid c. 14:0 acid d. 18:1 acid

19.19 cis, cis, cis-9, 12, 15-octadecatrienoic acid

19.21 there are more fatty acid residues present in a fat than an oil

19.23 triacylglycerols are triesters formed from the reaction between three fatty acid molecules and a glycerol molecule

19.25

$$\begin{array}{l} O \\ \| \\ H_2C-O-C-(CH_2)_{14}-CH_3 \\ \;| \quad\;\; O \\ \;| \quad\;\; \| \\ HC-O-C-(CH_2)_{14}-CH_3 \\ \;| \quad\;\; O \\ \;| \quad\;\; \| \\ H_2C-O-C-(CH_2)_{14}-CH_3 \end{array}$$

19.27

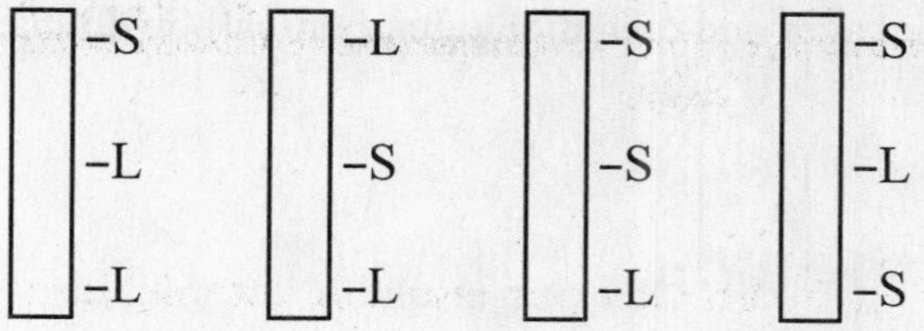

–S	–L	–S	–S
–L	–S	–S	–L
–L	–L	–L	–S

19.29 a. palmitic acid, myristic acid, oleic acid
b. oleic acid, palmitic acid, palmitoleic acid

19.31 a. glycerol and three fatty acids
b. glycerol and three fatty acid salts

19.33 $CH_2{-}CH{-}CH_2$ (each carbon bearing OH: OH OH OH), $CH_3{-}(CH_2)_{14}{-}COOH$, $CH_3{-}(CH_2)_{12}{-}COOH$,

$CH_3{-}(CH_2)_7{-}CH{=}CH{-}(CH_2)_7{-}COOH$

19.35 glycerol, palmitic acid, myristic acid, oleic acid

19.37 $CH_2{-}CH{-}CH_2$ (each carbon bearing OH: OH OH OH), $CH_3{-}(CH_2)_{14}{-}COO^-Na^+$, $CH_3{-}(CH_2)_{12}{-}COO^-Na^+$,

$CH_3{-}(CH_2)_7{-}CH{=}CH{-}(CH_2)_7{-}COO^-Na^+$

19.39 glycerol, sodium palmitate, sodium myristate, sodium oleate

19.41 hydrogenation requires the presence of double bonds

19.43 six

19.45 a.

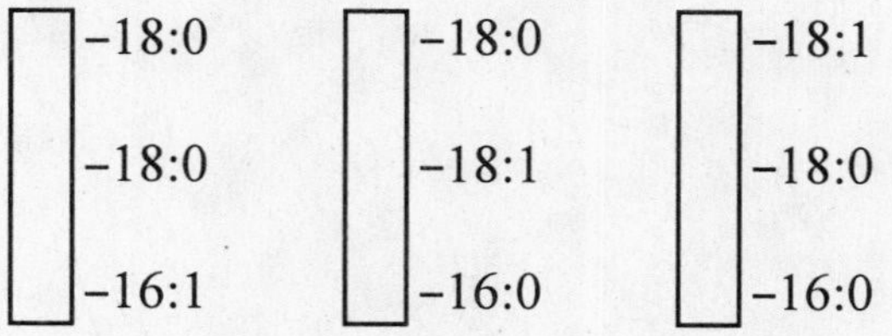

–18:0	–18:0	–18:1
–18:0	–18:1	–18:0
–16:1	–16:0	–16:0

b.

–18:0	–18:0	–18:0
–18:1A	–18:1B	–18:0
–16:0	–16:0	–16:1

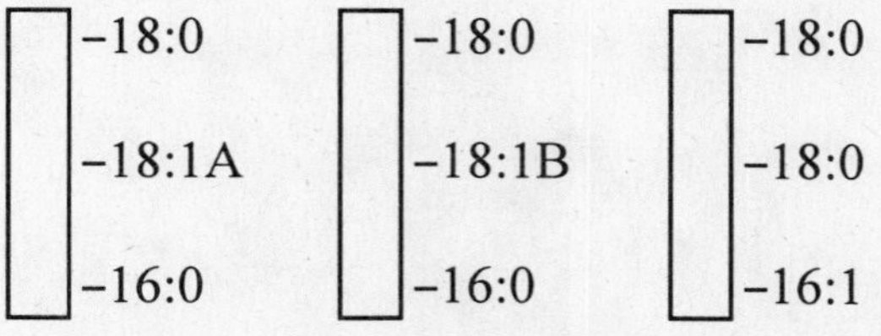

18:1A and 18:1B denote, respectively, the hydrogenation of the first and second double bonds in the 18:2 acid

19.47 rancidity results from hydrolysis of ester linkages and oxidation of carbon-carbon double bonds

19.49

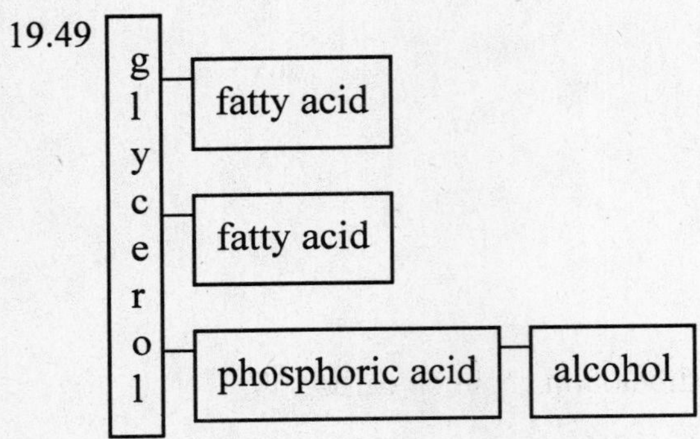

19.51 $HO\text{–}CH_2\text{–}CH_2\text{–}\overset{+}{N}\text{–}(CH_3)_3$, $HO\text{–}CH_2\text{–}CH_2\text{–}\overset{+}{N}H_3$, $HO\text{–}CH_2\text{–}\underset{\displaystyle COO^-}{\underset{|}{CH}}\text{–}\overset{+}{N}H_3$

19.53 the head is the combined glycerol-phosphoric acid-amino alcohol part and the tails are the two fatty acids

19.55 | fatty acid | – | long-chain alcohol |

19.57 $CH_3\text{–}(CH_2)_{14}\text{–}\overset{\displaystyle O}{\overset{\|}{C}}\text{–}O\text{–}(CH_2)_{15}\text{–}CH_3$

19.59 a wax is a monoester (monoalcohol and fatty acid) and a triacylglycerol is a triester (trialcohol and 3 fatty acids)

19.61

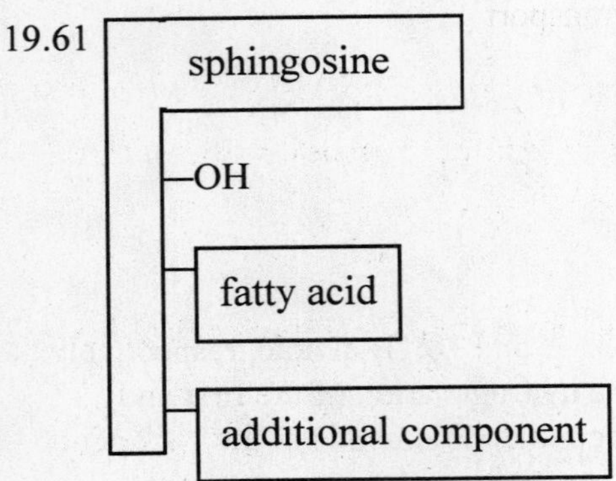

19.63 at carbon 2 of the carbon chain

19.65 the head is the additional component, the fatty acid is one tail, and the sphingosine chain is the other tail

19.67 the additional component in phosphoric acid-choline is sphingomyelins and it is a monosaccharide in cerebrosides

19.69

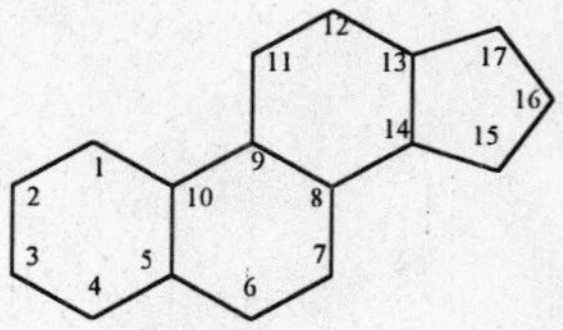

19.71 –OH on carbon 3, $-CH_3$ on carbons 10 and 13, hydrocarbon chain on carbon 17

19.73 solubilize lipids in an aqueous environment

19.75 sex hormones and adrenocortical hormones

19.77 estradiol has an –OH group on carbon 3 and testosterone has a ketone group on carbon 3, testosterone has an extra $-CH_3$ group at carbon 10

19.79 prostaglandins have a bond between carbons 8 and 12, which creates a cyclopentane ring

19.81 inflammatory process, pain and fever production, blood pressure regulation, induction of blood clotting, control of some reproductive functions, regulation of sleep/wake cycle

19.83 phosphoacylglycerols, sphingolipids

19.85 a two-layer thick structure of lipid molecules with nonpolar "tails" in the interior and polar "heads" on the exterior

19.87 creates "open" areas in the lipid bilayer

19.89 protein help is required in facilitated transport but not in passive transport

19.91 a. active transport b. facilitated transport c. active transport
d. passive transport and facilitated transport

19.93 a. neither b. glycerol-based c. neither d. sphingosine-based e. neither f. neither

19.94 a. no b. no c. no d. yes e. yes f. no

19.95 a. sphingomyelins b. triacylglycerols c. steroids d. leukotrienes e. prostaglandins
f. cerebrosides

19.96 a. storage lipid, nonpolar lipid b. storage lipid, nonpolar lipid
c. membrane lipid, polar lipid d. nonpolar lipid
e. membrane lipid, polar lipid f. membrane lipid, polar lipid

19.97 a. yes b. yes c. yes d. yes e. yes f. yes

19.98 a. phosphoacylglycerols and sphingomyelins b. sphingomyelins and cerebrosides c. cerebrosides d. sphingomyelins

19.99 a. 3, 0, 0 b. 4, 0, 0 c. 2, 1, 0 d. 1, 0, 0 e. 0, 1, 1 f. 4, 0, 0

19.100 a. no b. yes c. yes d. yes e. yes f. no

19.101 a. 2 b. 1, 2 c. 1, 2, 5, 6 d. 2, 3, 4

19.102 a. all b. 1, 2, 3, 4 c. all d. 5

19.103 a. 1, 2, 3, 5 b. 1, 5 c. 3 d. 3, 5

19.104 a. 2, 4, 5 b. 1, 6 c. 2, 4, 5 d. 1, 2, 4, 5, 6

Proteins | Chapter 20

Solutions to Exercises

20.1 a. yes b. no c. no yes

20.3 the identity of the R group (side chain)

20.5 a. phenylalanine, tyrosine, tryptophan b. methionine, cysteine
c. aspartic acid, glutamic acid d. serine, threonine, tyrosine

20.7 an amino group is part of the side chain

20.9 the side chain is part of a cyclic structure

20.11 a. alanine b. leucine c. methionine d. tryptophan

20.13 asparagine, glutamine, isoleucine, tryptophan

20.15 a. polar neutral b. polar acidic c. nonpolar d. polar neutral

20.17 L-family

20.19 a. $COOH$ (top); H_2N—C—H; CH_2—OH (bottom)

b. $COOH$ (top); H—C—NH_2; CH_2—OH (bottom)

c. $COOH$ (top); H—C—NH_2; CH_3 (bottom)

d. $COOH$ (top); H_2N—C—H; CH_2—$CH(CH_3)$—CH_3 (bottom)

20.21 they exist as charged species (zwitterions)

20.23 a. COO^- (top); $H_3\overset{+}{N}$—C—H; CH_2—$CH(CH_3)$—CH_3 (bottom)

b. COO^- (top); $H_3\overset{+}{N}$—C—H; $CH(CH_3)$—CH_2—CH_3 (bottom)

c. COO^- (top); $H_3\overset{+}{N}$—C—H; CH_2—SH (bottom)

d. COO^- (top); $H_3\overset{+}{N}$—C—H; H (bottom)

20.25 a. COO^- (top); $H_3\overset{+}{N}$—C—H; CH_2—OH (bottom)

b. $COOH$ (top); $H_3\overset{+}{N}$—C—H; CH_2—OH (bottom)

c. COO^- (top); H_2N—C—H; CH_2—OH (bottom)

d. $COOH$ (top); $H_3\overset{+}{N}$—C—H; CH_2—OH (bottom)

20.27 the pH at which zwitterion concentration in a solution is maximized

20.29 two –COOH groups are present, which deprotonate at different times

20.31 a. towards positive electrode b. isoelectric
c. towards negative electrode d. towards positive electrode

20.33 aspartic acid migrates toward the positive electrode, histidine migrates toward the negative electrode, and valine does not migrate

20.35 a short sequence of amino acids linked through peptide bonds

20.37 carboxyl group and amino group

20.39 $H_3\overset{+}{N}-\underset{\underset{CH_3}{|}}{\underset{CH-CH_3}{\underset{|}{CH}}}-\overset{O}{\overset{\|}{C}}-\overset{H}{\overset{|}{N}}-\underset{\underset{C_6H_5}{|}}{\underset{CH_2}{\underset{|}{CH}}}-\overset{O}{\overset{\|}{C}}-\overset{H}{\overset{|}{N}}-\underset{\underset{SH}{|}}{\underset{CH_2}{\underset{|}{CH}}}-COO^-$

20.41 Ser is the N-terminal end in Ser-Cys and Cys is the N-terminal end in Cys-Ser

20.43 Ser–Val–Gly Val–Ser–Gly Gly–Val–Ser
Ser–Gly–Val Val–Gly–Ser Gly–Ser–Val

20.45 a. Ser–Ala–Cys b. Asp–Thr–Asn

20.47 a. two b. two

20.49 nothing; they are the same

20.51 the sequence of amino acids in the protein chain

20.53 α-helix, β-pleated sheet, triple helix

20.55 intermolecular involves two separate chains and intramolecular involves a single chain bending back on itself

20.57 yes, both α-helix and β-pleated sheet can occur at different locations in the same molecule

20.59 secondary structure hydrogen bonding involves C=O· · · H–N interactions; tertiary structure hydrogen bonding involves R-group interactions

20.61 a. hydrophobic b. electrostatic c. disulfide bond d. hydrogen bonding

20.63 a multi-chain protein

20.65 a. fibrous protein: long, thin, fibrous shape; globular protein; a roughly spherical or globular shape
b. fibrous protein: water-insoluble; globular protein: dissolves in water or forms a stable suspension in water

20.67 a simple protein contains only amino acids; a conjugated protein has one or more other chemical components besides amino acids

20.69 yes, both Ala and Val are products in each case

20.71 drug hydrolysis would occur in the stomach

20.73 Ala–Gly–Met–His–Val–Arg

20.75 five: Ala–Gly–Ser, Gly–Ser–Tyr, Ala–Gly, Gly–Ser, Ser–Tyr

20.77 secondary, tertiary, and quaternary

20.79 primary structure is the same

20.81 triple helix

20.83 bonded to 5-hydroxylysine residues

20.85 an antigen is a substance foreign to the human body and an antibody is a substance that defends against an invading antigen

20.87 four polypeptide chains which have constant and variable amino acid regions, two chains are longer than the other two; 1 to 12% carbohydrate present by mass; long and short chains are connected through disulfide linkages

20.89 suspend and transport lipids in the bloodstream

20.91 a. tertiary b. tertiary c. secondary d. primary

20.92 a. alanine b. leucine c. threonine d. aspartic acid

20.93 a. +1 b. +1 c. +1 d. +3

20.94 a. –1 b. –1 c. –4 d. –1

20.95

$$\begin{array}{ccc} & COOH & \\ & | & \\ H_2N - & + & - H \\ H_3C - & + & - H \\ & | & \\ & CH-CH_3 & \\ & | & \\ & CH_3 & \end{array} \qquad \begin{array}{ccc} & COOH & \\ & | & \\ H - & + & - NH_2 \\ H - & + & - CH_3 \\ & | & \\ & CH-CH_3 & \\ & | & \\ & CH_3 & \end{array} \qquad \begin{array}{ccc} & COOH & \\ & | & \\ H_2N - & + & - H \\ H - & + & - CH_3 \\ & | & \\ & CH-CH_3 & \\ & | & \\ & CH_3 & \end{array} \qquad \begin{array}{ccc} & COOH & \\ & | & \\ H - & + & - NH_2 \\ H_3C - & + & - H \\ & | & \\ & CH-CH_3 & \\ & | & \\ & CH_3 & \end{array}$$

20.96 a. 24 b. 36 c. 20

20.97 collagen has a triple-helix structure; alpha-keratin has a double-double helix structure

20.98 a. $H_3\overset{+}{N}-CH(CH_2OH)-COOH$; $H_3\overset{+}{N}-CH(CH_3)-COOH$; $H_3\overset{+}{N}-CH(CH_2SH)-COOH$

b. $H_2\overset{+}{N}-CH(CH_2OH)-COO^-$; $H_2\overset{+}{N}-CH(CH_3)-COO^-$; $H_2\overset{+}{N}-CH(CH_2SH)-COO^-$

20.99 a. all choices b. 4 c. 1 d. 2, 4

20.100 a. 1, 5, 6 b. 3, 4 c. 3, 4 d. 1, 3, 4, 5, 6

20.101 a. 4, 5 b. 1, 2, 3 c. 6 d. 1, 2, 3, 4, 5

20.102 a. 1, 4, 5, 6 b. 3 c. 3, 6 d. (1, 5), (1, 6), (5, 6)

Enzymes — Chapter 21

Solutions to Exercises

21.1 catalyst

21.3 more efficient, more specific

21.5 a. yes b. no c. yes d. yes

21.7 a. add a carboxyl group to pyruvate b. remove H_2 from an alcohol
c. reduce an L-amino acid d. hydrolyze maltose

21.9 a. sucrase b. pyruvate decarboxylase
c. glucose isomerase d. lactate dehydrogenase

21.11 a. pyruvate b. galactose
c. an alcohol d. an L-amino acid

21.13 a. isomerase b. lyase c. ligase d. transferase

21.15 a. isomerase b. lyase c. transferase d. hydrolase

21.17 a. decarboxylase b. lipase c. phosphatase d. dehydrogenase

21.19 a. conjugated b. conjugated c. simple d. conjugated

21.21 A cofactor may be inorganic or organic; a coenzyme is a organic cofactor

21.23 to provide additional functional groups

21.25 the portion of an enzyme actually involved in the catalysis process

21.27 the substrate must have the same shape as the active site

21.29 interactions with amino acid R groups

21.31 a. accepts only one substrate
b. accepts substrate with a particular type of bond

21.33 absolute specificity and stereochemical specificity

21.35 a. absolute b. stereochemical

21.37 rate increases until enzyme denaturation occurs

21.39 enzymes vary in the number of acidic and basic groups present

21.41

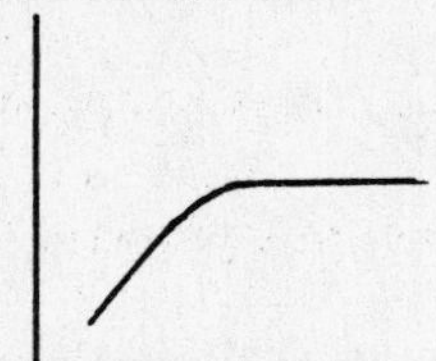

21.43 nothing; the rate remains the same

21.45 no; only one molecule may occupy the active site at a given time

21.47 a. reversible competitive b. reversible noncompetitive, irreversible
c. reversible noncompetitive, irreversible d. reversible noncompetitive

21.49 enzyme that has quaternary structure and more than one binding site

21.51 the product of a subsequent reaction in a series of reactions inhibits a prior reaction

21.53 inactive precursor of an enzyme

21.55 so that the enzymes do not "digest" the glands that produce them

21.57 competitive inhibition of the enzyme necessary for conversion of PABA to folic acid

21.59 it has absolute specificity for bacterial transpeptidase

21.61 dietary organic compounds needed by the body in trace amounts

21.63 13

21.65 a. fat-soluble b. water-soluble c. water-soluble d. water-soluble

21.67 a. likely b. unlikely c. unlikely d. unlikely

21.69 serves as a cosubstrate in the formation of collagen

21.71 coenzymes

21.73 a. all B vitamins b. B_{12} c. B_{12} d. thiamin and biotin

21.75 alcohol, aldehyde, acid

21.77 vitamin K is necessary for the formation of blood clotting agents; vitamin E functions as an antioxidant

21.79 cofactors, bone components, regulation of fluid movement

21.81 a. P, Cl b. Na, K, Ca, Mg, Cl c. Ca, P d. Ca, K, Mg

21.83 a. transport of oxygen b. formation of the hormone thyroxin

21.85 a. 0.15 mg b. 0.070 mg (male), 0.055 mg (female)

21.87 a. an apoenzyme is the protein portion of a conjugated enzyme; a proenzyme is an inactive precursor of an enzyme
b. a simple enzyme contains only protein; an allosteric enzyme has two or more protein chains and two binding sites
c. a coenzyme is an organic cofactor; an isoenzyme is one of several similar forms of an enzyme
d. a conjugated enzyme has both a protein and a nonprotein portion; holoenzyme is another name for a conjugated enzyme

21.88 a. alcohol, ketone b. double bond, alcohol
c. double bond, alcohol d. double bond, ketone

21.89 a. no b. no c. no d. yes e. yes f. yes

21.90 a. major mineral b. major mineral c. trace mineral d. trace mineral e. trace mineral
f. major mineral

21.91 a. oxidation-reduction reactions
b. addition of a group to or removal of a group from a double bond in a manner that does not involve hydrolysis or oxidation/reduction
c. conversion of a compound into another isomeric with it
d. bonding together of two molecules with the involvement of ATP
e. hydrolysis reactions
f. transfer of functional groups between two molecules

21.92 enzyme plus substrate produces an enzyme-substrate complex which breaks apart to regenerate the enzyme and a product molecule

21.93 a. ethanol b. zinc ion c. protein molecule d. alcohol dehydrogenase

21.94 a. tissue plasminogen activator b. lactate dehydrogenase c. creatine phosphokinase
d. aspartate transaminase

21.95 a. 1, 2, 3 b. 4 c. 1 d. 6

21.96 a. 2, 3 b. 2, 3, 5 c. 1, 2, 4, 5, 6 d. (1, 4)

21.97 a. 1, 3, 6 b. 2 c. 1, 3, 4, 5, 6 d. 1, 3, 6

21.98 a. 4, 5, 6 b. 1, 2, 3 c. 4, 6 d. 2, 3

Nucleic Acids — *Chapter 22*

Solutions to Exercises

22.1 at carbon 2 ribose has an –H atom and –OH group and deoxyribose has 2 –H atoms

22.3 a. pyrimidine b. pyrimidine c. purine d. purine

22.5 a. one b. four c. one

22.7 a. adenine b. guanine c. thymine d. uracil

22.9 a. ribose b. 2-deoxyribose c. 2-deoxyribose d. ribose

22.11 a. false b. false c. false d. false

22.13

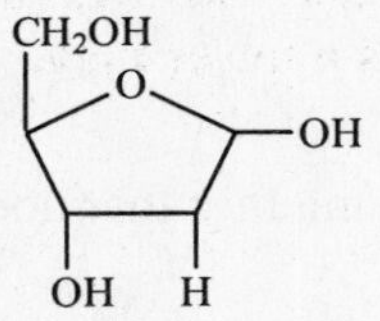

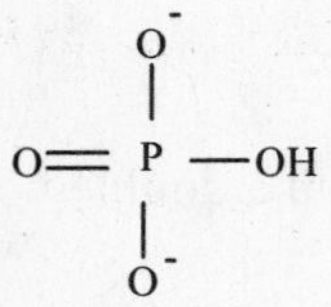

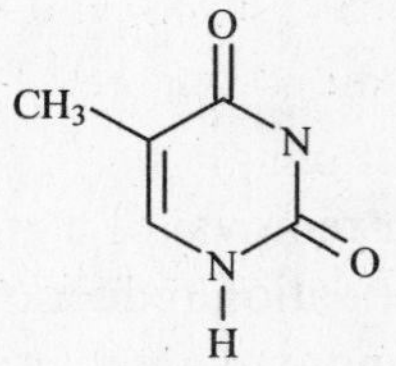

CH₂OH, O, OH, OH, H; O^-, O= P —OH, O^-; O, CH₃, N, N, O, H

22.15 sequence of alternating pentose and phosphate units

22.17 base sequence

22.19 5' end has a phosphate group attached to the 5' carbon; 3' end has a hydroxyl group attached to the 3' carbon

22.21

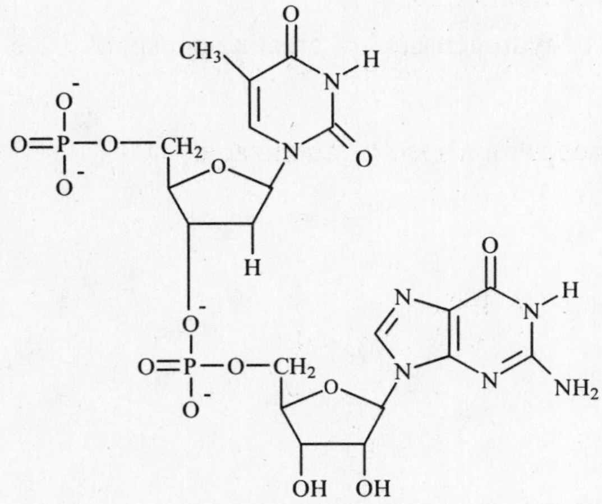

22.23 two polynucleotide chains coiled around each other in a helical fashion

22.25 a. 36% b. 14% c. 14%

22.27 G–C pairing involves 3 hydrogen bonds and A–T pairing involves 2 hydrogen bonds

22.29 5'–TAGCC–3'

22.31 a. 3'–TGCATA–5' b. 3'–AATGGC–5' c. 5'–CGTATT–3' d. 3'–TTGACC–5'

22.33 20 hydrogen bonds; 3 per G–C and 2 per A–T

22.35 catalyzes the unwinding of the double helix structure

22.37 a. 3'–TGAATC–5' b. 3'–TGAATC–5' c. 5'–ACTTAG–3'

22.39 the two strands are antiparallel (5' → 3' and 3' → 5') and only the 5' → 3' strand can grow continuously

22.41 a DNA molecule bound to a group of proteins

22.43 (1) RNA contains ribose instead of deoxyribose, (2) RNA contains the base U instead of T, (3) RNA is single-stranded rather than double-stranded, and (4) RNA has a lower molecular mass

22.45 a. ptRNA b. tRNA c. tRNA d. ptRNA

22.47 DNA unwinding, base pairing, ribonucleotide linkage, RNA release

22.49 A–U, C–G, G–C, T–A

22.51 3'–UACGAAU–5'

22.53 3'–AAGCGTC–5'

22.55 exons are the parts of genes that convey genetic information

22.57 3'–AGUCAAGU–5'

22.59 a three-nucleotide sequence in mRNA that codes for a specific amino acid

22.61 a. Leu b. Asn c. Ser d. Gly

22.63 a. CUC, CUA, CUG, UUA, UUG b. AAC
c. AGC, UCU, UCC, UCA, UCG d. GGU, GGC, GGA

22.65 the base T cannot be present in a codon

22.67 Met–Lys–Glu–Asp–Leu

22.69 a cloverleaf shape with three hairpin loops and one open side

22.71 covalent bond

22.73 a. UCU b. GCA c. AAA d. GUU

22.75 a. Thr b. Leu c. Pro d. Ser

22.77 (1) activation of tRNA, (2) initiation, (3) elongation, (4) termination, and (5) post-translational processing

22.79 A site

22.81 gly: GGU, GGC, GGA or GGG; ala: GCU, GCC, GCA or GCG; cys: UGU or UGC; val: GUU, GUC, GUA or GUG; tyr: UAU or UAC

22.83 a. Gly–Tyr–Ser–Ser–Pro b. Gly–Tyr–Ser–Ser–Pro c. Gly–Tyr–Ser–Ser–Thr

22.85 a DNA or RNA molecule with a protein coating

22.87 (1) attaches itself to cell membrane, (2) opens a hole in the membrane, and (3) injects itself into the cell

22.89 contains a "foreign" gene

22.91 host for a "foreign" gene

22.93 recombinant DNA is incorporated into a host cell

22.95 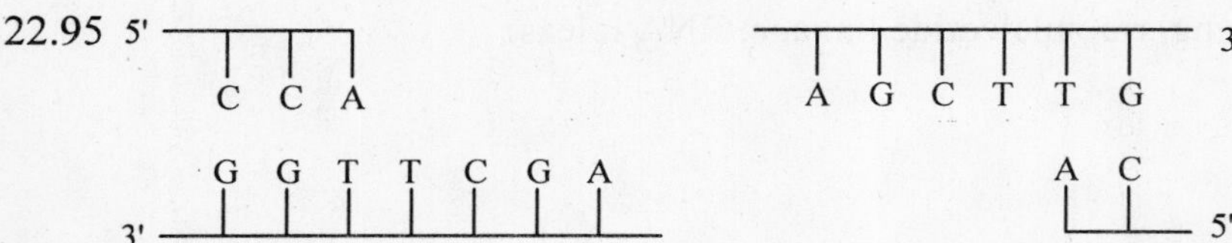

22.97 a. thymine is a methyl uracil; the methyl group is on C–5'
b. adenine is the 6-amino derivative of purine, and guanine is the 2-amino-6-oxo derivative

22.98 a. 3'-GTATGTCGGACCTTCGAT-5'
b. 3'–GUAUGUCGGACCUUCGAU-5'
c. 5'–GUA–UGU–CGG–ACC–UUC–GAU–5'
d. 5'–CAU–ACA–GCC–UGG–AAG–CUA–3'
e. Val–Cys–Arg–Thr–Phe–Asp

22.99 a. 1 b. 4 c. 2 d. 4

22.100 a. 3 b. 2 c. 1 d. 3

22.101 a. 3 b. 1 c. 4 d. 1, 2, 3, and 4

22.102 a. 1 b. 2 c. 4 d. 3

22.103 A

22.104 3.3 million base pairs spread over 23 pairs of chromosomes and 50,000 to 100,000 genes

22.105 a. 3, 5, 6 b. 1, 2, 4 c. (1, 2), (1, 4), (2, 4), (3, 5), (3, 6), (5, 6) d. (2, 5)

22.106 a. 1, 2 b. 4, 6 c. 1, 2, 4, 6 d. 1, 2

22.107 a. 1, 6 b. 3, 5, 6 c. 2, 4, 5, 6 d. 3, 5, 6

22.108 a. all b. 1 c. 5 d. (2, 6), (3, 4)

Biochemical Energy Production — Chapter 23

Solutions to Exercises

23.1 anabolism – synthetic; catabolism – degradative

23.3 a series of consecutive biochemical reactions

23.5 large molecules are broken down to smaller molecules, energy is produced

23.7 prokaryotic cells have no nucleus and the DNA is usually a single circular molecule; eukaryotic cells have their DNA in a membrane-enclosed nucleus

23.9 a small structure within the cell cytoplasm that carries out a specific cellular function

23.11 inner membrane

23.13 region between the inner and outer membranes

23.15 adenosine triphosphate

23.17

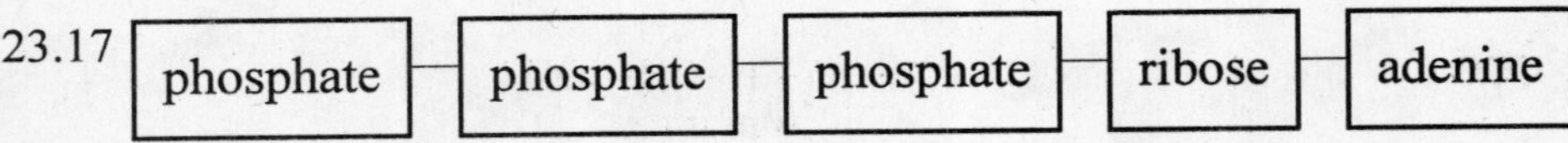

23.19 three phosphates versus one phosphate

23.21 adenine versus guanine

23.23 $\text{ATP} \xrightarrow{H_2O} \text{ADP} + P_i$

23.25 flavin adenine dinucleotide

23.27

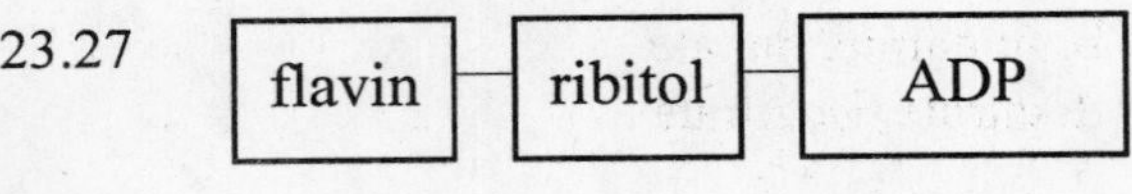

23.29

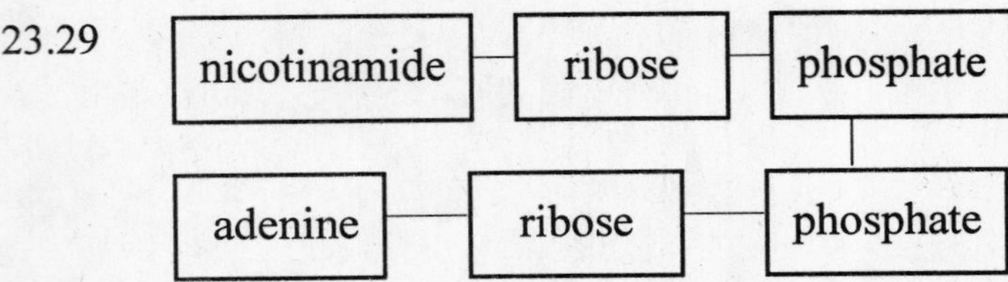

23.31 nicotinamide subunit

23.33 a. $FADH_2$ b. NAD^+

23.35 a. nicotinamide b. riboflavin

23.37 [2-aminoethanethiol] — [pantothenic acid] — [phosphorylated ADP]

23.39 a compound with a greater free energy of hydrolysis than is typical for a compound

23.41 free monophosphate species

23.43 a. phosphoenolpyruvate b. creatine phosphate c. 1,3-diphosphoglycerate d. AMP

23.45 (1) digestion, (2) acetyl group formation, (3) citric acid cycle, (4) electron transport chain and oxidative phosphorylation

23.47 tricarboxylic acid cycle, Kreb's cycle

23.49 acetyl CoA

23.51 a. 2 (steps 3, 4) b. 1 (step 6) c. 2 (steps 3, 8) d. 2 (steps 2, 7)

23.53 a. steps 3, 4, 6, 8 b. step 2 c. step 7

23.55 $^{-}OOC–CH_2–CH_2–COO^{-}$, succinate; $^{-}OOC–CH{=}CH–COO^{-}$, fumarate;

$^{-}OOC–CH(OH)–CH_2–COO^{-}$, malate; $^{-}OOC–C(=O)–CH_2–COO^{-}$ oxaloacetate

23.57 oxidation and decarboxylation

23.59 a. NAD^{+} b. FAD

23.61 a. isocitrate, α-ketoglutarate b. fumarate, malate
c. malate, oxaloacetate d. citrate, isocitrate

23.63 respiratory chain

23.65 O_2

23.67 a. oxidized form of flavin mononucleotide
b. cytochrome
c. iron/sulfur protein
d. reduced form of nicotinamide adenine dinucleotide

23.69 a. mobile b. fixed c. fixed d. mobile

23.71 NADH, FeSP, cyt c_1, cyt a_3

23.73 a. oxidation b. reduction c. oxidation d. reduction

23.75 a. $FADH_2$, CoQ, $2Fe^{3+}$ b. $FMNH_2$, 2Fe(II)SP, $CoQH_2$

23.77 a. FMN + 2Fe(II)SP b. CoQ, $CoQH_2$

23.79 ATP synthesis from ADP using energy from the electron transport chain

23.81 protons (H^+ ions)

23.83 intermembrane space

23.85 ATP synthase

23.87 protons flow through the ATP synthase complex

23.89 three

23.91 they enter the ETC at different stages

23.93 a. (1) b. (2) c. (2) d. (1) e. (1) f. (2)

23.94 a. (1) b. (2) c. (1) d. (3) e. (1) f. (2)

23.95 a. 4 b. 2, 3, and 4 c. 1, 2, 3, and 4 d. 3

23.96 a. 1 b. 4 c. 4 d. 2

23.97 a. 4 b. 1 c. 3 d. 1

23.98 the products from the CAS, FAD and NAD^+, are the starting reactants for the ETC

23.99 a. inside mitochondria b. inside mitochondria

23.100 FAD is the oxidizing agent for carbon-carbon double bond formation

23.101 a. 1, 2, 3, 4, 6 b. 1, 2, 4 c. 1, 5 d. 1

23.102 a. 2, 3, 4 b. 5, 6 c. 2 d. 3

23.103 a. 1, 2, 4, 6 b. 2, 4, 5, 6 c. 1 d. 6

23.104 a. 3, 6 b. 2, 6 c. 1, 4, 6 d. 3, 6

Carbohydrate Metabolism Chapter 24

Solutions to Exercises

24.1 mouth; salivary α-amylase

24.3 small intestine; pancreas

24.5 outer membranes of intestinal mucosal cells; hydrolysis of sucrose

24.7 glucose, galactose, fructose

24.9 glucose

24.11 NAD^+

24.13 formation of glucose 6-phosphate, a species that cannot cross cell membranes

24.15 dihydroxyacetone phosphate, glyceraldehyde 3-phosphate

24.17 two

24.19 two

24.21 steps 1, 3 and 6

24.23 cytoplasm

24.25 a. glucose 6-phosphate b. 2-phosphoglycerate c. phosphoglyceromutase d. ADP

24.27 a. step 10 b. step 1 c. step 8 d. step 6

24.29 a. +2 b. +4

24.31 a.

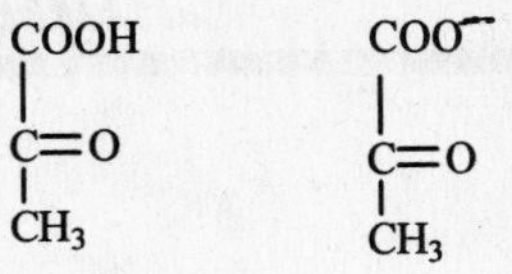

COOH
|
C=O
|
CH_3

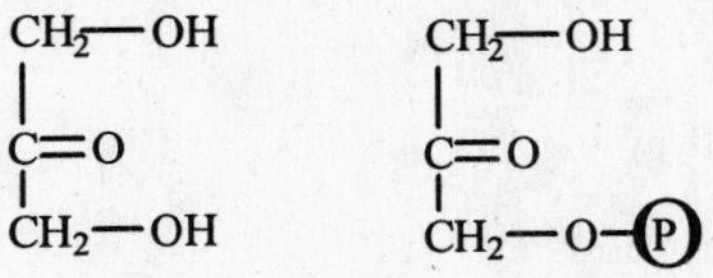

COO^-
|
C=O
|
CH_3

b.

CH_2—OH
|
C=O
|
CH_2—OH

CH_2—OH
|
C=O
|
CH_2—O—Ⓟ

c.

Ⓟ—O—CH_2 O CH_2—OH
HO
OH
OH

Ⓟ—O—CH_2 O CH_2—O—Ⓟ
HO
OH
OH

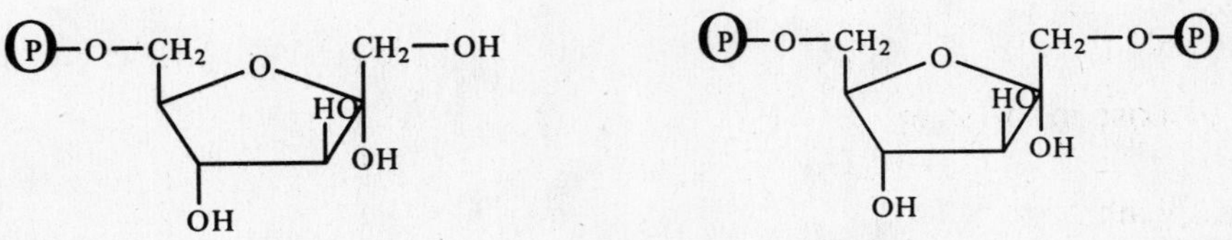

d.

COOH
|
CH—OH
|
CH_2—OH

CHO
|
CH—OH
|
CH_2—OH

24.33

1 CH_2—O—Ⓟ
|
2 C=O
|
3 CH_2—OH

4 CHO
|
5 CH—OH
|
6 CH_2—O—Ⓟ

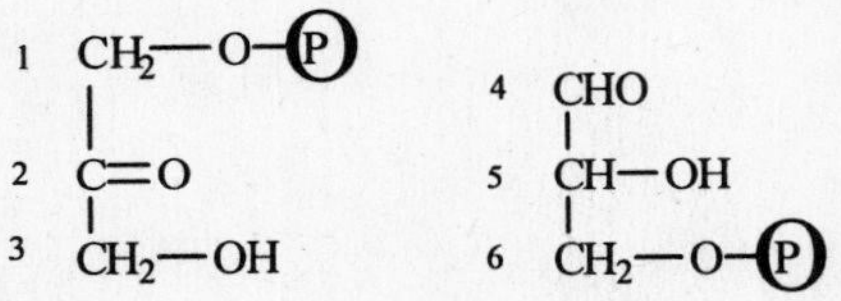

24.35 acetyl CoA, lactate, ethanol

24.37 pyruvate + CoA + NAD^+ → acetyl CoA + NADH + CO_2

24.39 NADH is oxidized to NAD^+, a substance needed for glycolysis

24.41 CO_2

24.43 glucose + 2ADP + $2P_i$ → 2 lactate + 2ATP

24.45 decreases ATP production by 2

24.47 36 ATP versus 2 ATP

24.49 two

24.51 glycogenesis converts glucose to glycogen and glycogenolysis is the reverse process

24.53 glucose 6-phosphate

24.55 UTP

24.57 UDP + ATP → UTP + ADP

24.59 step 2

24.61 in liver cells the product is glucose and in muscle cells it is glucose 6-phosphate

24.63 as glucose 6-phosphate

24.65 the liver

24.67 2-step pathway for step 10; different enzymes for steps 1 and 3

24.69 oxaloacetate

24.71 converted to glucose in the liver

24.73 glucose 6-phosphate

24.75 NADPH is consumed in its reduced form; NADH is consumed in its oxidized form (NAD^+)

24.77 glucose 6-phosphate + $2NADP^+$ + H_2O → ribulose 5-phosphate + CO_2 + 2NADPH + $2H^+$

24.79 CO_2

24.81 increases rate of glucogen synthesis

24.83 increases blood glucose levels

24.85 pancreas

24.87 epinephrine attaches to cell membrane and stimulates the production of cAMP

24.89 glucagon (liver cells) and epinephrine (muscle cells)

24.91 a. all four
b. glycogenesis and glycogenolysis
c. glycolysis and gluconeogenesis
d. gluconeogenesis

24.92 a. glycolysis
b. glycolysis
c. glycolysis, gluconeogenesis, glycogenesis
d. glycogenesis

24.93 a. glycolysis b. glycogenesis c. glycogenolysis d. glyconeogenesis

24.94 a. 2 ATP b. 36 ATP c. 2 ATP d. 2 ATP

24.95 a. 2 b. 2 c. 3 d. 4

24.96 a. when the body requires free glucose
b. anaerobic conditions in muscle; red blood cells
c. when the body requires energy
d. anaerobic conditions in yeast

24.97 a. 12 moles b. 4 moles c. 4 moles d. 72 moles

24.98 a. glucose supply is adequate, body doesn't need energy
b. glucose supply is adequate, body needs energy
c. ribose 5-phosphate or NADPH are needed
d. free glucose supply is not adequate

24.99 a. 1, 2, 4, 5, 6 b. 1, 4, 5, 6 c. 5 d. 2, 3

24.100 a. 1 b. 4 c. 6 d. 6

24.101 a. 2 b. 6 c. 2 d. 2

24.102 a. 4, 5, 6 b. 1, 2, 3, 4, 5 c. 1, 2 d. 3, 6

Lipid Metabolism — *Chapter 25*

Solutions to Exercises

25.1 98%

25.3 no effect

25.5 because lipids have a long residence time in the stomach

25.7 acts as an emulsifier

25.9 hydrolysis results in release of only two of the three fatty acid units, producing a monoglyceride + two free fatty acids

25.11 reassembled into triacylglycerols; converted to chylomicrons

25.13 they have a large storage capacity for triacylglycerols

25.15 hydrolysis of triacylglycerols in adipose tissue; entry of hydrolysis products into bloodstream

25.17 activates hormone-sensitive lipase

25.19 glycerol 3-phosphate, dihydroxyacetone phosphate

25.21 one

25.23 outer mitochondrial membrane

25.25 ATP is converted to AMP and $2P_i$

25.27 shuttles acyl groups across the inner mitochondrial membrane

25.29 a. alkane to alkene b. alkene to 2° alcohol c. 2° alcohol to ketone

25.31 trans-isomer

25.33 a. step 3, turn 1 b. step 2, turn 2 c. step 4, turn 2 d. step 1, turn 1

25.35 compounds a and d

25.37 a. 7 turns b. 5 turns

25.39 a cis-trans isomerase converts a cis-(3, 4) double bond to a trans-(2, 3) double bond

25.41 a. glucose b. fatty acids

25.43 a. 4 turns b. 5 acetyl CoA c. 4 NADH d. 4 $FADH_2$ e. 2 high-energy bonds

25.45 78 ATP

25.47 they yield the same amount of $FADH_2$; the additional enzymes needed to process unsaturated fatty acids do not require FADH

25.49 4 kcal versus 9 kcal

25.51 (1) dietary intakes high in fat and low in carbohydrates, (2) inadequate processing of glucose present, and (3) prolonged fasting

25.53 ketone body formation occurs when oxaloacetate concentrations are low

25.55 $CH_3{-}\overset{\overset{\displaystyle O}{\|}}{C}{-}CH_2{-}\overset{\overset{\displaystyle O}{\|}}{C}{-}O^-$, $CH_3{-}\overset{\overset{\displaystyle OH}{|}}{CH}{-}CH_2{-}\overset{\overset{\displaystyle O}{\|}}{C}{-}O^-$, $CH_3{-}\overset{\overset{\displaystyle O}{\|}}{C}{-}CH_3$

25.57 liver mitochondria

25.59 formation of acetoacetyl CoA

25.61 accumulation of ketone bodies in blood and urine

25.63 cytoplasm versus mitochondrial matrix

25.65 acyl carrier protein; polypeptide chain replaces phosphorylated ADP

25.67 liver, adipose tissue, mammary glands

25.69 it is involved in the citrate shuttle system

25.71 carrier of C_2 units for a growing fatty acid chain

25.73 (1) condensation, (2) hydrogenation, (3) dehydration, and (4) hydrogenation

25.75 a. step 1, cycle 1 b. step 2, cycle 2 c. step 3, cycle 2 d. step 4, cycle 1

25.77 compounds b and d

25.79 C_{16} fatty acid

25.81 it is needed to convert saturated fatty acids to unsaturated fatty acids

25.83 a. 6 rounds b. 6 malonyl ACP c. 6 ATP bonds d. 12 NADPH

25.85 a. 13%–17% b. 83%–87%

25.87 a. mevalonate b. isopentenyl pyrophosphate c. squalene

25.89 a. fewer than b. fewer than c. same as

25.91 a. fatty acid spiral b. fatty acid spiral c. lipogenesis
d. glycerol catabolism e. ketogenesis f. ketogenesis

25.92 a. fatty acid catabolism b. lipogenesis
c. lipogenesis d. ketogenesis
e. consumption of molecular O_2 e. consumption of molecular O_2

25.93 a. step 1 b. step 1 c. step 3 d. step 4

25.94 (1), (3), (4), and (2)

25.95 a. true b. false c. false d. false

25.96 a. incorrect b. correct c. incorrect d. correct

25.97 a. incorrect b. correct c. correct d. correct

25.98 glucose, C_8 fatty acid, sucrose, C_{14} fatty acid

25.99 a. 1, 5 b. 4 c. 6 d. 3

25.100 a. 2, 3, 4 b. 1, 5 c. 1, 5, 6 d. 2, 5, 6

25.101 a. 1, 2, 4, 5 b. 6 c. 2, 4, 5, 6 d. 1, 3

25.102 a. 2, 3, 5, 6 b. 4, 5, 6 c. 1, 3, 4, 5, 6 d. 5, 6

Protein Metabolism

Chapter 26

Solutions to Exercises

26.1 occurs in the stomach with gastric juice as the denaturant

26.3 pepsinogen is the inactive precursor of pepsin

26.5 gastric juice is acidic (1.5–2.0 pH) and pancreatic juice is basic (7–8 pH)

26.7 shuttle molecules facilitate the passage of amino acids through the intestinal wall

26.9 total supply of free amino acids available for use

26.11 cyclic process of protein degradation and resynthesis

26.13 nitrogen intake exceeding nitrogen output produces a positive nitrogen balance; nitrogen output exceeding nitrogen intake produces a negative nitrogen balance

26.15 it becomes negative, proteins are degraded to get the needed amino acid

26.17 protein synthesis, synthesis of nonprotein nitrogen-containing compounds, nonessential amino acid synthesis, and energy production

26.19 a. essential b. nonessential c. nonessential d. essential

26.21 b and c

26.23 an amino acid and an α-keto acid

26.25 a. $HO{-}CH(CH_3){-}CH(\overset{+}{N}H_3){-}COO^- + CH_3{-}C(=O){-}COO^- \rightarrow HO{-}CH(CH_3){-}C(=O){-}COO^- + CH_3{-}CH(NH_3){-}COO^-$

b. $CH_3{-}CH(\overset{+}{N}H_3){-}COO^- + {}^-OOC{-}CH_2{-}C(=O){-}COO^- \rightarrow CH_3{-}C(=O){-}COO^- + {}^-OOC{-}CH_2{-}CH(\overset{+}{N}H_3){-}COO^-$

c. $H{-}CH(\overset{+}{N}H_3){-}COO^- + {}^-OOC{-}CH_2{-}CH_2{-}C(=O){-}COO^- \rightarrow H{-}C(=O){-}COO^- + {}^-OOC{-}CH_2{-}CH_2{-}CH(\overset{+}{N}H_3){-}COO^-$

d. $HO{-}CH(CH_3){-}CH(\overset{+}{N}H_3){-}COO^- + {}^-OOC{-}CH_2{-}CH_2{-}C(=O){-}COO^- \rightarrow$

$HO{-}CH(CH_3){-}C(=O){-}COO^- + {}^-OOC{-}CH_2{-}CH_2{-}CH(\overset{+}{N}H_3){-}COO^-$

26.27 pyruvate, α-ketoglutarate, oxaloacetate

26.29 coenzyme that participates in the amino group transfer

26.31 conversion of an amino acid into a keto acid with the release of ammonium ion

26.33 oxidative deamination produces ammonium ion and transamination produces an amino acid

26.35 a. ${}^-OOC{-}CH_2{-}CH_2{-}C(=O){-}COO^-$

b. $HS{-}CH_2{-}C(=O){-}COO^-$

c. $CH_3{-}C(=O){-}COO^-$

d. $C_6H_5{-}CH_2{-}C(=O){-}COO^-$

26.37 transamination of the α-keto acid produces the amino acid (leucine)

$$CH_3-\underset{\displaystyle CH_3}{\overset{|}{C}}H-CH_2-\underset{}{C}H(\overset{+}{N}H_3)-COO^-$$

26.39 a. aspartate b. glutamate c. pyruvate d. α-ketoglutarate

26.41 $H_2N-\overset{O}{\overset{\|}{C}}-NH_2$

26.43 carbamoyl phosphate

26.45 an amide group

26.47 $H_3\overset{+}{N}-$, $H_2N-\overset{O}{\overset{\|}{C}}-NH-$, $H_2N-\overset{\overset{+}{N}H_2}{\overset{\|}{C}}-NH-$

26.49 carbamoyl phosphate

26.51 ornithine

26.53 conversion of citrulline to argininosuccinate

26.55 a. citrulline b. ornithine c. argininosuccinate d. carbamoyl phosphate

26.57 equivalent of four ATP molecules

26.59 goes to the citric acid cycle where it is converted to oxaloacetate, which can then be reconverted to aspartate

26.61 α-ketoglutarate, succinyl CoA, fumarate, oxaloacetate

26.63 a. acetoacetyl CoA and acetyl CoA b. succinyl CoA and acetyl CoA c. fumarate and oxaloacetate d. α-ketoglutarate

26.65 degradation products can be used to make glucose

26.67 glutamate

26.69 pyruvate, α-ketoglutarate, 3-phosphoglycerate, oxaloacetate, and phenylalanine

26.71 hydrolyzed to amino acids

26.73 in biliverdin the heme ring has been opened and one carbon atom has been lost (as CO)

26.75 biliverdin, bilirubin, bilirubin diglucuronide, urobilin

26.77 urobilin

26.79 excess bilirubin

26.81 pyruvate, acetyl CoA, acetoacetate

26.83 amino acids are converted to acetyl CoA; lipogenesis converts acetyl CoA to fatty acids; fatty acids are converted to triacylglycerols

26.85 converted to body fat stores

26.87 a. (1) b. (2) c. (3) d. (2) e. (1) and (3) f. (3)

26.88 a. (1) b. (3) c. (3) d. (2) e. (1) f. (2)

26.89 (1), (3), (2), and (4)

26.90 a. 3 b. 2 c. 1 d. 3

26.91 a. transamination b. deamination c. deamination d. transamination

26.92 a. 1 b. 4 c. 2 d. 3

26.93 a. true b. false c. true d. true

26.94 a. 2 b. 1 c. 2 d. 3

26.95 a. 2, 3, 4, 5 b. 2, 3, 4, 6 c. 1, 6 d. 1, 2, 3

26.96 a. all choices b. 2, 3, 4, 6 c. 1 d. 5

26.97 a. all choices b. 2, 3, 4, 5, 6 c. 1 d. 1

26.98 a. 1 b. 1, 2, 4, 5 c. 1, 2, 4, 5 d. 2, 4, 5